RESEARCH PROGRESS IN MATERIALS SCIENCE

RESEARCH PROGRESS IN MATERIALS SCIENCE

WILLIAM OLSSON
AND
FILIP LINDBERG
EDITORS

Nova Science Publishers, Inc.
New York

For permission to use material from this book please contact us:
Telephone 631-231-7269; Fax 631-231-8175
Web Site: http://www.novapublishers.com

LIBRARY OF CONGRESS CATALOGING-IN-PUBLICATION DATA

Research progress in materials science / William Olsson and Filip Lindberg (editors).
p. cm.
Includes bibliographical references and index.
ISBN 978-1-60456-629-1 (hardcover)
1. Materials science. I. Olsson, William. II. Lindberg, Filip.
TA403.8.R47 2008
620.1'1--dc22
2008013833

Published by Nova Science Publishers, Inc. ✦ New York

CONTENTS

PREFACE

Materials science includes those parts of chemistry and physics that deal with the properties of materials. It encompasses four classes of materials, the study of each of which may be considered a separate field: metals; ceramics; polymers and composites. Materials science is often referred to as materials science and engineering because it has many applications. Industrial applications of materials science include processing techniques (casting, rolling, welding, ion implantation, crystal growth, thin-film deposition, sintering, glassblowing, etc.), analytical techniques (electron microscopy, x-ray diffraction, calorimetry, nuclear microscopy (HEFIB) etc.), materials design, and cost/benefit tradeoffs in industrial production of materials. This book examines new research in this frontier field.

As well known, the borate crystals are an important laser material and nonlinear optical crystal. For examples, the double borate crystals of Nd^{3+}:$YAl_3(BO_3)_4$ and Nd^{3+}:$GdAl_3(BO_3)_4$ are an excellent self-doubling frequency laser crystal; the borate crystal of β-BaB_2O_4 (β-BBO) and LiB3O5 (LBO) are an excellent nonlinear optical crystal. After an examination of known fluxes and growth, this article reviews recent advances in the development of the rare earth and Cr^{3+}-doped $RX_3(BO_3)_4$ crystal growth as well as the spectral characterizations. After an examination of the relative phase diagram of β-BBO, Chapter 1 reviews recent advance in the developments of β-BBO and LBO crystal growth. Except above, this article also describes recent development in the growth of $M_3Ln_2(BO_3)_4$ and $M_3Ln(BO_3)_3$ crystals.

Some hazardous materials show characteristic features in their reaction with transition metals; they are unstabilized, and sometimes decompose when in contact with transition metals. Their chemical structure indicates that they have a high tendency to interact with transition metals since they include electronegative atoms. Transition metal atoms attract lone pairs of electronegative atoms. This tendency leads to the formation of a complex of transition metals and the molecules. However, some complexes, especially those with unstable molecules, can dissociate immediately along with decomposition of unstable chemicals, or they can be unstabilized by heat, friction and other stimulation because bond strength is affected and weakened by electronic condition change through complex formation. The effect sometimes results in a reaction beyond control and explosion. Chapter 2 provides an overview of current research on the reactivity of transition metals and unstable chemicals, and the decomposition behavior of unstable chemicals interacting with metals. The subject is related through three viewpoints: the stability change caused by the interactions, the reaction features from the viewpoint of heat release, and the interaction features of oxygen and nitrogen with Fe^{3+}.

Recently, several novel autotrophic biological nitrogen removal processes have been developed, including partial nitritation, denitrification by nitrifier, anaerobic ammonium oxidation and its combination. As compared with the classical nitrogen removal process via nitrification and denitrification, these novel processes have various advantages in economical and environmental aspects, such as no need for external carbon addition, negligible sludge production and less energy and oxygen requirement.

This chapter deals with a brief review and recent laboratory scale experimental developments in lithoautotrophic nitrogen removal under an anaerobic environment: anaerobic ammonium oxidation by *Planctomycetes*-like bacteria and *Nitrosomonas*-like bacteria. Chapter 3 addresses the specifics of process, microbial diversity, performance characteristics and the future challenges for application. In laboratory experiments, sustainable autotrophic nitrogen elimination from ammonium-rich wastewaters such as synthetic wastewater (100~200 mg/L NH_4-N) and high strength slurry-type piggery waste (56 g COD/L and 4.3 g/L NH_4-N) was investigated by using an upflow anaerobic granular sludge bed reactor. In the research, two types of anaerobic autotrophic nitrogen elimination, namely by *Planctomycetes* and *Nitrosomonas* species, were observed according to the type of the substrate. In the experiment using synthetic wastewater, the microbial community, based on cloned 16S rDNA analysis, indicated that an anoxic ammonium oxidizer (autotrophic denitrifying nitrifier) such as *Nitrosomonas eutropha* and *Nitrosomonas europaea* and other autotrophic denitrifiers mainly contributed to nitrogen elimination. The reactor showed quite low nitrogen conversion (0.02~0.03 kg N/m^3-d). FISH (Fluorescence *in situ* Hybridization) analysis of the biomass treating piggery waste demonstrated that *Planctomycetes*-like bacteria (especially, Candidatus "*Kuenenia stuttgartiensis*") played a significant role in nitrogen removal, contrary to the foregoing experiments for ammonium-rich synthetic wastewater. Average nitrogen conversion was 0.59~0.66 kg N/m^3-d (0.3~0.32 kg NH_4-N/m^3-d). The specific microbial activity in the batch experiment was 0.1 g NH_4-N/g VSS/day (0.77 g T-N/g VSS/day), indicating that potential denitritation is also quite high. Based on mass balance, the result revealed that the Anammox reaction was influenced less by the degree of partial denitritation, and the Anammox bacteria did not compete with denitritation bacteria. The autotrophic nitrogen-removing granular sludge showed quite favorable physical characteristics in biomass settleability and size.

Several earthquakes have demonstrated that many existing bridge columns are seismically vulnerable. A number of retrofitting and upgrading techniques including steel and concrete jacketing have been developed in order to reduce the seismic vulnerability of the highway system. Recently, researchers have been considering utilizing fiber reinforced polymers to replace traditionally used steel/concrete jackets. Extensive research has been conducted worldwide by a number of researchers to verify enhancements made by FRP retrofitting system, as well as to develop design guidelines for future implementations. Chapter 4 presents common failure modes of existing reinforced concrete columns as well as a brief discussion of the traditional retrofitting solutions. Developments in retrofitting of rectangular bridge columns using FRP are presented as well. Different international codes and guidelines for retrofitting of rectangular columns using FRP are reviewed as well. Investigations to date have demonstrated that the shear and flexural capacity of rectangular columns can be adequately enhanced by externally bonded FRP; whereas FRP retrofitting was less effective in improving the behavior of columns having deficient lap splices. In terms of design, several codes and guidelines present design equation for shear and confinement design. However, for rectangular columns having deficient lap splice, retrofitting using FRP is not recommended by several codes.

The formation process of carbon nanotubes (CNTs) on a thermally treated 6H-SiC(000-1) carbon face substrate was investigated using X-ray photoemission spectroscopy (XPS), atomic-resolution ultrahigh-vacuum scanning tunneling microscopy (UHV-STM) and transmission electron microscopy (TEM). STM observations showed that carbon networks formed on the surface of carbon nanocap structures, which appear in the initial stage of CNT growth. This indicates that the carbon nanocaps are crystalline when they are formed on the SiC surface. By suppressing surface oxidation during heating, details of the formation process of carbon nanocaps were successfully observed using UHV-STM. Based on the experimental results, we proposed a model that describes the formation of CNT, and also a model that explains why the chirality of the CNTs is the zigzag-type. In Chapter 5 the authors also investigated the effect of ambient oxygen gas on the SiC surface and the phase diagram of nanocap formation was obtained. Our results show that both active and passive oxidation transitions occurred depending on the oxygen pressure. They also show that carbon nanocaps formed at about 1200°C, irrespective of the ambient pressure, when the oxygen pressure was under 10^{-2} Pa.

It is well known that the system of Cooper's pair is described by boson symmetric wave functions, but Cooper's pair operators are bosons only when the moments **k** are different and they are fermions for equal **k**. The analysis of trilinear commutation relations for the Cooper pair (pairon) operators reveals that they correspond to the modified parafermi statistics of rank $p = 1$. Two different expressions for the Cooper pair number operator are presented in Chapter 6. The authors demonstrate that the calculations with a Hamiltonian expressed via pairon operators is more convenient using the commutation properties of these operators without presenting them as a product of fermion operators. This allows to study problems in which the interactions between Cooper's pairs are also included. The problem with two interacting Cooper's pairs is resolved and its generalization in the case of large systems is discussed.

It is shown that in site representation, the hole-pair operators obey the same commutation relations (paulion) as the Cooper pair operators in impulse representation, although the latter describe delocalized quasiparticles. In quasi-impulse representation, the hole-pair operators are also delocalized and their exact commutation relations correspond to a modified parafermi statistics of rank M($M$is the number of sites in a "superlattice" formed by the centers of mass of each hole pair). From this follows that one state can be occupied by up to Mpairs. Even in the absence of dynamic interaction, the system of hole pairs is characterized by some immanent interaction, named after Dyson as kinematic interaction. This interaction appears because of the deviation of the quasiparticle statistics from the Bose (Fermi) statistics and its magnitude depends on the concentration of hole pairs. In spite of the non-bosonic behavior, there is no statistical prohibition on the Bose-Einstein condensation of coupled hole pairs.

The development and application of catalytic processes is one of the fundamental issues for fine chemicals industry where many transformations are still based on stoichiometric reactions from the area of organic chemistry, and often large amounts of waste are formed. Therefore there is a need for more selective, especially catalytic processes. Major activities in industrial process research and development are directed to environmental aspects, raw material costs, technology, and selectivity of a reaction. In Chapter 7, new industrially important catalytic processes for the application to the synthesis of the vitamin α-tocopherol and its acetate derivative are presented. Protocols with rare earth catalysts provide the advantages of high selectivity and low catalyst loading for a variety of reactions. Here the authors report on state of the art in this area, and on recent results obtained in our laboratories. For example, the known processes for the synthesis of α-tocopherol suffer from disadvantages like formation of by-products and waste-water problems. To overcome these drawbacks new types of catalysts have been developed for the synthesis of α-tocopherol and its acetate derivative, starting from trimethylhydroquinone or its 1-acetate, and phytol, isophytol or derivatives, e. g., halides or esters like alkanoates. Catalysts used are various rare earth metal triflates or triflimides. This new approach to vitamin E or derivatives thereof has several advantages, based on high yields and selectivities. Furthermore the reactions can be carried out in continuous or batch-wise modes. When using catalyst-to-substrate ratios from 1:100 to 1:10'000, yields of above 94% can be achieved.

In Chapter 8, photoluminescence, excitation and absorption spectra are investigated for thin films lightly and heavily doped with TPD, $Ir(ppy)_3$, Alq_3, PtOEP and mCP and for these neat films at various temperatures between 10 and 290 K. The photoluminescence spectra are compared with the electroluminescence spectra of organic light emitting diodes with emitting layers of $Ir(ppy)_3$ and PtOEP. Aggregates such as dimer and trimer are observed in the neat films of $Ir(ppy)_3$ and Alq_3. On the other hand, dimer and trimer are formed in TPD, PtOEP- and mCP-doped films and these neat films, together with monomer. In the case of PtOEP, large size aggregates which consist of many PtOEP molecules are also formed even in lightly doped films. Such a difference between the group of $Ir(ppy)_3$ and Alq_3 and the group of TPD, PtOEP and mCP is caused by difference of competition and balance among solid state solvation effect, dipole-dipole interactions between dopant molecules, and intermolecular covalent bonding. It is suggested that molecular geometrical structure gives rise to the difference.

In Chapter 9, the main aspects of the application of the fractal analysis for the description of the behavior of macromolecular coils in the diluted solutions are considered. The coil structure is quantitatively described with the help of its fractal (Hausdorff) dimension characterizing the distribution of coil elements in space. A number of the experimental methods of the determination this dimension is offered. The interrelation of the classical and fractal (structural) characteristics of macromolecular coils is shown. The fractal model is used for the description of such phenomena as film formation from polymer solutions, flocculation and self-diffusion of macromolecular coils. The example of the application of the theory of fractional derivatives for the description of macromolecular coil the main parameters of a in solution is give.

In: Research Progress in Materials Science
Editors: W. Olsson and F. Lindberg, pp. 1-23

ISBN: 978-1-60456-629-1

Chapter 1

RECENT PROGRESS OF BORATE CRYSTAL GROWTH

Guofu Wang
Fujian Institute of Research on the Structure of Matter,
Chinese Academy of Sciences,
Fuzhou, Fujian 350002, China

Abstract

As well known, the borate crystals are an important laser material and nonlinear optical crystal. For examples, the double borate crystals of Nd^{3+}:$YAl_3(BO_3)_4$ and Nd^{3+}:$GdAl_3(BO_3)_4$ are an excellent self-doubling frequency laser crystal; the borate crystal of β-BaB_2O_4 (β-BBO) and LiB3O5 (LBO) are an excellent nonlinear optical crystal. After an examination of known fluxes and growth, this article reviews recent advances in the development of the rare earth and Cr^{3+}-doped $RX_3(BO_3)_4$ crystal growth as well as the spectral characterizations. After an examination of the relative phase diagram of β-BBO, this article reviews recent advance in the developments of β-BBO and LBO crystal growth. Except above, this article also describes recent development in the growth of $M_3Ln_2(BO_3)_4$ and $M_3Ln(BO_3)_3$ crystals.

2. Crystal Growth and Spectroscopic Characterizations of Rare-Earth-Doped Rx3(Bo3)4 Crystals

2.1. Structural Characteristic of $RX_3(BO_3)_4$ Crystals

The double borate with general formula $RAl_3(BO_3)_4$ (where R=Y, Nd, Gd, Ru, Tb, Dy, Ho, Er, Yb) and $RCr_3(BO_3)_4$ (where R=Sm, Gd) were firstly reported by Ballman [1], which were synthesized by a flux method from K_2SO_4-$3MoO_3$ and PbF_2-$3B_2O_3$ fluxes. These compounds are isostructural with the mineral huntite $CaMg_3(CO_3)_4$ with space group R32 [2]. Hong et al [3] firstly reported the structural details of $NdAl_3(BO_3)$ $_4$ crystal. The $NdAl_3(BO_3)_4$ crystal belongs to hexagonal with space group R32 and the unit cell dimensions of a= 9.341Å, c=7.3066 Å, Z=3[3]. The structure of $NdAl_3(BO_3)_4$ crystal consists of NdO_6 distorted octahedral, AlO_6 distorted octahedral and BO_3 planar triangle. The structure of $NdAl_3(BO_3)_4$ crystal is illustrated in Figure 1 [3]. However, in 1979, Jarchow reported that the

$NdAl_3(BO_3)_4$ crystal was monoclinic with two space groups C2/c and C2 [4]. Lutz and Zwiter also reported in 1981 and 1983 that $NdAl_3(BO_3)_4$ crystal is monoclinic with two space group C2/c and C2 [5,6]. In 1983 Belokoneva reported that $NdAl_3(BO_3)_4$ crystal was monoclinic with space group C2/c [7]. Finally, author discovered that $NdAl_3(BO_3)_4$ crystal is a polymorphous compound and has two-phase transition points and three phase structure, i.e. R32, C2/c and C2 space groups [8]. In fact, The other $RX_3(BO_3)_4$ (R=Y, Lanthanum; X= Al, Sc, Cr) crystal also exist the polymorphous phenomena. For example, $GdAl_3(BO_3)_4$ crystal has a space group C2 except the space group R32 [9]. $LaSc_3(BO_3)_4$ crystal also exists three phase structures of R32, C2/c and Cc [10-12]. However, no matter which phase structure in $RX_3(BO_3)_4$ crystal, it always consist of the RO_6 distorted octahedral, XO_6 distorted octahedral and BO_3 planar triangle, as shown in Figures 1. The RO_6 polyhedra are isolated with R ion and R ion joined by means of R–O–X–O– R. The R-R distance between is from 5.596 Å to 6.623 Å. This structural characteristic results in a weaker interaction between laser upper-levels when the active ions, such as Nd^{3+}, Er^{3+}, were doped into the $RX_3(BO_3)_4$ crystals. As a result, the rare earth –doped $RX_3(BO_3)_4$ crystals have a low fluorescence concentration quenching effect [3, 8,10,11].

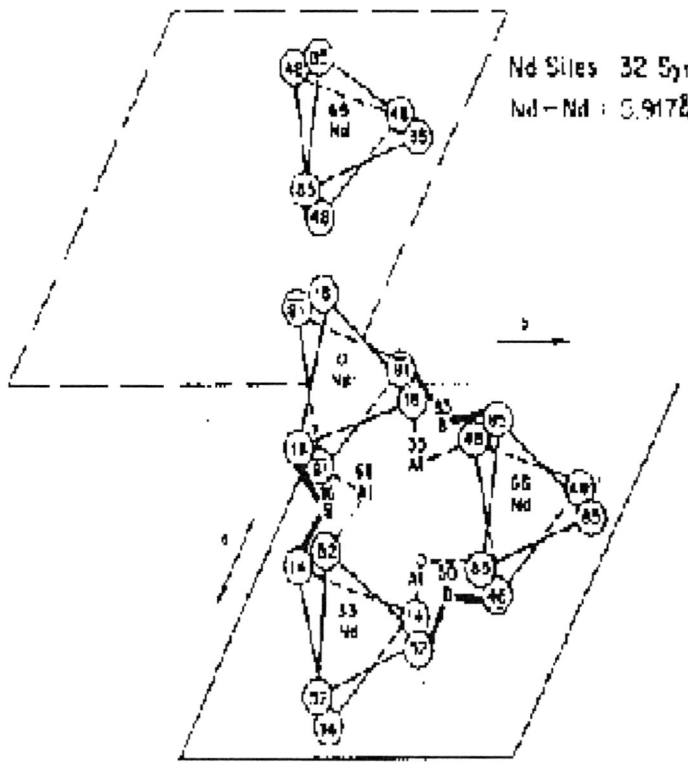

Figure 1. A structure fragment of $NdAl_3(BO_3)_4$ crystal with space group R32 [3].

2.2. Selection and Characterization of Fluxes

Since $RAl_3(BO_3)_4$ crystals melt incongruently at high temperature and decompose into their constituent phases of RBO_3 and $AlBO_3$, inconsequence, $RX_3(BO_3)_4$ crystals are only grown from the high temperature solutions except $LaSc_3(BO_3)_4$ crystal. Up to now, a number fluxes were reported for the growth of $RAl_3(BO_3)_4$ crystals, such as PbF_2-B_2O_3, $Li_2B_4O_7$, $Na_2B_4O_7$, BaO-B_2O_3, $K_2Mo_3O_{10}$ and $K_2Mo_3O_{10}$–KF(K_2SO_4 or PbF_2)[13-34]. However, the growth of $RX_3(BO_3)_4$ crystals with large size and good quality is extremely difficult. All attempts to grow large rare earth aluminoborate crystals with sufficiently good optical quality for laser application have met with little success. Since the number of possible fluxes is vast and detailed principles that may be used in the selection of the optimum solvent for the growth of a given material have not yet been established, the most difficult aspect of crystallization from high-temperature solution is the choice of an appropriate flux. The essential requirements of a flux are that: Firstly, the solubility of the desired crystal is high without forming a stable compound; Secondly, it has a low volatility at growth temperature; Thirdly, it also has a low viscosity. Mixed fluxes are commonly used since the composition may be varied so as achieve a compromise between, for example, solubility and volatility and hence optimize the properties of the flux.

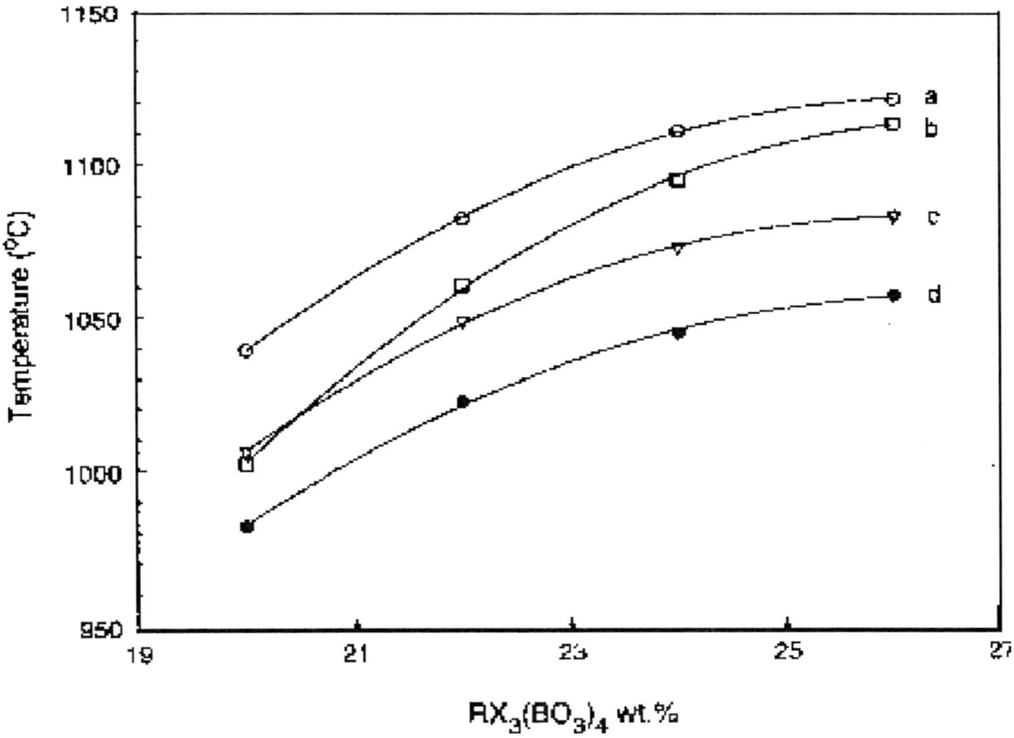

Figure 2. Solubility curve of (a) $YAl_3(BO_3)_4$ in $K_2Mo_3O_{10}$; (b) $YAl_3(BO_3)_4$ in $K_2Mo_3O_{10}$-B_2O_3; (c) $GdAl_3(BO_3)_4$ in $K_2Mo_3O_{10}$; (d) $GdAl_3(BO_3)_4$ in $K_2Mo_3O_{10}$-B_2O_3 [38].

The potassium trimolybdate, $K_2Mo_3O_{10}$, was firstly reported as a flux from which to grow the $RAl_3(BO_3)_4$ crystals [34]. Although $K_2Mo_3O_{10}$ is well known as a successful flux for the growth of $RAl_3(BO_3)_4$ crystals, it has two potential disadvantages. Firstly, the solubility of the desired crystal may be quite low, so that the solute may not dissolve completely and homogeneously, even though heated to a temperature above the saturation temperature and retained there for long time. Secondly, it is highly volatile at high temperature. In the preparation of YAB crystals from a $K_2Mo_3O_{10}$ flux, losses of up to 50wt.% of the solution occurred during growth at temperature between 1000~1100^{0}C over a 4-5 day period [22]. Inevitably, poor quality crystals containing significant solution inclusions were obtained due to rapid, uncontrolled fluctuations in growth rate [35]. In consequence, it is necessary to improve the solubility of potassium trimolybdate and to reduce evaporation of the solution.

In 1995 author and S. T. Jung et al used the $K_2Mo_3O_{10}$–B_2O_3 mixed flux to grow Cr^{3+} or Nd^{3+}-doped $YAl_3(BO_3)_4$ crystals, respectively [36,37]. Results shown $K_2Mo_3O_{10}$–B_2O_3 mixed flux is available to grow $RAl_3(BO_3)_4$ crystals. Author further investigated the characterization of $K_2Mo_3O_{10}$–B_2O_3 mixed flux. The result shown that the addition of B_2O_3 to the $K_2Mo_3O_{10}$ flux increases the solubility of $YAl_3(BO_3)_4$ and $GdAl_3(BO_3)_4$ crystal in the solvent and reduces the saturation temperature (see Figure 2)[38]. The addition of B_2O_3 to the $K_2Mo_3O_{10}$ flux reduces the volatility of these solutions (see Figure 3) [38]. The $K_2Mo_3O_{10}$–B_2O_3 mixed flux is available to grown the $RAl_3(BO_3)_4$ crystals. The $K_2Mo_3O_{10}$–$Li_2B_4O_7$ was developed for the growth of $RSc_3(BO_3)_4$ crystals [38].

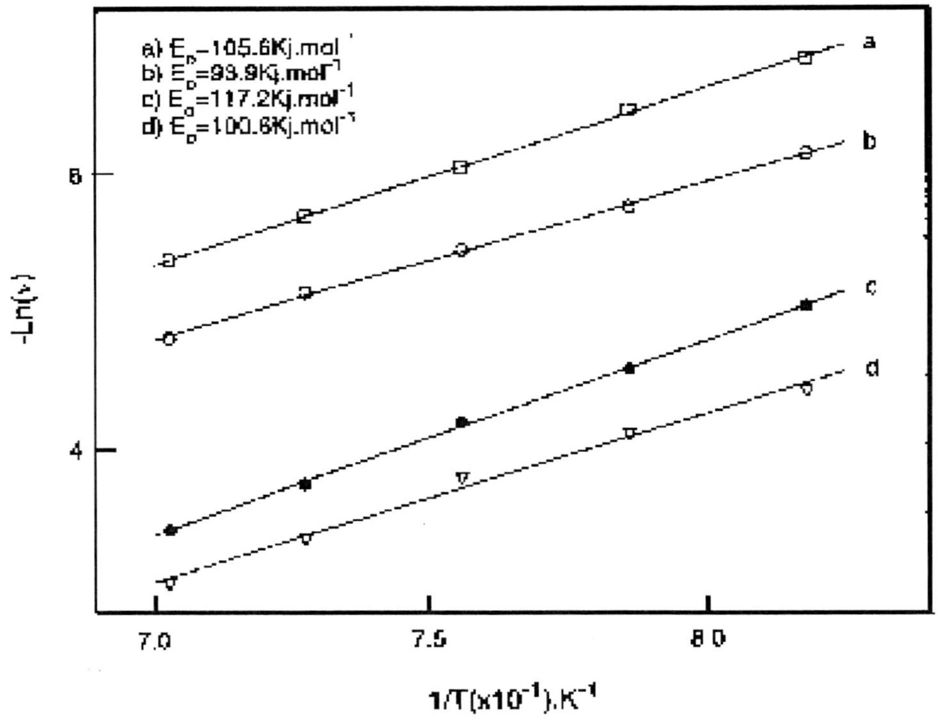

Figure 3. Arrhenius plot of evaporation rate versus temperature for solutions of $YAl_3(BO_3)_4$ in $K_2Mo_3O_{10}$-B_2O_3 ;(b) $YAl_3(BO_3)_4$ in $K_2Mo_3O_{10}$; (c) $GdAl_3(BO_3)_4$ in $K_2Mo_3O_{10}$-B_2O_3 (d) $GdAl_3(BO_3)_4$ in $K_2Mo_3O_{10}$ [38].

2.3. Crystal Growth

$LaSc_3(BO_3)_4$ and $NdSc_3(BO_3)_4$ can be only grown by the Czochralski method in $RX_3(BO_3)_4$ borate family. Nd^{3+}:$LaSc_3(BO_3)_4$crystals with dimension $\phi 20\times 50$ mm^3 were successfully grown by the Czochralski method [39]. Recently, Cr^{3+}-doped $LaSc_3(BO_3)_4$ with dimension $\phi 25\times 35$ mm^{3}has also has been grown at a pulling rate of 1.5mm/h and a rotating rate of 15rpm by the Czochralski method, as shown in Figure 4 [40]. Except above crystals, all $RX_3(BO_3)_4$ borate crystals were growth by top-seeded solution growth (TSSG) method. Figures 5-7 shows some rare-earth-doped $RAl_3(BO_3)_4$ crystal.

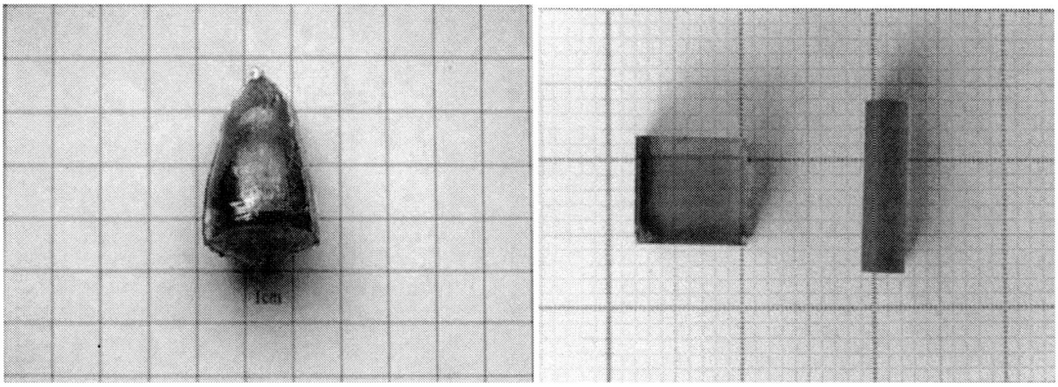

Figure 4. Cr^{3+}:$LaSc_3(BO_3)_4$ crystal grown by the Czochralski method [40].

Figure 5. Ti^{3+}:$YAl_3(BO_3)_4$ crystal with dimension $42\times 36\times 24$mm^3 grown by TSSG method [41].

Figure 6. Nd^{3+}:$YAl_3(BO_3)_4$ crystal with dimension 45×27×20mm^3 grown by A. D. Jiang et al using TSSG method in Fujian Institute of Research on the Structure of Matter, CAS.

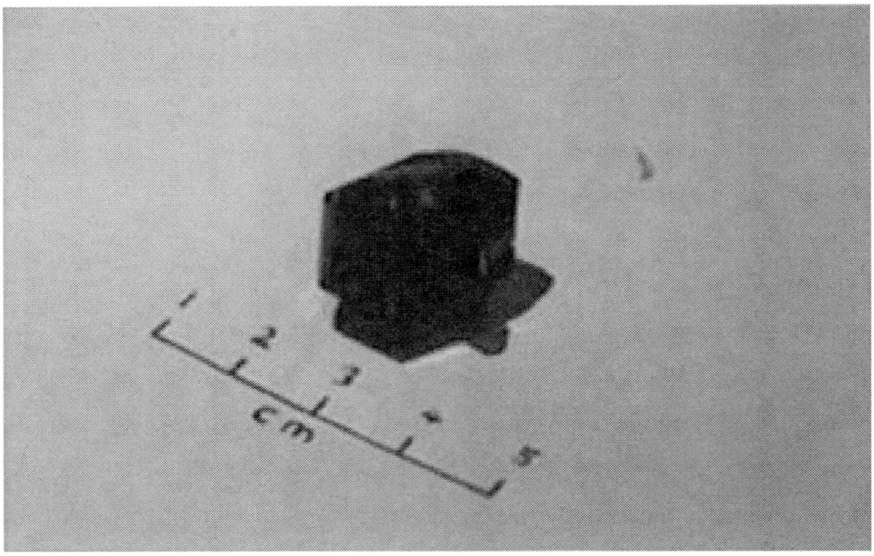

Figure 7. Cr^{3+}:$YAl_3(BO_3)_4$ crystal with dimension 12×16×22mm^3 grown by TSSG method [38].

Table 1. The appropriate flux for some doped $RX_3(BO_3)_4$ crystal

Crystals	Flux	Crystal Size	Ref.
$Nd^{3+}:YAl_3(BO_3)_4$	Unknown	Up to 45mm	[46]
$Nd^{3+}:GdAl_3(BO_3)_4$	$K_2Mo_3O_{10}/LiF$	20×18×15mm³	[42]
$Cr^{3+}:YAl_3(BO_3)_4$	$K_2Mo_3O_{10}/B_2O_3$	22×16×112mm³	[36,38]
$Yb^{3+}:YAl_3(BO_3)_4$	$K_2Mo_3O_{10}/B_2O_3$ $K_2Mo_3O_{10}$//PbF/PbO	15×10×10mm³ 15×15×30mm³	[43] [44]
$Yb^{3+}:GdAl_3(BO_3)_4$	$Yb^{3+}:YAl_3(BO_3)_4$	15×10×1omm³	[45]

Table 1 lists the grown rare-earth-doped $RX_3(BO_3)_4$ borate crystals. The grown $RX_3(BO_3)_4$ borate crystals generally contained some inclusion and crack. The quality of the growing crystals depends on seed orientation. For examples, $Nd^{3+}:YAl_3(BO_3)_4$ crystal was grown from the $K_2O/3MoO_3/0.5B_2O_3$ using a seed with <001> direction, the inclusion-free crystal was obtained [27]. Figure 6 shows $Nd^{3+}:YAl_3(BO_3)_4$ crystal with excellent quality and morphology, which was grown using a seed with [001] direction.

2.4. Optical Characterization of Rare-Earth-Doped $RX_3(BO_3)_4$ Crystals

$Nd^{3+}:YAl_3(BO_3)_4$ (NYAB) and $Nd^{3}:GdAl_3(BO_3)_4$ (NGAB) are an important laser crystal in $RX_3(BO_3)_4$ borate crystals. Since $RAl_3(BO_3)_4$ with space group has nonlinear optical property [47], they can be regarded as a self-frequency doubling laser material when Nd^{3+} is doped into $RAl_3(BO_3)_4$ crystal. Dorozhkin et al firstly realized the self-frequency doubling lasing performance from 1.32μm to 0.66 μm [48]. Then Lu et al reported the self-frequency doubling lasing experiment from 1.06μm to 0.53μm under dye pulse laser pumping [49]. Luo et al further reported experimental result that 78mJ of 0.53μm green laser output was obtained under xenon flash lamp pumping [46]. Now $Nd^{3+}:YAl_3(BO_3)_4$ has been became an excellent self-frequency doubling laser crystal for diode laser pumping.

$Nd^{3+}:GdAl_3(BO_3)_4$ is another self-frequency doubling laser crystal in $RX_3(BO_3)_4$ borate crystals [50,51]. The absorption and emission spectra of $Nd^{3+}:GdAl_3(BO_3)_4$ crystal. $Nd^{3+}:YAl_3(BO_3)_4$ and $Nd^{3+}:GdAl_3(BO_3)_4$ crystals have a common spectroscopic characteristic. Both have a broad absorption band at 810nm, which is ideal for the diode laser pumping, and have large emission cross-sections. The emission cross sections are $1.4\times10^{-19}cm^2$ for $Nd^{3+}:YAl_3(BO_3)_4$ and $6.3\times10^{-19}cm^2$ for $Nd^{3+}:GdAl_3(BO_3)_4$, respectively [52, 42].

2.5. Spectroscopic Characterization of Cr^{3+}-Doped $RX_3(BO_3)_4$ Crystals

Cr^{3+}-doepd $RX_3(BO_3)_4$ crystals are a potential tunable solid-state laser materials. Figures 8-14 show the fluorescence spectra of $Cr^{3+}:LuAl_3(BO_3)_4$ (Cr:LAB), $Cr^{3+}:YAl_3(BO_3)_4$ (Cr:YAB), $Cr^{3+}:GdAl_3(BO_3)_4$ (Cr:GAB), Cr^{3+}:Gd $Ga_3(BO_3)_4$ (Cr:GGB), $GdCr_3(BO_3)_4$ (Cr:GCB)

Cr^{3+}:$YSc_3(BO_3)_4$ (Cr:YSB), Cr^{3+}:$GdSc_3(BO_3)_4$ (Cr:GSB) crystals at room temperature. They demonstrated the different spectral features. As well known, the spectrum of Cr^{3+}-doped tunable solid-state laser is dependent on the crystal-field strength Dq and Racah parameter B and C. When the active ion Cr^{3+} occupies in the different crystal field sites which dependents the value of Dq/B, they exhibits the different spectral feature. When in the strong crystal field, the fluorescence spectrum exhibits only sharp line (R-line); when in medium crystal field, the fluorescence spectrum consists of R-line and broad band, as shown in Figures 8-9; when in the weak crystal field, the fluorescence spectrum is only a broad band, as shown in Figures 10-11. We believe that these spectral features are closely related to chemical composition of $RX_3(BO_3)_4$ double borate crystals. In the $RX_3(BO_3)_4$ double borate crystals there are two trivalent cation positions: R= Y^{3+} and rare-earth element occupy the distorted oxygen trigonal prisms; X= Al^{3+}, Sc^{3+}, Ga^{3+} or Cr^{3+} occupy the center of distorted oxygen octahedra. The ionic radii of R^{3+} and X^{3+} ions cover a wide range: La^{3+}(1.15Å) > Gd^{3+}(1.11 Å) > Y^{3+}(0.95 Å) > Lu^{3+}(0.86 Å); Sc^{3+}(0.83 Å) > Cr^{3+}(0.65 Å) > Ga^{3+}(0.62 Å) > Al^{3+}(0.55 Å). Table 1 shows the relationship between the crystal field strength Dq/B and the ionic radii of R^{3+} and X^{3+} ions of Cr^{3+}:$RX_3(BO_3)_4$ crystals. It is found that the crystal field strength Dq/B is decreased with the increasing of R^{3+} and X^{3+} radium. The larger size of R^{3+} and X^{3+} is, the weaker the crystal field strength in Cr^{3+}:$RX_3(BO_3)_4$ crystals is. In other words, the "size effect" of R^{3+} and X^{3+} strongly affects on crystal field strength of tunable laser crystals. According to the "size effect", we guessed that the Cr^{3+}:$LaSc_3(BO_3)_4$ crystal should have a weak crystal field since it has the largest ionic radii of R^{3+} and X^{3+} ions in $RX_3(BO_3)_4$ borate family. The spectroscopic characterizations of Cr^{3+}:$LaSc_3(BO_3)_4$ crystals have been investigated [53,54]. Figures 12-13 show the absorption and fluorescence spectrum of Cr^{3+}:$LaSc_3(BO_3)_4$ crystal. The Cr^{3+}:$LaSc_3(BO_3)_4$ crystal has a weak crystal field strength (Dq/B=2.27) and broad emission band extending from 740 to 1280 nm with a peak at 960 and a full-width at half-maximum (FWHM) of 194nm [53,54].

Table 2. The ions radius and crystal field strength in Cr^{3+}-doped $RX_3(BO_3)_4$ borate crystals

	Crystal	Dq/B	Crystal	Dq/B	Crystal	Dq/B	Crystal	Dq/B
Ion radius(Å)	(Al) 0.55		Ga 0.62		Cr^{3+} 0.65		Sc^{3+} 0.83	
(Lu) 0.86	$LuAl_3(BO_3)_4$							
(Y) 0.95	$YAl_3(BO_3)_4$	2.50	$YGa_3(BO_3)_4$	/	$YCr_3(BO_3)_4$	/	$YSc_3(BO_3)_4$	2.39
(Gd) 1.11	$GdAl_3(BO_3)_4$	2.52	$GdGa_3(BO_3)_4$	/	$GdCr_3(BO_3)_4$	/	$GdSc_3(BO_3)_4$	2.45
(La) 1.15	$LaAl_3(BO_3)_4$	/					$LaSc_3(BO_3)_4$	2.26

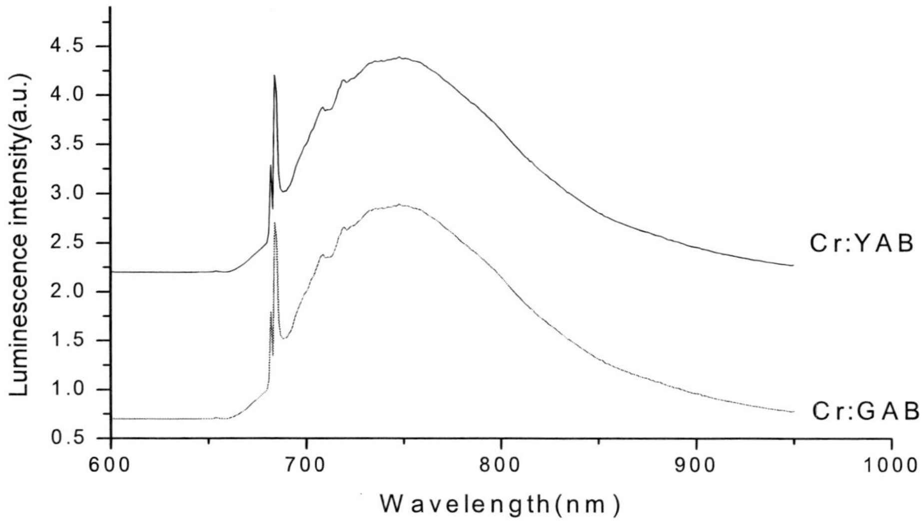

Figure 8. Fluorescence spectra of Cr^{3+}:$YAl_3(BO_3)_4$ and Cr^{3+}:$GdAl_3(BO_3)_4$ crystals [55,56].

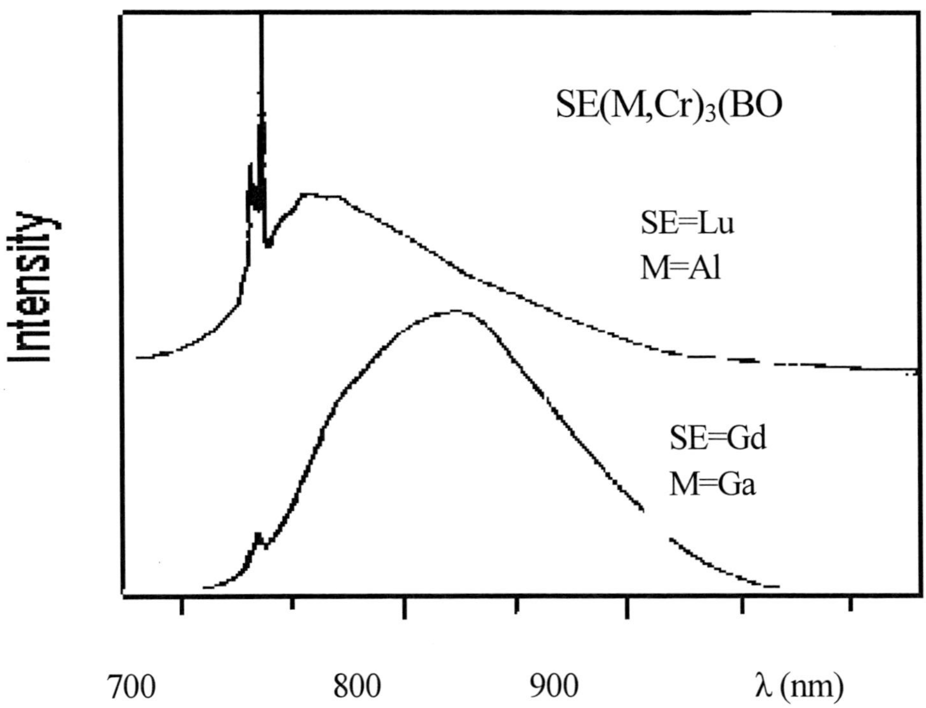

Figure 9. Fluorescence spectra of Cr^{3+}:$LuAl_3(BO_3)_4$ and Cr^{3+}:$GaGd_3(BO_3)_4$ crystals [57].

Up to now, the lasing output dost not achieved in Cr^{3+}-doped $RX_3(BO_3)_4$ crystals. Can the Cr^{3+}-doped $RX_3(BO_3)_4$ crystals become potential tunable solid state laser gain media? Recently the laser characterization of Cr^{3+}:$LaSc_3(BO_3)_4$ crystal has been investigated in our laboratory. The lasing output has been firstly achieved in Cr^{3+}:$LaSc_3(BO_3)_4$ crystal. The laser rod of 0.6at% Cr^{3+}:$LaSc_3(BO_3)_4$ with dimensions of $\phi 3.3\times 11mm^2$ was cut from the as-grown crystal. Laser rod was pumped by a single Xenon flash lamp with different transmittances of the output mirrors at 963 nm, which were 9.8%, 5.8%, 4.6%, 3.3%, 2.4%, respectively. Figure 14 shows the pulse output energy versus the total input energy to the flash lamp with T=3.3%, 5.8%, 9.84%. The results shows that the pulsed laser output energy of 4.05mJ at 963nm for Cr^{3+}:$LaSc_3(BO_3)_4$ crystal with the size of $\phi 3.3\times 11mm^3$ have been obtained, when pumped by a xenon flash-lamp, of which input energy 13.2J and the transmittance of the output mirror T=3.3% at 963nm. The over and slope efficiencies are η_o=0.036% and η_s=0.04%, respectively.

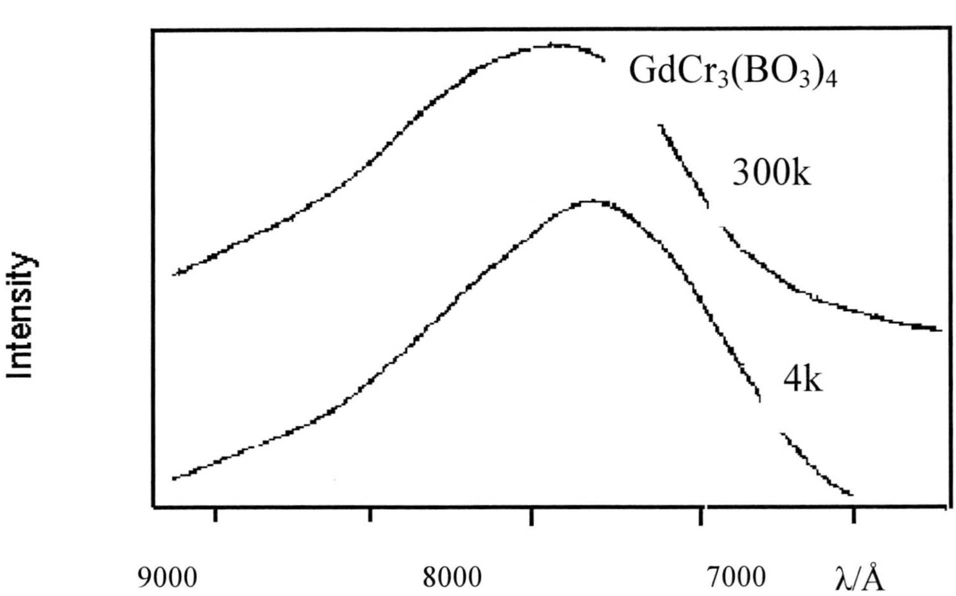

Figure 10. Fluorescence spectra of $GdCr_3(BO_3)_4$ crystals [57].

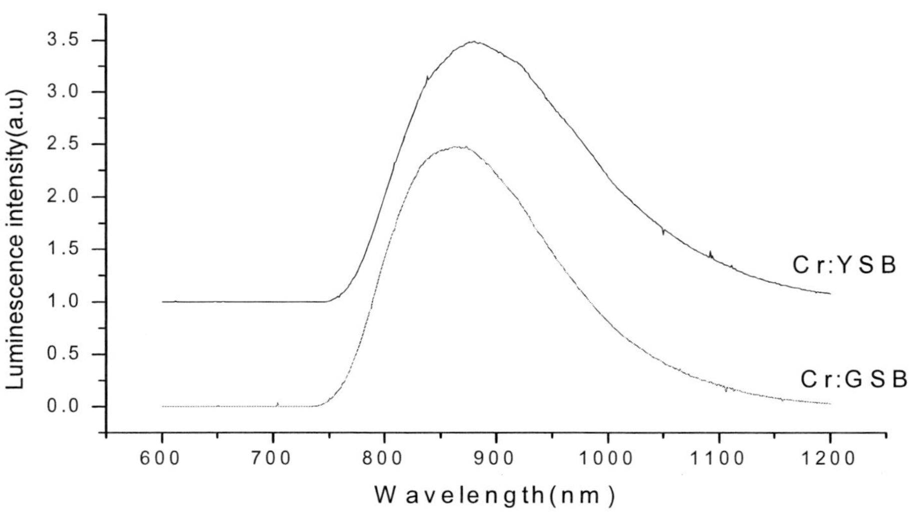

Figure 11. Fluorescence spectra of Cr^{3+}:$YSc_3(BO_3)_4$ and Cr^{3+}:$GdSc_3(BO_3)_4$ crystals [55,56].

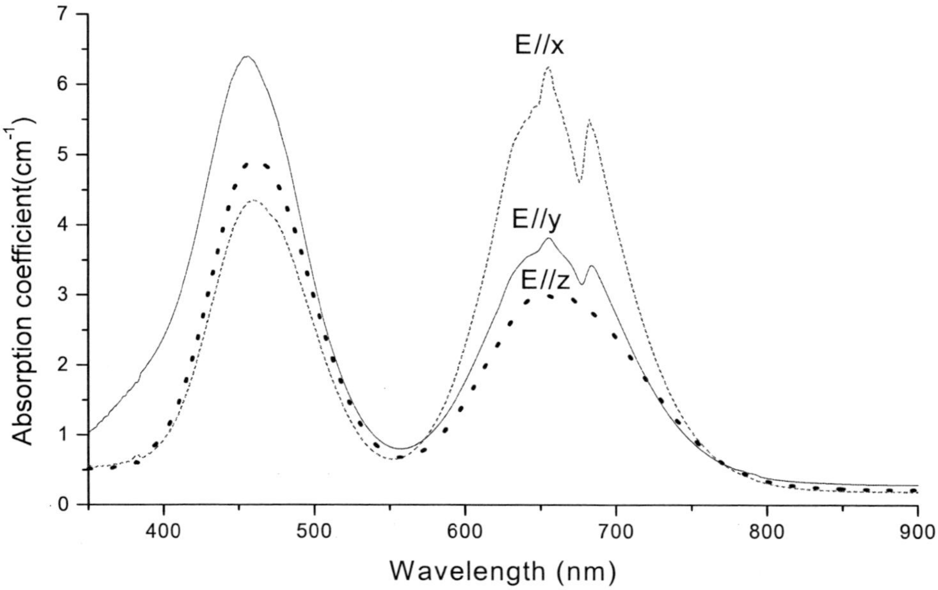

Figure 12. Polarized absorption spectra of Cr^{3+}: $LaSc_3(BO_3)_4$ at room temperature [54].

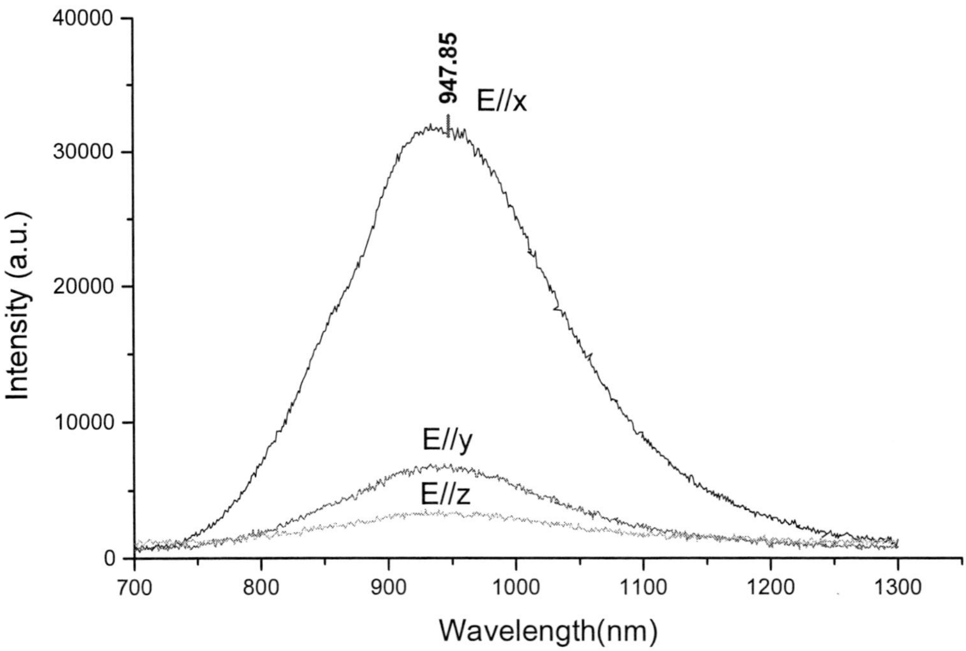

Figure 13. Polarized photoluminescence spectra of Cr^{3+}: $LaSc_3(BO_3)_4$ at room temperature [54].

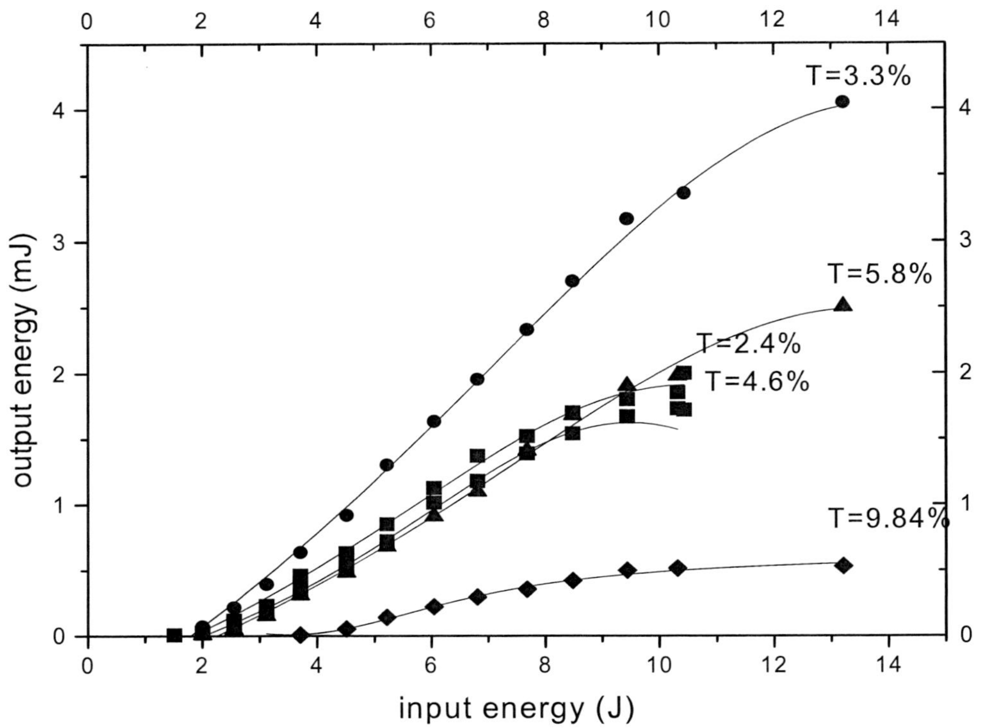

Figure 14. Laser output energy vs input energy as pumped by a flash lamp.

3. Crystal Growth of β-BBO and LBO Crystal

3.1. Phase Diagram with Relation to BaB_2O_4 Crystal

The low temperature phase BaB_2O_4 (β-BBO) is an excellent nonlinear optical crystal. β-BBO was firstly developed in 1985 [58]. It exhibits a number of important properties such as broad transparency range (190-3500nm), large efficient SHG coefficients (d_{eff}=6KDP at 10.6μm), large birefringence (n_0-n_e=0.115 at 10.6μm)[59], and high laser damage threshold (2.3GW/cm^2 for 14ns pulse at 10.6μm) [60]. Since BaB_2O_4 crystal has a phase transition at 925^0C [61], it was only grown by TSSG method. Then, before the growing β-BBO it is important to investigate the phase diagram with relation to β-BBO crystal.

The BaB_2O_4–Na_2O phase diagram was firstly developed in 1981[62]. According to this phase diagram, the β-BBO crystal was firstly grown by TSSG method in 1985[63]. Another phase diagrams with relation to β-BBO crystal were also investigated in this period, such as BaB_2O_4–Li_2O and BaB_2O_4–$Li_2B_2O_4$ [64], BaB_2O_4–K_2O and BaB_2O_4–$K_2B_2O_4$ [65], and BaB_2O_4–SrO and BaB_2O_4–SrB_2O_4 [65]. These phase diagram can be applied to grow the β-BBO crystal except the BaB_2O_4–SrO and BaB_2O_4–SrB_2O_4. In BaB_2O_4–SrO and BaB_2O_4–SrB_2O_4 systems the Sr^{2+} ions can stabilize the high temperature phase (α-BBO) to room temperature, and β-BBO crystal cannot be obtained.

3.2. Crystal Growth of β-BBO Crystal and α-BBO

Up to now, several growth techniques of β-BBO crystal were reported, such as the Czochralski technique [67-69], the LHPG (laser heated pedestal growth) method [70], the TSZM (traveling solvent-zone melting) method [71], the top-seeded solution growth (TSSG) technique [63, 72-77] and the immersion-seeded solution growth (ISSG) technique [78]. In these techniques, the TSSG method is only can be grown the large sized crystals of β-BBO. Previously, there are a number of fluxes used for the growth of β-BBO crystals, such as B_2O_3, BaF_2, $BaCl_2$, Li_2O, Na_2O [63], NaF [66, 67, 74], Na_2O [66, 67], NaCl [78], Na_2O-NaF [79], Na_2O- B_2O_3 [80], Na_2SO_4 and CaF_2[81]. However, these fluxes have different characterizations. For examples, used B_2O_3, Li_2O, $Na_2B_2O_4$, $Na_2B_4O_7$ as the flux, the growth of β-BBO crystals is quite difficult due to the high viscosity of solutions or narrow crystallizing rang. Used NaCl, Na_2SO_4, $BaCl_2$, BaF_2 and CaF_2 as the flux, the growth performance is difficult to be carried out because of the high volatility of solutions [79]. Na_2O is a typical flux for the growth of β-BBO crystals. In the BBO-Na_2O pseudo-binary diagram [62], there is a large growing rang from 925^0C to 755^0C for the growth of β-BBO crystals, which allows to yield large β-BBO crystal. M. Roth et al. determined BBO-NaF pseudo-binary diagram and investigated the viscosity and volatilities of BBO-NaF system [74]. Compared with the BBO-Na_2O system, in BBO-NaF system the slope of liquid curve is more smooth than that in BBO-Na_2O system, which signifies that yield of β-BBO crystal in BBO-NaF system would exceed that in BBO-Na_2O system.

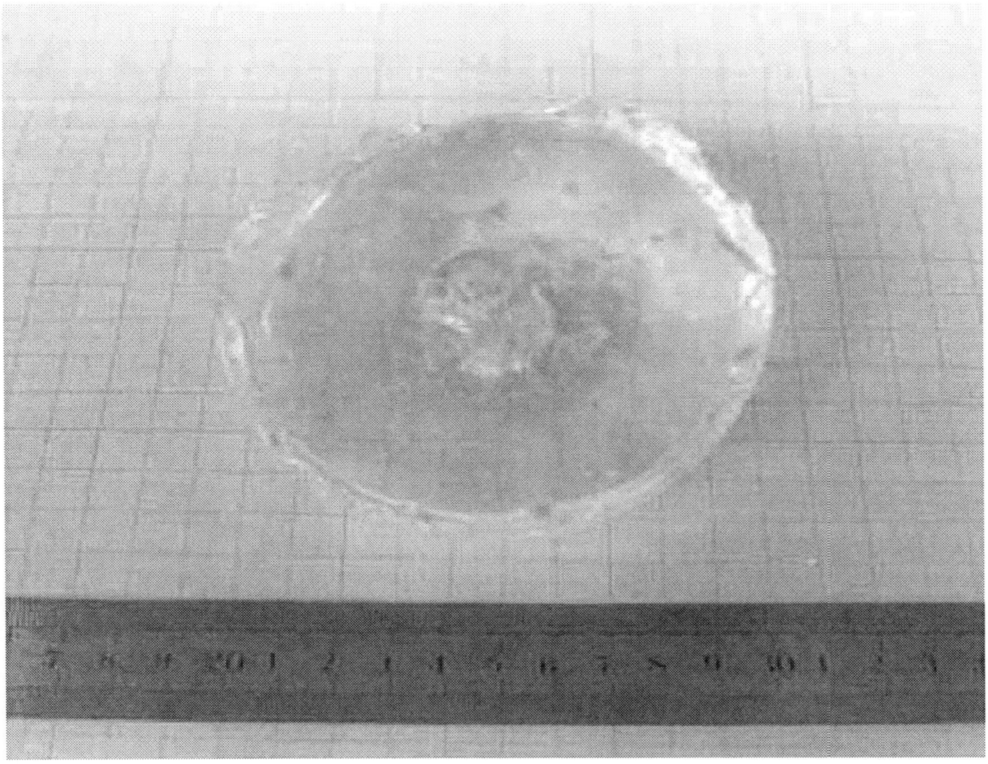

Figure 15. β-BBO crystal with dimensions of 100mm diameter and 40mm length (where grey spots are the solution stuck on surface of the crystal) [82].

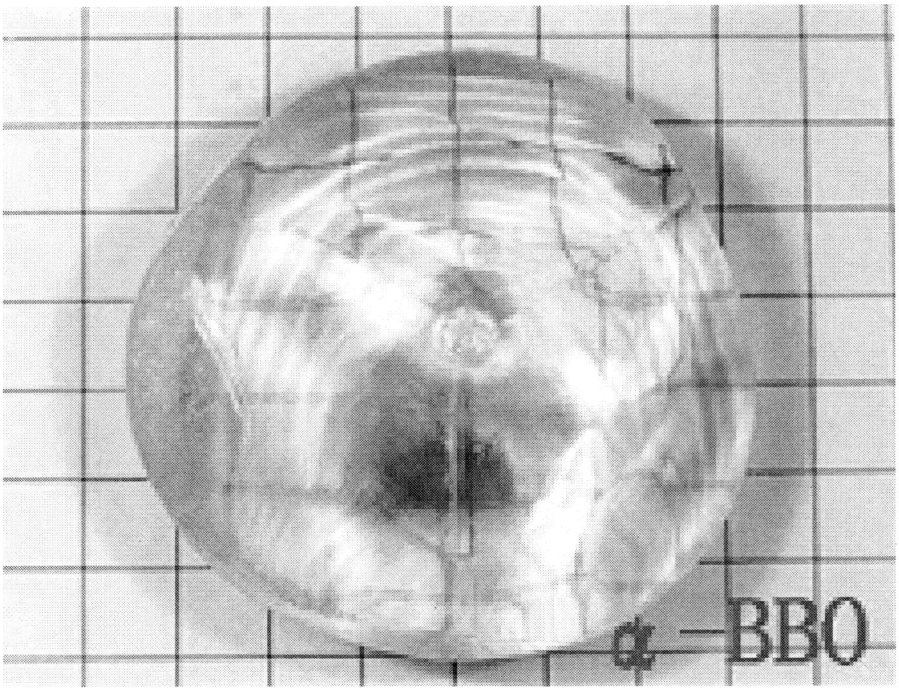

Figure 16. Sr^{3+}-doped BBO crystal grown by the Czochraskli method [83].

Recently, W. Chen and author reported the growth of the β-BBO crystals using TSSG method from the BBO-NaF system. The crystal was grown along [001] orientation. During the growth period, the crystal was slowly cooled at a rate of 0.05-0.1K/h and rotated at a rate of 5-20 rpm. When the growth process was ended, the crystal was pulled up surface of the solution and slowly cooled to room temperature at the rate of 50K/h. The crystal of β-BBO with dimensions up to 100mm diameter and 40mm length, 800g weight and free inclusion was obtained, as shown in Figure 15 [82].

The high temperature phase BaB_2O_4 (α-BBO) is an excellent highly birefringent crystal which is characterized by a wide transmission range of 189-3500nm and n_0=1.6776 and n_e=1.5359 at 533nm. Although α-BBO crystal can be grown by the Czochralski method, the perfect α-BBO crystal is difficult to obtain since it tends to crack due to the α→β phase transition at 925^0C. In fact, the α→β phase transition temperature can be controlled and lowered. Recently, according to the phase diagram of BaB_2O_4–SrO and BaB_2O_4–SrB_2O_4 [65], S. F. Wu and author doped a small amount Sr^{3+} into the α-BBO and grown the Sr^{3+}-doped α-BBO crystal [83]. Sr^{3+}-doped α-BBO crystal was grown by the Czochralski method at a pulling rate of 1.5~2.5mm/h and a rotating rate of 10rpm. An Sr^{3+}-doped α-BBO crystal with $\phi 70\times 30mm^2$ and free cracking has been successfully grown [83], as shown in Figure 16. Why the Sr^{3+} ion can stabilized the α-BBO crystal to room temperature? In 1991 author has described it that the Sr^{3+} ions play a role in the Sr^{3+}-doped α-BBO crystal [84]. When Sr^{3+} ions are doped into BBO crystal formed $Sr_xBa_{1-x}B_2O_4$ solid solution, where Sr^{3+} ions substitute parts of Ba^{2+} ions to decrease the free energy of the system and lower the α→β phase transition temperature. As a result, the α-BBO crystal can be stabilized to room temperature [84]. The results of transmission spectra and birefractive indices show that Sr^{2+} ions do not affect the optical quality and properties of α-BBO crystal [83,85].

3.3. Crystal Growth of LBO

LiB_3O_5 (LBO) crystal was firstly developed by C. T. Chuang et al in 1989 [86]. LBO crystal is one of the most important crystals for nonlinear optical application, such as second harmonic generation, optical parametric and planner waveguiding. The LBO crystals were grown by TSSG method employing boron oxide-rich charges [87-91]. Figure 17 shows a crystal of LBO with dimension $\phi 80\times 40mm^3$ using a boron oxide-rich solution. Recently the growth of LBO crystal has been further developed. V. I. Kosyakov et al investigated the topolopy of liquidus surface in B_2O_3–$Li_2O{\cdot}B_2O_3$–$Li_2O{\cdot}MoO_3$– MoO_3 system, and succeeded to define the region of LBO primary crystallization on the phase diagram and to grow optically good-quality LBO crystal [92]. LBO single crystal with 290g weight and good quality has been successfully grown at a cooling rate of 0.5~1.0^{0}C/day using a seed oriented parallel to [001][92]. Figure 18 shows a LBO single crystal with dimension $49\times 48\times 47mm^3$ grown in our laboratory, which was grown at a cooling rate of 0.5~1.0^{0}c/day using a seed with [100] orientation from B_2O_3–$Li_2O{\cdot}B_2O_3$–$Li_2O{\cdot}MoO_3$– MoO_3 system.

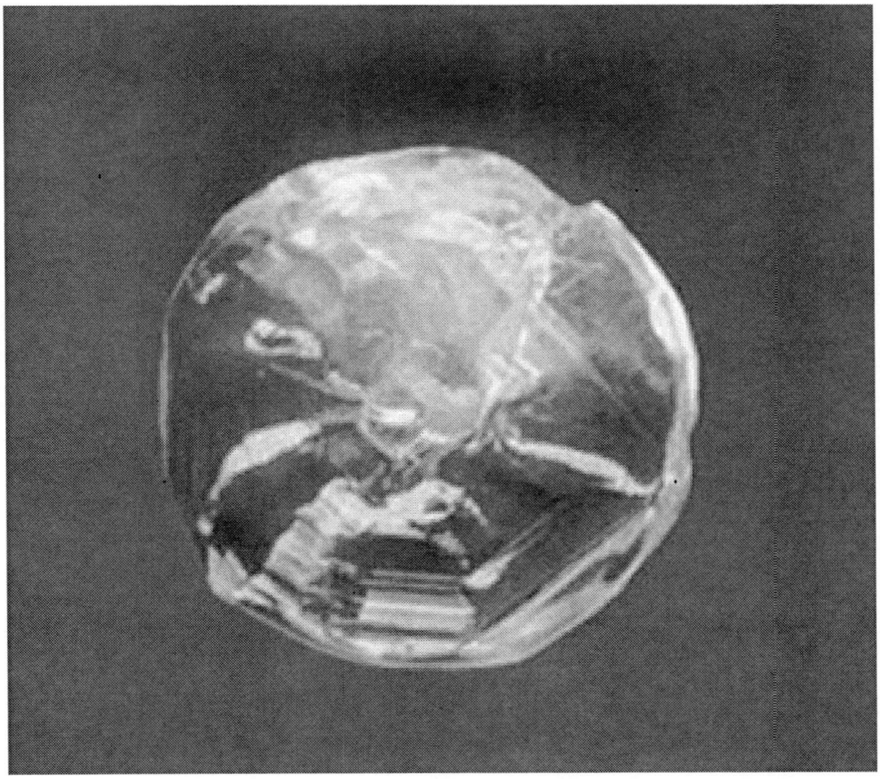

Figure 17. LBO crystal with dimension $\phi 80\times 25mm^3$ grown by A. D. Jiang et al using TSSG method in Fujian Institute of Research on the Structure of Matter, CAS.

Figure 18. LBO crystal with dimension $49\times48\times47mm^3$ grown from B_2O_3–$Li_2O\cdot B_2O_3$–$Li_2O\cdot MoO_3$– MoO_3 system.

4. Crystal Growth of $M_3Ln_2(BO_3)_4$ and $M_3Ln(BO_3)_3$

Another double borate with formula $M_3Ln_2(BO_3)_4$ (M=Ca, Sr Ba; Ln= La~Lu and Y) and $M_3Ln(BO_3)_3$ (M= Ba, Sr; Ln=La~Lu, Y, Sc) were recently reported as a new laser gain medium[93-100]. These borate crystals are different from the $RX_3(BO_3)_4$, the most of $M_3Ln_2(BO_3)_4$ and $M_3Ln(BO_3)_3$ crystals can be grown by the Czochralski method. Some of these crystals were successfully grown by the Czochralski method., as shown in Figures 19-22 [101-104]. The pulling rate was 1~2mm/h. The rotating rate was 10~30 rpm.

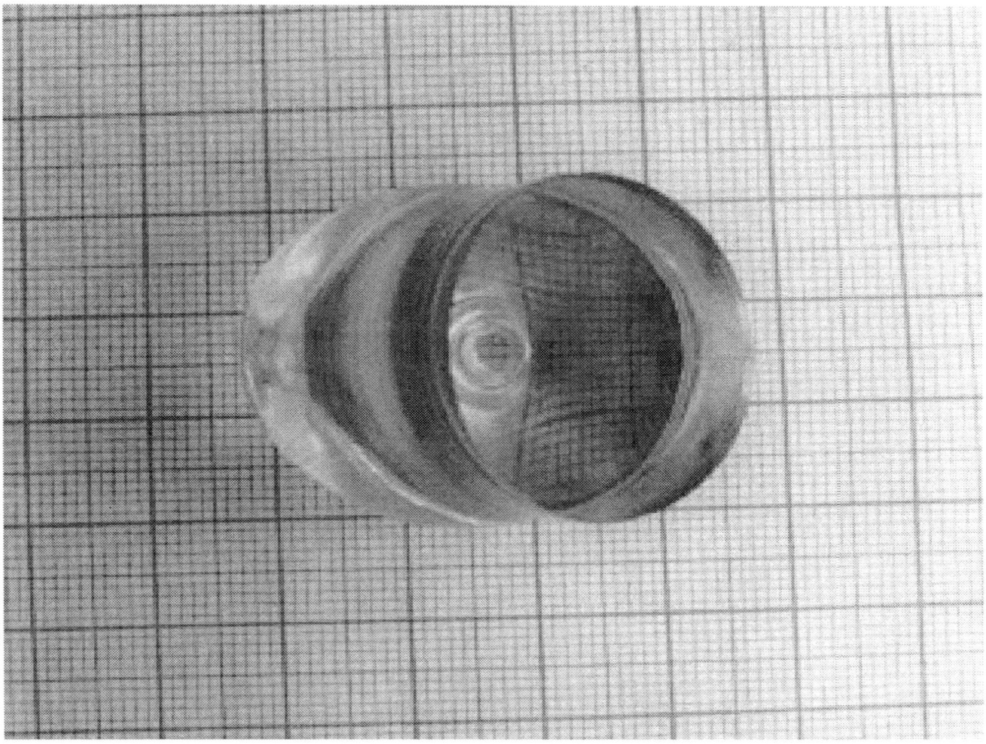

Figure 19. Crystal of Nd^{3+}: $Sr_3Y(BO_3)_3$ grown by Czochralski method [101].

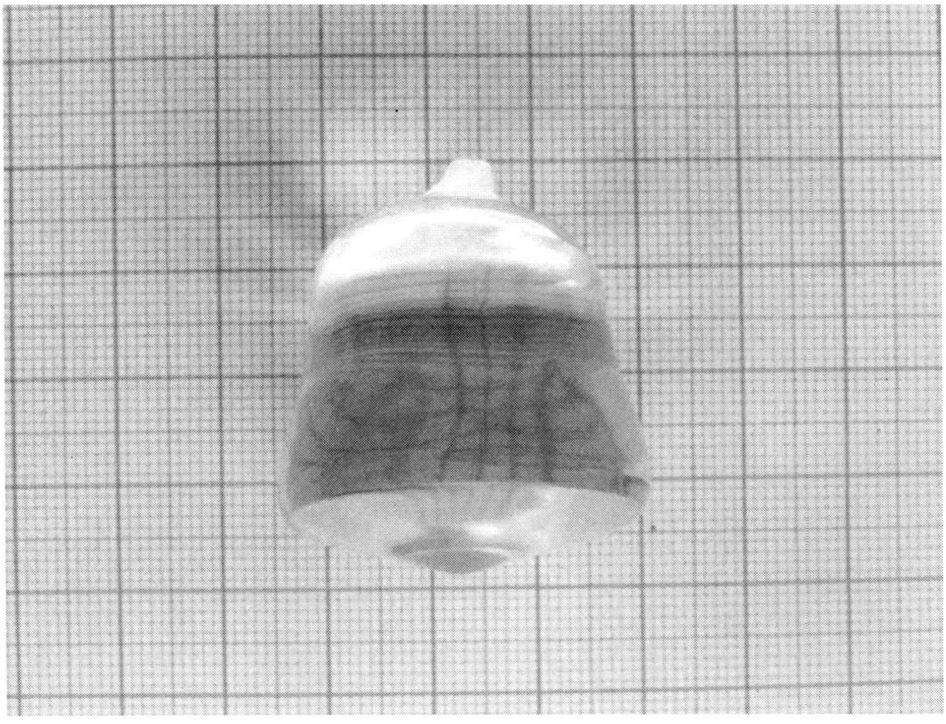

Figure 20. Yb^{3+}:$Sr_3Gd(BO_3)_3$ crystal grown by Czochralski method [102].

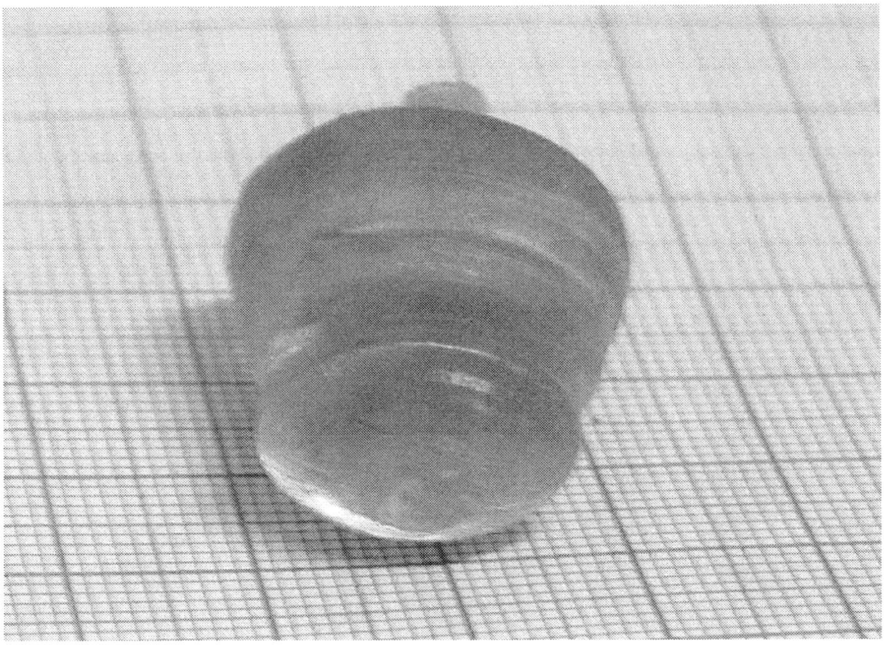

Figure 21. Nd^{3+}:$Sr_3Y_2(BO_3)_4$ crystal grown by Czochralski method [103].

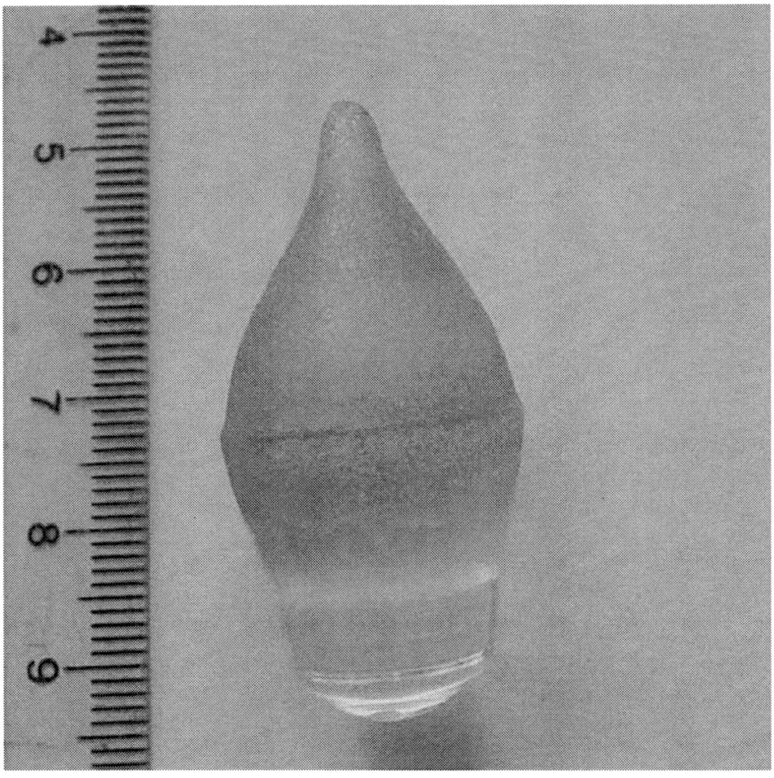

Figure 22. Er^{3+}/Yb^{3+}: $Ca_3Y_2(BO_3)_4$ crystal grown by Czochralski method [104].

5. Summary

In past three decades, there are a number of publications on various $RX_3(BO_3)_4$ crystals doped Nd^{3+}, Cr^{3+}, Yb^{3+} and other rare earth ion. In particular, Nd^{3+}: $YAl_3(BO_3)_4$ and Nd^{3+}: $YAl_3(BO_3)_4$ crystals have been become self-frequency doubling laser crystal. The Nd^{3+}: $LaSc_3(BO_3)_4$ crystal has been became a laser crystal with high efficiency for diode laser pumping [105]. The growth technoligic of these crystals has been greatly improved. The $K_2Mo_3O_{10}$ flux plays an important role in TSSG method, since the $K_2Mo_3O_{10}$ and $K_2Mo_3O_{10}/B_2O_3$ fluxes are useful to improve the quality and size of $RX_3(BO_3)_4$ crystals. The β-BBO and LBO crystals are two of the most important crystals for nonlinear optical application and are widely applied. The β-BBO and LBO crystals have become the commercial nonlinear optical crystal. In past two decades, there are a number of publications on various β-BBO and LBO crystal growth. The application of new system B_2O_3–$Li_2O{\cdot}B_2O_3$–$Li_2O{\cdot}MoO_3$– MoO_3 for LBO and NaF for β-BBO would further improve the grown size and quality of β-BBO and LBO crystals.

Acknowledgements

Author is indebted to Prof. A. D. Jiang for him supporting photographs (see Figure 9 and Figure 25). Some works in this article were supported by the National Natural Science Foundation of China (50272066 and 60378031), Natural Science Foundation of Fujian Province (E9810015) and Project of Science and Technology of Fujian Province (2001H107), respectively.

References

[1] A. Ballman, *Amer. Mineral.* **47**(1962)1380-1383.
[2] D. Mills. *Inorg. Chem.* **1**(1962)960.
[3] H. H-P, Hong, K. Dwight, *Mat. Res. Bull.* **9**(1974)1661.
[4] O. Jarchow, F. Lutz, K. H. Klaska, *Z. Krist.* **149**(1981)162.
[5] F. Lutz, G, Huber, *J. Crystal Growth*, **52**(1981)646.
[6] W. Zwiker, S. Colak, *AD-A119457(US)* (1983).
[7] E. L. Belokoneva, T. I. Timchenko, *Kristall.* **28**(1983)1118.
[8] G. F. Wang, M. Y. He, Z. D. Luo, *Mat Res. Bull.* **26**(1991)1085.
[9] E. L. Belokoneva, T. I. Timchenko, *Spv. Phys. Crystallorgr* **28**(1983)658.
[10] G. F. Wang, M. Y. He, W. Z. Chen. Z. B. Lin, S. F. Lu, Q. J. Wu, *Mat. Res. Innovat.* **2**(1999)341.
[11] M. Y. He, G. F. Wang, Z. B. Lin, W. Z. Chen. S. F. Lu, Q. J. Wu, *Mat. Res. Innovat.* **2**(1999)345.

[12] A. V. Goryunov, G. M. Kuzmicheva, B. V. Mukhin, E. V. Zharikov, A. Yu. Ageev, *Zhurnal Neorgnicheskoi Khimii*, **41**(1996)1605.
[13] N. I. Leonyuk, *Izv. Aka. Nauk SSSR, Neorg. Matter.*, **12**(1976)554.
[14] S. F. Akhmetov, G. L. Akhmetova, L. V. S. Kovalenko, N. I. Leoyuk, A. V. Pashkova, *Kristtolografiya*, **23** (1978)230.
[15] A. V. Azizov, N. I. Leonyuk, S. T. I. Timchenko, A. N. V. Belov, *Sov. Phys. Dokpl*, **24**, (1979), 313.
[16] N. I. Leonyuk, A.V. Pashkova, T I.Timchenko, *Sov. Phys. Dokpl*, **24**(1979)233.
[17] A.V. Azizov, N. I. Leonyuk, V. R. Rezvyi, T. I. Timchenko, A. N. V. Belov, *Sov. Phys. Dokpl*, **27**(1982)95.
[18] J. Wang, B. S. Lu, H. F. Pan, M. H. Jiang, *J. Shandong University (Chinese)*, **21**(1986)14.
[19] V. N. Kolov, P. Peshev, *J. Crystal Growth*, **144**(1994)187.
[20] S. Lu, H. F. Pan, *J. Shandong University (Chinese)*, **21**(1986)31.
[21] N. I. Leonyuk, A. V. Azizov, A. N. V. Belov, *Sov. Phys. Dokpl, Sov. Phys. Dokpl*, **27**(1978)374.
[22] A.V. Azizov, N. I. Leonyuk, V. R. Rezvyi, T. I. Timchenko, A. N.V.Belov, *Sov. Phys. Dokpl*, **27**(1982) 95.
[23] N. I. Leonyuk, *Izv. Aka. Nauk SSSR, Neorg. Matter.*, **12**(1976)554.
[24] F. Kellendonk, G. Blasse, *J. Chem. Phys.*, **75**(1981)511.
[25] E. M. Dianov, M.V. Dmitruk, A.Y. Karasik, E. O. Kirpichenkova, V. V. Osiko, V. G. Ostroumov, M. I. Timosheckkin, I. A. Shcherbakov, *Sov. Quantum Electron.*, **10**(1980)1222.
[26] V. I. Chani, K. Shimamura, K. Inoue, T. Fukuda, K.Sugiyama, *J. Crystal Growth*, **132**(1993)173.
[27] S. T. Jung, D. Y Choi, J. K. Kaug, S. T. Chung, *J. Crystal Growth*, **148**(1995)207.
[28] S. T. Jung, J. K. Kang, S. J. Chung, *J. Crystal Growth*, **149**(1995)207.
[29] S. R. Chinn, H.Y-P. Hong, *Opt. Commun.*, **15**(1975)345.
[30] Chen, *J. Crystal Growth*, **89**(1988)295.
[31] T. I. Timchenko, N. I. Leonyuk and G. S. Butzuova, *Soviet Phys. Cryst.*, **25**(1980)515
[32] F. Lutz, M. Ieiss, J. Muller, *J. Crystal Growth*, **47**(1979)130.
[33] F. Lutz, D. Ruppel, M. Leiss, *J. Crystal Growth*, **48**(1980)41.
[34] N. I. Leonyuk, A. V. Pashkova, T. D. L. Semenova, *Izv. Aka. Nauk SSSR, Neorg. Matter.*, **11**(1975)181.
[35] Elwell, H. J. Scheel, *Crystal Growth from High Temperature Solutions,* (Academic Press, New York, 1974), pp86-107.
[36] G. F. Wang, H. G. Gallagher, T. P. J. Han, B. Henderson, *J. Crystal Growth*, **153**(1995)169.
[37] S. T. Jung, D. Y. Choi, J. K. Kang, S. J. Chung, *J. Crystal Growth*, **148**(1995)207.
[38] G. F. Wang, H. G. Gallagher, T. P. J. Han, B. Henderson, *J. Crystal Growth*, **163**(1996)272.
[39] J-P. Meyn, T. Jensen, G. Huber, *IEEE J. Quant. Electron*. **30**(1994)913.

[40] X. F. Long, G. F. Wang, T. P. J. Han, *J. Crystal Growth*, **249**(2003)191.
[41] G. F. Wang, T. P. J. Han, H. G. Gallagher, B. Henderson, *J. Crystal Growth*, **181**(1997)48.
[42] G. F. Wang, Z. B. Lin, Z. S. HU, T. P. J. Han, H. G. Gallagher, J-P. R. Wells, *J. Crystal Growth*, **233**(2001)755.
[43] J. S. Liao, Y. F. Lin, Y. J. Chen, Z. D. Luo, Y. D. Huang, *J. Crystal Growth*, **267**(2004)134.
[44] H. D. Jiang, J. Li, J. Y. Wang, X. B. Hu, H. Liu, B. Teng, C. Q. Zhang, P. Dekker, P. Wang, *J. Crystal Growth*, **233**(2001) 248.
[45] J. S. Liao, Y. F. Lin, Y. J. Chen, Z. D. Luo, Y. D. Huang, *J. Crystal Growth*, **269**(2004)484.
[46] Z. D. Luo, A. D. Jiang, Y. C. Huang, M. W. Qiu, *Chinese Phys. Lett*, **6**(1989)440.
[47] A. A. Filimonov, N. L. Leonyuk, L. B. meissner, T. I. Timchenko, L. S. Rez, *Kristall. Technich*, **9**(1974)63.
[48] L. M. Dorozhkin et al, Sov. Tech. *Phys. Lett*, **7**(1981)555.
[49] S. Lu et al, Chinese *Phys. Lett*, **3**(1986)411.
[50] A. Brenier, C. Tu, M. Qiu, A. Jiang, J. Li, B. Wu, *J. Opt. Soc. Am. B*, **18**(2001)1104.
[51] C. Tu, M. Qiu, Y Huang, X Chen, A. Jiang, Z. Luo, *J. Crystal Growth*, **208**(2001)487.
[52] D. Jaquel, J. Capmany, Z. D. Luo, J. Garefa Solé, *J. Phys: Condens. Matter*, **9**(1997)9715.
[53] X. F. Long, Z. B. Lin, Z. S. Hu, G. F. Wang, T. P. J. Han, *J. Alloys and Compounds*, **347**(2002)52.
[54] X. F. Long, Z. B. Lin, Z. S. Hu, G. F. Wang, *Chem. Phys. Lett.* **392**(192).
[55] G. F. Wang, H. G. Gallagher, T. P. J. Han, B. Henderson, *J. Crystal Growth*, **163**(1996)272.
[56] G. F. Wang, T. P. J. Han, H. G. Gallagher, B. Henderson, *Appl. Phys. Lett*, **67**(1995)3906.
[57] J. P. Meyu, T. Jensen, G. Huber, *IEEE J. Quantum Electrom*. **30**(1994)913.
[58] Chen, B. Wu, A. Jiang. G. You, *Science Sin. B*, **28**(1985)235.
[59] R. C. Eckardt, H. Masuda, Y. X. Fan, R. L. Byer, *IEEE J. Quantum Electron*, 26 (1990) 922.
[60] H. Nakatani, W. R. Bosenberg, L. K. Cheng, C. L. Tang, *Appl. Phys. Lett*, **53**(1988) 2587.
[61] K. H. Hübner, *Neues Jahrb. Mineral. Monatsh*, **111**(1969)325.
[62] Q. Z. Huang J. K. Liang, *Acta Physica Sinica*, **30**(1981)559.
[63] A. Jiang, F. Cheng, Q. Lin, Z. Cheng, Y. Zheng, *J. Crystal Growth*, **79**(1986)963.
[64] Q. Z. Huang, G. F. Wang, J. K. Liang, *Acta Physica Sinica*, **33**(1984)76.
[65] G. F. Wang, Q. Z. Huang, *Acta Physica Sinica*, **34**(1985)562.
[66] G. F. Wang, Q. Z. Huang, J. K. Liang, *Acta Chimica Sinica*, **42** (1984)41.
[67] K.Ito, F.Marumo, Y. Kuwano, *J. Crystal Growth*, **106**(1990)728.
[68] H. Kouta, Y. Kuwano, K. Ito and F. Marumo, *J. Crystal Growth*, **114**(1991)676.
[69] H. Kouta, S. Imoto and Y. Kuwano, *J. Crystal Growth*, **128**(1993)938.
[70] Y. Tang, R. K. Route, R. S. Felgelson, *J. Crystal Growth*, **91**(1988)81.

[71] R. O. Hengel, F. Fischer, *J. Crystal Growth*, **114**(1991)656.
[72] W. R. Bosenberg, R. J. Lane and C. L. Tang, *J. Crystal Growth*, **108**(1991)394.
[73] Perlov, M. Roth, *J. Crystal Growth*, **137**(1994)123.
[74] M. Roth, D. Perlov, *J. Crystal Growth* **169**(1996)734.
[75] A. Jiang, *China Patent No. 85 1 01617.0*, April 1,1985.
[76] Jiang, *China Patent No.ZL92 1 12921.1*, November 4,1992.
[77] H. Hong, T. Zhao, S. Huan, W. Zhong, *J. Synthetic Crystals* **20**(1991)221 (Chinese).
[78] P. F. Bordui, G. D. Calvert and R. Blachman, *J. Crystal Growth*, **129**(1993)371.
[79] Y. S. Oseledchik, V. V. Osadchuk, A. L. Prosvirnin, A. F. Selevich, *J. Crystal Growth*, **131** (1993) 199.
[80] V. Nikolov, P. Peshev, K. Khubanov, *J. Solid State Chemistry*, **97**(1992)36.
[81] Q. Huang, Z. Liang, *J. Crystal Growth*, **97**(1989)720.
[82] W. Chen, A. D. Jiang, G. F. Wang, *J. Crystal Growth*, **256**(2003)383.
[83] S. F. Wu, G. F. Wang, J. L. Xie, X. Q. Wu, Y. F. Zhang, X. Lin, *J. Crystal Growth*, **245**(2002)84.
[84] F. Wang, *J. Inorg. Matter*, **6**(1991)326 (in Chinese).
[85] J. F. Liu, X. M. He, J. Xu, G. Q. Zhou, S. M. Zhou, G. J. Zhao, S. Z. Li, *J. Crystal Growth*, **260**(2004)486.
[86] T. Chen, Y. C. Wu, A. D. Jiang, B C. Wu, G. M. You, R K. Li, S. Lin, *J. Opt. Soc. Am. B*, **6**(1989)616.
[87] S. A. Markgraf, Y. Furukawa, M. Sato, *J. Crystal Growth*, **140**(1994)343.
[88] P. Shumov, V. S. Nikolov, A. T. Nenov, *J. Crystal Growth*, **144**(1994)218.
[89] Bruck, R. J. Raymakers, R. K.Route, R. S. Feigelson, *J. Crystal Growth*, **128**(1993)993
[90] S. A. Guretskii, A. P. Ges, D. J. Zhigunov, A. A. Innatenko, N. A. Kalanda, L.A. Kurnevich, A. M. Luginets, A. S. Milovanov, P. V. Molchan, *J. Crystal Growth*, **156** (1995) 410.
[91] J. W. Kim, C. S. Yoon, H. G. Gallagher, *J. Crystal Growth*, **222**(2001)760.
[92] V. I. Kosyakov, N. A. Pylneva, Z. G. Bazarova, A. M. Yurkin, *Mat. Res. Bull.*, **36**(2001)573.
[93] K. I. Schaffers, T. Alekel III, P. D. Thompson, J. R. Cox, D. A. Keszler, *J. Am. Chem. Soc.*, **112**(1990)7068.
[94] P. D. Thompson, D. A. Keszler, *Chem. Mater.*,**6** (1994) 2005.
[95] J. R. Cox, D. A. Keszler, J. Huang, *Chem. Mater.*, **6** (1994) 2008.
[96] K. I. Schaffers, P. D. Thompson, T. Alekel III, J. R. Cox, D. A. Keszler, *Chem Mater.*,**6**(1994) 2014.
[97] T. N. Khamaganova, N. M. Kuperman, Zh. G. Bazarova, *J. Solid State Chem.*, **145** (1999) 35.
[98] P-H. Haumesser, R. Gaumé, J-M. Benitez, B. Viana, B. Ferrand, G. Aka, A. Vivien. *J. Crystal Growth*, **233** (2001) 233.
[99] Druon, S. Chénais, P. Raybaut, F. Balembois, P. Georges, R. Gaumé, G. Aka, B. Viana, S. Mohr, D. Kopf, *Optics Letters*,**27** (2002) 197.
[100] S. Chénais, F. Druon, F. Balembois, P. Georges, R.Gaumé, P. H. Haumesser, B. Viana, G. P. Aka,D. Vivien, *J. Opt. Soc. Am. B*, **19**(2002) 1083.
[101] J. G. Pan, Z. S. Hu. Z. B. Lin, G. F. Wang, *J. Crystal Growth*, **260**(2004)456.
[102] J. G. Pan, G. F> Wang, *J. Crystal Growth*, **262**(2004)527.
[103] Y. Zhang, Z. B. Lin, Z. S. Hu, G. F. Wagng, *J. Solid State Chem.* **177**(2004) 3184.

[104] B. Wei, Z. S. Hu, Z. B. Lin, L. Z. Zhang, G. F. Wang, *J. Crystal Growth*, **273** (2004) 190.

[105] J-P. Meyn, T. Jensen, G. Huber, *IEEE J. Quantum Electron*. **30**(1994)913.

In: Research Progress in Materials Science
Editors: W. Olsson and F. Lindberg, pp. 25-53

ISBN: 978-1-60456-629-1

Chapter 2

Reactivity of Unstable Chemicals in the Presence of Transition Metals

Mieko Kumasaki
National Institute of Occupational Safety and Health

Abstract

Some hazardous materials show characteristic features in their reaction with transition metals; they are unstabilized, and sometimes decompose when in contact with transition metals. Their chemical structure indicates that they have a high tendency to interact with transition metals since they include electronegative atoms. Transition metal atoms attract lone pairs of electronegative atoms. This tendency leads to the formation of a complex of transition metals and the molecules. However, some complexes, especially those with unstable molecules, can dissociate immediately along with decomposition of unstable chemicals, or they can be unstabilized by heat, friction and other stimulation because bond strength is affected and weakened by electronic condition change through complex formation. The effect sometimes results in a reaction beyond control and explosion. This article provides an overview of current research on the reactivity of transition metals and unstable chemicals, and the decomposition behavior of unstable chemicals interacting with metals. The subject is related through three viewpoints: the stability change caused by the interactions, the reaction features from the viewpoint of heat release, and the interaction features of oxygen and nitrogen with Fe^{3+}.

1. Introduction

Hazardous materials are those materials which represent characteristic features or have potential to pose damages: physical damages by explosion and runaway reaction, or biological damage such as mutation or acute toxicity [1,2]. The damages impact basic reliance on chemical industries even though they have provided demanded chemicals, and contributed to an affluent society.

The physical damage, in particular, can produce widespread as well as long-term consequences. It sometimes triggers biological damage, also. In terms of wide-spread consequence, the chemical plant explosion occurring in 1991 in Japan is taken for example.

The explosion killed and injured personnel working inside the premises, even far from the explosion center. The deadly fragments flew about 100m to break through iron bars on the window. Other fragments flew more than 250m to crash through a ceiling. As for a long-term consequence, destruction of facility allowed chemicals on site to spill into rivers, ocean, soil, and air. The chemicals can remain long term in the environment and may cause a biological impact. For this reason, knowledge of the reactivity of hazardous materials is essential to retain them under control, and to prevent any disaster for the sustainable development of modern society.

One of the characteristic reactions of some hazardous materials is found in their reaction with transition metals [1]. Considering the chemical structures of hazardous materials, most of them include electronegative atoms, especially for unstable and reactive molecules. Electronegative atoms can easily interact with transition metals by donating lone pairs to vacant atomic orbit of transition metals. The interaction results in the formation of a complex: species of a central metal atom or ion surrounded by a set of ligands [3]. A complex consists of coordination linkage whose bond electrons are donated by ligands. Various kinds of complexes are found in nature such as hemoglobin, Vitamin B12, and so on. These complexes are stabile to serve in a body or natural environment when ligands are stable enough.

On the contrary, complex formation has an intensified impact on unstable molecules as ligands. Transition metals provide perturbation to the electronic condition of a molecule, and this can change molecular form and bond strength. Strength of bond affects molecular stability since chemicals generally start decomposition with a break in weak bonds. Therefore unstable molecules can change bond strength and be unstabilized through complex formation.

Furthermore, some complexes dissociate immediately after complex formation, along with rearrangement within ligands or decomposition of ligands. While this kind of function is utilized as catalysts to generate valuable chemicals [4], it sometimes leads a reaction beyond control (runaway reaction) and explosion. Unstable chemicals as ligands decompose easily, and have the potential to lead to explosion because of their fragility.

This article provides an overview of current research on the interaction between transition metals and unstable chemicals. It can be divided into three parts: the stability change caused by interactions, the reaction features from the viewpoint of heat release, and the interaction features of oxygen and nitrogen with Fe^{3+}.

The first part describes cases where the kind of interaction changes sensitivity to heat, friction, electronic spark, and mechanical impact. Tetrazole and triazole (azoles) are subjected to investigation. The chemicals are utilized in pyrotechnic devices as energetic materials: that means they are categorized in reactive materials. By formation of complexes, they change sensitivity as well as energy release behavior.

The second part focuses on heat release behavior of hydrogen peroxide (H_2O_2), hydrazine (NH_2NH_2), and hydroxylamine (NH_2OH) which are the simplest molecules containing electronegative atoms. Simple chemical structure enables us to eliminate structural hindrance to metal-electronegative atom interaction which frequently occurs in a large molecule. Additionally, tracking heat release behavior provides information about reactions since heat generation directly corresponds to the enthalpy difference between reactant and stable products. The results can provide some insight into reactivity difference between nitrogen and oxygen, as well as their individual reactivity.

Contrary to the second part, the third part relates the reaction between NH_2OH and Fe^{3+} affected by steric hindrance. This part focuses on the reaction of substituted NH_2OH by the methyl group. The Steric hindrance by the methyl group can restrict the reaction center in NH_2OH. This part also relates the protected iron ion by ligands, acidity of solutions, and charge distribution on NH_2OH by way of ab initio molecular orbital calculation.

2. The Stability Change Caused by Interactions: The Stability Change of Tetrazole and Triazole through the Interaction with Transition Metal [5]

This section provides an overview of the effect of the interaction on the sensitivity to azoles by way of synthesized azole-metal complexes. Azoles are energetic materials which are frequently used as an ingredient of pyrotechnic devices. In the field of energetic materials, azole complexes with transition metals, however, had less attraction compared to a mixture of azoles and transition metal compounds. For instance, transition metal oxides have been added to propellants and other energetic devices so as to enhance reactivity by releasing oxygen [7,8]. Some oxides achieved better heat conduction and promotion of the electron-transfer process as well as improvement of burning rate.

In that sense, the evaluation of azole-metal complexes started originally from the attempt to enhance the reactivity of azoles by means of effective interaction between azoles and transition metal as catalysts [5]. Since the distance between transition metal and azole becomes shorter in complex, the interaction was expected to be stronger. Through the experiments and analysis, the interaction has been recognized as perturbation on azole's stability.

This section describes the evaluation of 1H-tetrazole-metal complex and 1H-1,2,4-triazole-metal complex in terms of sensitivity (Figure 1). 1H-tetrazole (abbreviated as 1HT) and 1H-1,2,4-triazole (abbreviated as 1HTRI) are basic azoles which can easily coordinate to transition metals through nitrogen atoms. Previous studies reported that azoles exist as bidentate ligands in many azole-metal complexes [8, 9]. Many transition metals are able to form complexes with azoles. This section is concerned only with complexes including copper, apart from other reports which have assessed their sensitivity and reactivity [10-13].

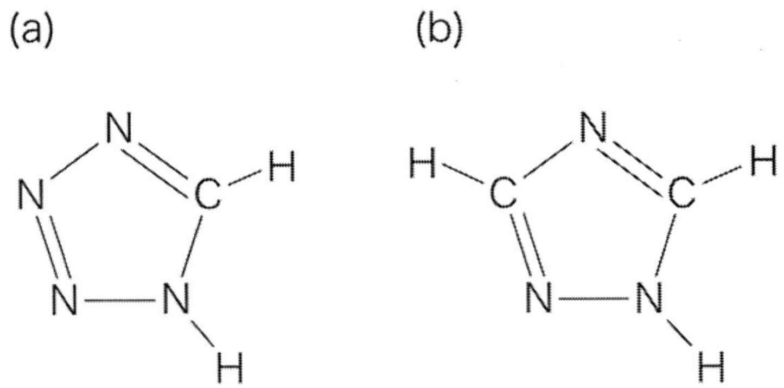

Figure 1. The chemical structures of 1H-tetrazole(a) and 1H-1,2,4-triazole(b).

2.1. Synthesis

On of the sample was prepared by mixing aqueous solutions of 1HT and $Cu(NO_3)_2$ $3H_2O$. Pale blue precipitate immediately generated. The precipitate was subsequently washed with water, methanol, ether, and dried under reduced pressure. Their molecular formula were determined to be $[Cu(CN_4H)_2]H_2O$ (abbreviated 1HTCu) based on an elemental analysis. One proton was liberated from 1HT since the parent molecule has a pKa value of about 4.8, which is similar to that of acetic acid [14]. All attempts to dissolve the complexes in common solvents failed because of poor solubility.

The other complex of 1HTRI was synthesized in the aqueous solution as well as one of 1HT. Purple precipitate was determined as $[Cu(C_2N_3H_3)_2](NO_3)_2$ (abbreviated 1HTRICu) based on elemental analysis. All attempts to dissolve the complexes in common solvents failed because of poor solubility, either. The poor solubility precluded recrystalization.

2.2. Sensitivity Tests to Friction, Mechanical Impact, and Electronic Spark

The drop Hammer Test is a popular method of assessing the sensitivity of explosives to mechanical impact [15]. There exist a few kinds of the tests for mechanical impact developed at Swiss Chemicals Industry, Bundesanstalt für Materialforschung und prüfung (Federal Institute for Materials Research and Testing in English, BAM), the U.S. Bureau of Mines, and Japan Industrial Standards (JIS). In this study, experiments carried out in the manner of JIS. The test sample is sandwiched between two cylinders and placed on anvil in a tester so that 5kg of a dropping hammer free-fall onto it. The experimental results provide sensitivity to mechanical impact as a function of energies calculated from dropping heights.

Based on the up-and-down method, the 50% explosion height and the corresponding energy were determined as is presented in Table 1 [16]. For the complexes, the test was carried out 6 times with a hammer height setting of 50cm. Both complexes did not explode in this test and were categorized as grade 8 according to JIS. The test revealed that complex is more stable to mechanical impact than azoles. Although this test is strongly influenced by physical properties such as grain size and humidity, the interaction between azole and transition metal appear to stabilize azole to mechanical impact.

Friction sensitivity test was carried out in accordance with JIS K 4810 by using of BAM friction test apparatus [17]. Substance is spread onto a porcelain plate, and the ceramic place is dragged under weighed porcelain stick. Sensitivity of substance to friction is evaluated from an amount of load at which one explosion occurs at least out of 6 trials. The load is utilized as indexes for 1/6 explosion point. From the friction sensitivity test, the 1/6 explosion point was equal to be 25.2kgf for 1HT and 24.2-25.2kgf for 1HTRI. On the other hand, 1HTCu and 1HTRI did not explode at the highest load of 36.0kgf. Therefore, they are too insensitive to measure using the BAM friction sensitivity test. These results are summarized in Table 2. This test is strongly influenced by physical properties such as grain size and humidity as well as drop hammer test. However, the azole-metal complexes can be considered to be more stable to friction than azoles.

Table 1. Summary of the results of Drop Hammer Test

Samples	H_{50}[cm]	E_{50}[J]
1HT	12	5.9
1HTRI	3.4	1.7
1HTCu	>50	>24.5
1HTRICu	>50	>24.5

Table 2. Summary of the results of BAM friction sensitivity test

Samples	$M_{1/6}$[kgf]
1HT	25.2
1HTRI	24.2-25.2
1HTCu	>36.0
1HTRICu	>36.0

Meanwhile, they showed different tendency of the sensitivity to electronic spark from other stimulation. Sensitivity to electronic spark was evaluated according to Electrostatic Sensitivity Test in Japan Explosives Society Standard (IV) [18]. The sample was sandwiched between two electrodes connected to condensers and the electrical spark applied by discharge. The sensitivity was estimated by the 50% explosion point using an up-and-down method.

The results of the electric spark ignition are shown in Table 3. The test revealed that the complexes were sensitive compared to the pure azoles. The sensitivity of 1HTCu was close to the mixture of Al and $KClO_4$ ($\log E_{50}$ is -0.76) which is utilized as spark source in fireworks [19, 20]. As for 1HTRICu, the sensitivity was close to the mixture of B and $KClO_3$ which is utilized as igniter for propellants and airbag inflator (0.06).

Sakata reported that the sodium and potassium salts of 1HT were more insensitive to impact, friction and electric spark than 1HT [21]. The copper complex showed the same trend except for the sensitivity to static electricity.

2.3. Thermal Analysis for Sensitivity to Heat

Sealed Cell-Differential Scanning Calorimetry (SC-DSC) was used for surveying behavior under heating conditions. In the analysis, the calorimeter heats sample and reference under the programmed temperature control. The heat release or absorption data as a function of temperature enable to estimate sensitivity of the sample to heat. The sample weight for the investigation was about 1.0mg. The heating rate was 10K/min.

Table 3. Summary of the results of sensitivity test to static electricity

Samples	$\log E_{50}$	E_{50}[mJ]
1HT	0.52	3.29
1HTRI	0.04	1.10
1HTCu	-0.88	0.13
1HTRICu	-0.08	0.84

Figures 2 and 3 show the SC-DSC charts of the azoles, azole-metal complexes and tetrazole-metal nitrate mixtures. Horizontal axis indicates temperature. Upward peaks indicate heat release (exothermic) while downwards do heat absorption (endothermic). The exothermic peaks of the complex stretched sharply compared with the pure azole and azole-metal nitrate mixture. The 1HT in complexes decompose faster than as pure azole. These results confirmed that an azole definitely interacts with a metal and coordination to the metal effects on the decomposition velocity.

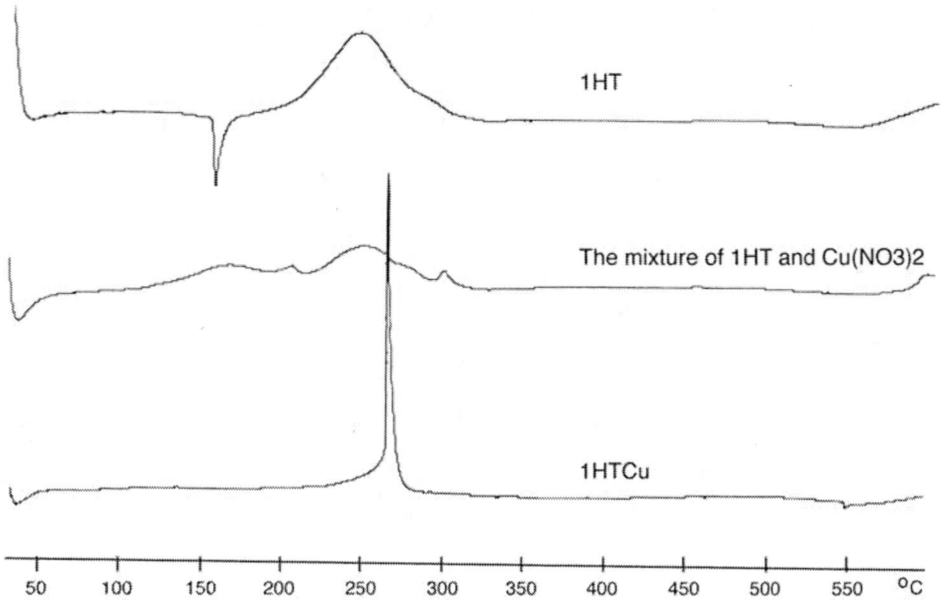

Figure 2.

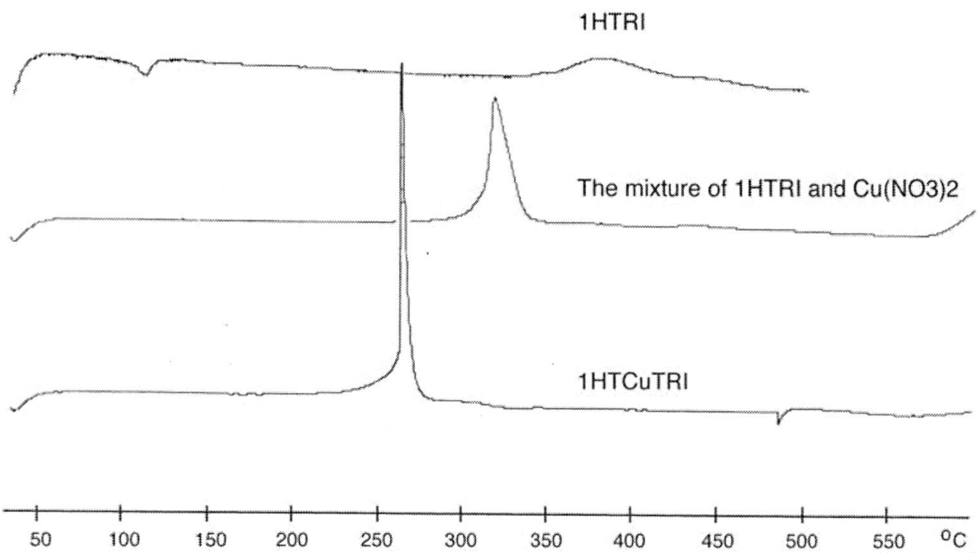

Figure 3.

In figures 2 and 3, the azoles exhibit endothermic behavior corresponding to melting below their decomposition temperatures at which heat release starts. For the complexes, however, the endothermic peaks disappeared. This disappearance is considered to be due to a change in their electronic properties. Neutral molecules link together via a Van der Waals force in the pure azole crystal. In 1HTCu, however, tetrazole negatively charged because of proton liberation. In this way, 1HT molecules and copper link via a Coulombic force in the complex solid because copper positively charged. Contrary to 1HT in complex, 1HTRI can be considered to slightly charge positively in complex because 1HTRI donate electrons to metal to form coordination linkage. The electron-donation also generates Coulombic force. Since the Coulombic force is stronger than the Van der Waals force, 1HTCu and 1HTRICu as ionic crystal are difficult to melt compared molecular crystals of 1HT and 1HTRI.

The values of T_{DSC} and Q_{DSC} are summarized in Table 4. T_{DSC} is an onset temperature of exothermic peak which indicates the temperature explosion or decomposition starts. Q_{DSC} is the heat of reaction per unit mass calculated based on peak integral. Since there are heavy copper atoms (and nitrate anion for 1HTRICu) in the complex, the Q_{DSC} of the complexes should decrease. As for 1HTCu, the heat of reaction per unit mass was naturally smaller than 1HT. However, the heat of reaction per azole molecule calculated based on the elemental analysis results is almost the same as 1HT, but larger than the potassium and sodium salts. Meanwhile, Q_{DSC} of 1HTRICu was larger than that of the 1H-1,2,4-triazole. This is considered to be due to fragmentation easily progresses in 1HTRICu. Williams et al. reported that triazoles are able to form polymers [22]. As a result of complex formation, a high reactivity achieved fragmentation during decomposition and a more exothermic reaction along with generation of small stable molecules.

As for T_{DSC}, the thermal stability of 1HTCu was higher compared to 1HT. In the study on substituted tetrazoles, Ohno concluded that tetrazoles were thermally stabilized by electron donating substituents because of an increase of aromaticity [23]. In 1HTCu, the tetrazole valence was −1 and the electronic state of triazole ring is similar to one in the tetrazole with electron-donating groups which make tetrazole ring slightly anionic.

Figure 4 indicates the optimized structures of the anionic and neutral 1H-tetrazoles by using a molecular orbital calculation at the HF/6-31G* level of theory with GAUSSIAN94 [24]. In the optimized structures, the uniformity of the bond lengths and an increase in the aromaticity were found in the anionic 1H-tetrazole. The calculation results indicate the anionic character of 1H-tetrazole influences the thermal stabilities. Furthermore, the N1-C5 and N3-N4 bonds are shorter in the anionic state than in the neutral state. Considering that the decomposition of 1H-tetrazole starts from breaking these bonds under the stated condition [25], it can be said that the anionic state is more stable than the neutral state.

For 1HTRICu, the triazole ring has a slight cationic character due to electron donation. The cationic state of the molecule was optimized and is shown in Figure 5. In this calculation, triazole valence was +1 so as to simulate slightly charged condition. Consequently, the standard deviation of the average bond length became greater in the cationic state than in the neutral state. This might indicate a decrease in aromaticity. That can explain the stability of 1HTRI decreased.

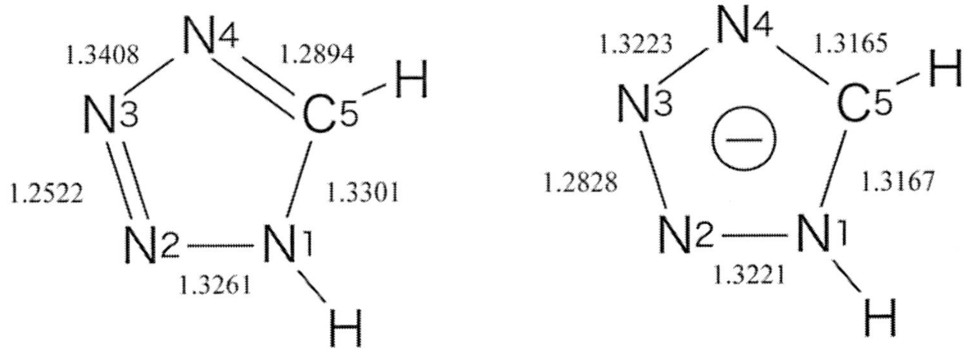

Figure 4. Optimized structures of neutral and negatively charged 1HT.

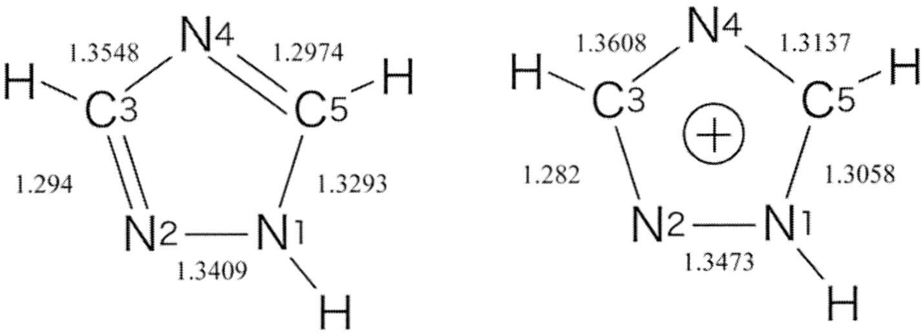

Figure 5. Optimized structures of Neutral and positively charged 1HTRI.

The concept of aromaticity can explain well the change in thermal stability for copper complex. However, some complex of 1HT and 1HTRI behave differently. For instance, complex of nickel and 1HTRI exhibited the similar thermal stability to 1HTRI [10]. To reveal principle for stability of the complexes, analysis will be required for structural information and coordination manner.

Table 4. Summary of the results under the heating condition

Sample	T_{DSC}[°C]	Q_{DSC}[J/g]	Q_{DSC}[kJ/mol]	Q_{DSC}[kJ/azole]
1HT	204	4257	298	
1HTCu	264	2516	552	276
1HTNa[21]	324	1700	157	
1HTK[21]	311	1620	175	
1HTRI	342	1506	104	
1HTRICu	312	1987	647	324

3. The Reaction Features from the Viewpoint of Heat Release: The Reaction between Transition Metal and Simple Molecules with Electronegative Atoms [26-29]

In chemical structures of reactive hazardous materials, nitrogen and oxygen atoms appear most frequently. Both of them have a capability to interact with transition metals by donating electrons used in coordination bond formation. Their affinity for transition metals must be one of the subjects of research to reveal the nature of the reactivity.

Starting research from simpler molecule is common line of reasoning for research on chemistry. This section describes the research on three chemicals with simple molecular structures: hydrazine (NH_2NH_2), hydroxylamine (NH_2OH), and hydrogen peroxide(H_2O_2). They have similar chemical structure composed of two electronegative atoms and hydrogen atoms. All of them exhibit high solubility into water. Furthermore, they have triggered several explosion accidents due to high reactivity. As for NH_2NH_2, it is easy to understand to explode since this compound is utilized in a rocket propellant ingredient. NH_2OH and its salts are widely available commercially and frequently used industrially in chemical synthesis and the manufacturing of semiconductors. NH_2OH, however, killed five people in the state of Pennsylvania in the U.S.A. in 1999, and four people in Japan, in 2000. The accidents occurred by H_2O_2 is too numerous to enumerate. The numerous number of the accident of H_2O_2 reflects its utility in chemical industry, as environmentally-friendly breaching agent and bactericide.

Besides, similar properties are found in some regards: redox behavior. They are available as both of reducing agent and oxidizing agent.

In general, H_2O_2, NH_2OH and NH_2NH_2 act as reducing agent in acidic solutions and as oxidizing agents in neutral or alkaline media. Based on the potential of half-reactions, H_2O_2 serves as oxidizing agent, NH_2NH_2 does as reducing agent, and NH_2OH is considered as the intermediate compound. However, by varying the conditions, H_2O_2 has a capacity as reducing agent in contact with stronger oxidizers [30, 31]

In acid solution:

$2H_2O \rightarrow H_2O_2 + 2H^+ + 2e^-$ E=1.77V as oxidizer (1)

$H_2O_2 \rightarrow O_2 + 2H^+ + 2e^-$ E=0.68 as reducer (2)

. $N_2H_5^+ \rightarrow N_2 + 5H^+ + 4e^-$ E=-1.16 as reducer [31] (3)

$NH_4^+ + H_2O \rightarrow HONH_3^+ + 2H^+ + 2e^-$ E=1.35 as oxidizer (4)

$HONH_3^+ \rightarrow 0.5N_2 + 2H^+ + H_2O + e-$ E=-1.87 as reducer (5)

$2HONH_3^+ \rightarrow N_2O + 6H^+ + H_2O + 4e^-$ E=-0.05 as reducer (6)

$2HONH_3^+ \rightarrow H_2N_2O_2 + 6H^+ + 4e^-$ E=-0.44 as reducer (7)

In basic solution:

$3OH^- \rightarrow HO_2^- + H_2O + 2e^-$ E=0.87 as oxidizer (8)

$N_2H_4 + 4OH^- \rightarrow N_2 + 4H_2O + 4e^-$ E=-1.16 as reducer [31] (9)

$2NH_2OH + 2OH^- \rightarrow N_2 + 4H_2O + 2e^-$ E=-3.04 as reducer (10)

$2NH_2OH + 4OH^- \rightarrow N_2O + 5H_2O + 4e^-$ E=-1.05 as reducer (11)

$2NH_2OH + 6OH^- \rightarrow N_2O_2^{2-} + 6H_2O + 4e-$ E=-0.73 as reducer (12)

Meanwhile, transition metal generally works as oxidizer. They can have several oxidation states and transition easily from higher oxidation state to lower oxidation state companied with oxidizing neighboring reducer.

Furthermore, some transition metals are able to reverse the direction of the transition, and be oxidized under some conditions. For example, Fe^{2+} is easily oxidized into Fe^{3+} in a basic solution. This kind of metal functions as catalysts through the process of exchange electrons between metal and substrates as is shown below. In the process, the metal can supply chain reaction carrier such as radicals and lead hazardous materials to react and decompose even in a small amount existence. In particular, it can trigger a severe accident if a substance behaves as both of oxidizer and reducer.

$$P\ (oxidizer) + M^{n+} \rightarrow P \quad + M^{(n+1)+}$$

$$R \quad (reducer) + M^{(n+1)+} \rightarrow R + M^{n+}$$

Through this section, the difference in reactivity of three compounds is observed in terms of heat release behavior.

3.1. An Experimental Apparatus for Tracking Heat Release Behavior: SuperCRC

Figure 6 shows the schematic view of the Super-CRC. Super-CRC is one of the reaction calorimeters. Among several reaction calorimeters on the market, Super-CRC is characterized by small-scaled (16ml at a maximum), differential system and built-in magnetic stirrer.

In this study, all experiments were carried out in an isothermal mode. A glass vial containing 1ml of substrates (aqueous solution of H_2O_2, NH_2OH and NH_2NH_2) was placed in the heat sink. An amount of 1ml of water was used as a reference. After the baseline got stabilized, 0.1g of the metal aqueous solutions(prepared to give 0.2mmol/g) were injected respectively to vials at once through Teflon tubes instead of needle so as to prevent metal contamination and completely dose a small amount of solutions. Since the Super-CRC is a differential calorimeter, the temperature difference between substrates and iron solution can be canceled as long as the same amount of solution is injected into both reactors. Therefore the heat flow curve is supposed to be independent of the differences. The data collection carried out under the continuous stirring. A heat-flow datum point was obtained every 3 seconds during the course of the reaction. The limitation of a maximum of 15,000 data points defined the maximum test duration as 12.5 hours.

If the reactions were complete within 12.5h, the measurements were stopped after the heat flow curves reached their baseline levels. When the chemicals slowly reacted and continuously generated heat, the measurement continued until the maximum test duration. Three trials were carried out for each sample, and average values of the peak and overall heat of the reaction were calculated.

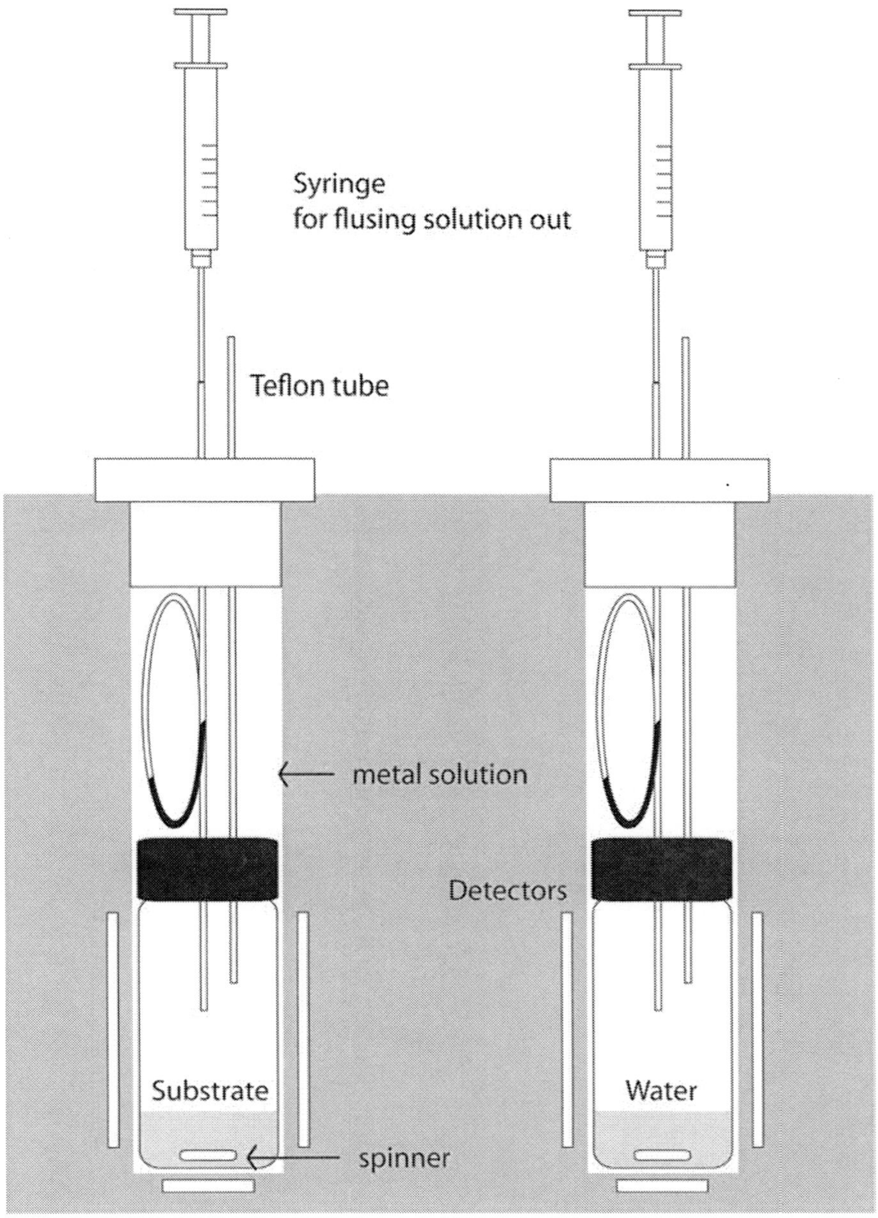

Figure 6. Schematic view of the Super-CRC.

3.2. Reactivity of H_2O_2 and NH_2NH_2 with Fe^{3+}

Figures 7, 8, and 9 show the reaction at 35°C with three Fe^{3+} compounds, $Fe(NO_3)_3$, $Fe_2(SO_4)_3$ and $FeCl_3$. The concentration of H_2O_2 and NH_2NH_2 is 30wt.%. The mixing with Fe^{3+} aqueous solution showed moderate heat release for NH_2NH_2 and completed to give dark precipitate regardless of the different anions. It remained unconfirmed whether any reduced iron species generated such as oxides of Fe^{2+}, complex like $[Fe(NH_2NH_2)_2Cl_2]$ $2H_2O$, $[Fe(NH_2NH_2)_2SO_4]$ H_2O, $[Fe(NH_2NH_2)_2(NO_3)_2]$ H_2O reported in previous study [32], and so forth. The precipitation, however, appeared $Fe(OH)_3$ due to the basicity of the NH_2NH_2 aqueous solution. Even under the reductive condition due to NH_2NH_2, it is possible that Fe persist in $Fe(OH)_3$ since basic condition lets reduction of Fe^{3+} more difficult rather than acidic one [33].

Meanwhile, the reactivity of H_2O_2 exhibited quite different feature. H_2O_2 released heat immediately after injection and gives transparent solution without precipitation. Considering the concentration of H_2O_2 ($8.8x10^{-3}$mol in a vial) and Fe(III)($2x10^{-5}$mol), Fe(III) species works as catalyst for H_2O_2.

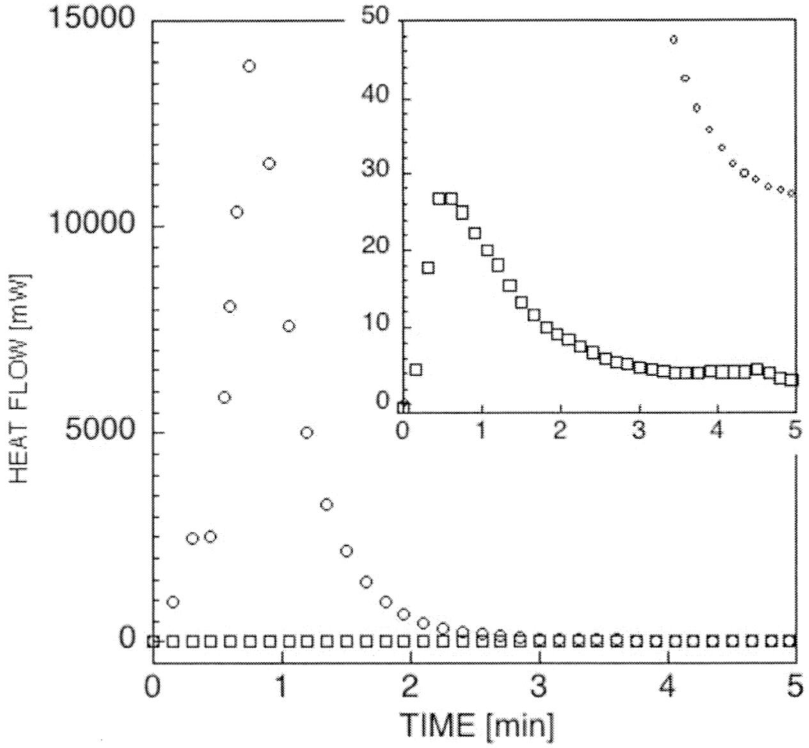

Figure 7. Heat release behavior in the reaction of H202 (circle) and NH2NH2 (square) with FeCI3. inset: the behavior of NH2NH2.

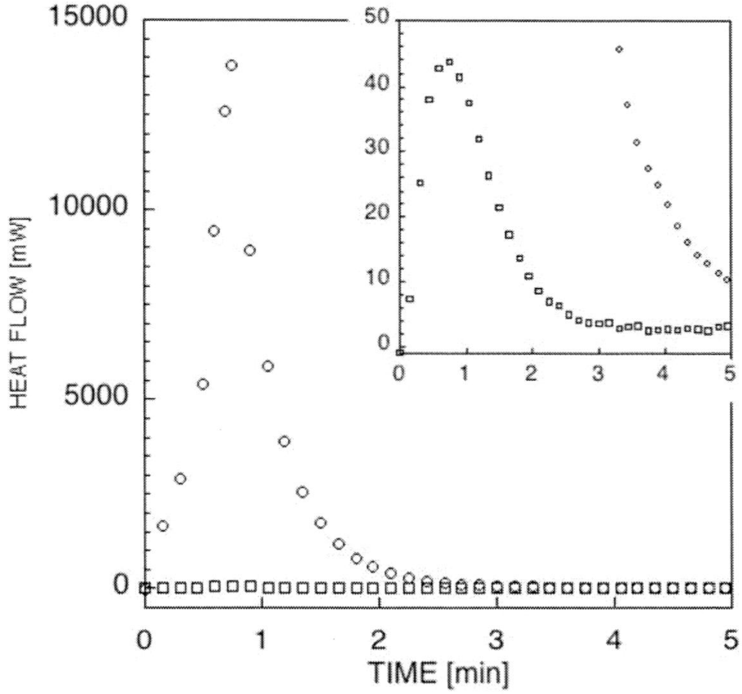

Figure 8. Heat release behavior in the reaction of H202 (circle) and NH2NH2 (square) with Fe(NO3)3. inset: the behavior of NH2NH2.

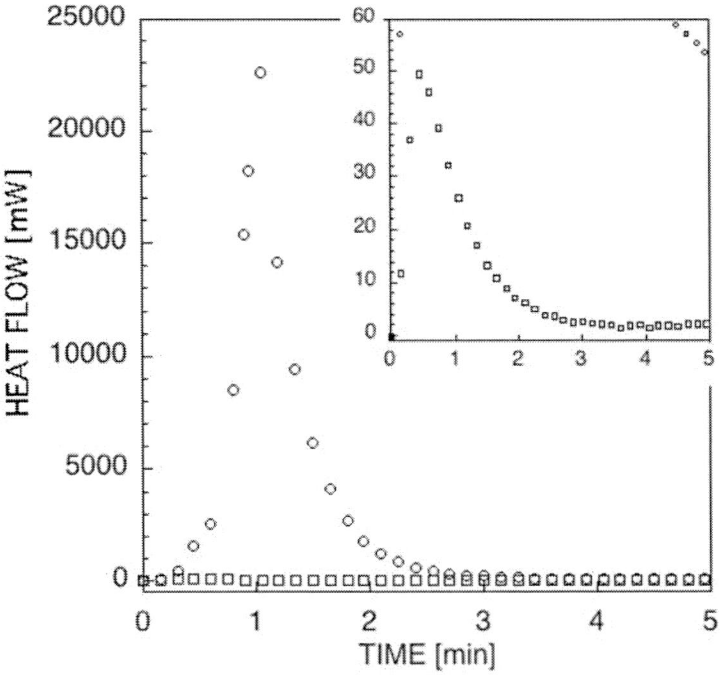

Figure 9. Heat release behavior in the reaction of H202(cicrle) and NH2NH (square) with Fe2(SO4)3/ inset: the behavior of NH2NH2.

The mechanism of the reaction between Fe(III) and H_2O_2 has not been focused intensively in spite of numerous researches on the reaction H_2O_2 and Fe(II) referred as the "Fenton reaction" [34, 35]. Among the research on Fe(III), Kremer mentioned the reaction of Fe^{3+} associated with oxidation of Fe^{2+} in a report. The idea that Fe^{2+} ions participate in the Fe^{3+} ion reaction, however, has not been proved [36] while the main process of the reaction considered as the fission of the O-O bond within complex with Fe^{3+} ion.

3.3. Reactivity of NH_2OH with Transition Metals

NH_2OH is intermediate compounds of H_2O_2 and NH_2NH_2. It has capability to oxidize or reduce depending on counterpart compounds. Reactivity with various kind of transition metal is overviewed both in basic and acidic conditions by way of calorimetric measurements [26,27].

An aqueous solution of NH_2OH (abbreviated as HA, basic solution), NH_2OH HCl (abbreviated as HACl, acidic solution), were used, and carefully prepared to give 2mol/l (about 6.4w.t.%). Solutions of the transition metals were prepared to give 0.2mmol/g of solutions by dissolving $Fe(NH_4)(SO_4)_2$ $12H_2O(Fe^{3+})$, $Cr(NH_4)(SO_4)_2$ $12H_2O(Cr^{3+})$, $K_2Cr_2O_7(Cr^{6+})$, $KMnO_7(Mn^{7+})$, $Co(NH_4)_2(SO_4)_2$ $6H_2O(Co^{2+})$, $[Co(NH_4)_6]Cl_3(Co^{3+})$, and $Cu(NH_4)_2(SO_4)$ $6H_2O(Cu^{2+})$.

UV-Vis absorption spectra were collected before and after the reactions in order to follow the redox reaction from the aspect of the valence of metals. In case of precipitation, fine precipitate leads Tyndall phenomenon and light scattering. That inhibited collection of UV-Vis absorption spectra.

3.3.1. Reactivity of NH_2OH with Fe^{3+}

Figure 10 shows a view of the exothermic behavior of HA and HACl caused by Fe^{3+}. The average overall heat of the reaction was 3.39J for HACl, and one for HA was more than 237.82J. The reaction did not allow to calculate exact overall heat of the reaction since the reaction released heat to the maximum test duration. HA was consumed during the reaction despite the HA/Fe^{3+} molar ratio of 100/1. Therefore, Fe^{3+} was supposed to act as a catalyst in the decomposition of HA.

After the injection of Fe^{3+}, HA released heat immediately. In the reaction mixture, red-blown precipitate considered as $Fe(OH)_3$ and the evolution of gas was observed. The gas was recognized as NH_3 from its odor and the results of a previous study [37]. The following reaction schemes have been suggested in previous studies under basic condition [30]:

$$NH_2OH \rightarrow 1/3NH_3 + 1/3N_2 + H_2O$$

In contrast, the reaction of HACl was complete within one hour. The addition of free Fe^{3+} to HACl provided a colorless mixture and no precipitation. The disappearance of the orange color in the Fe^{3+} meant that HACl had turned Fe^{3+} into Fe^{2+}. No further reaction due to Fe^{2+} was supposed to occur in an acidic condition.

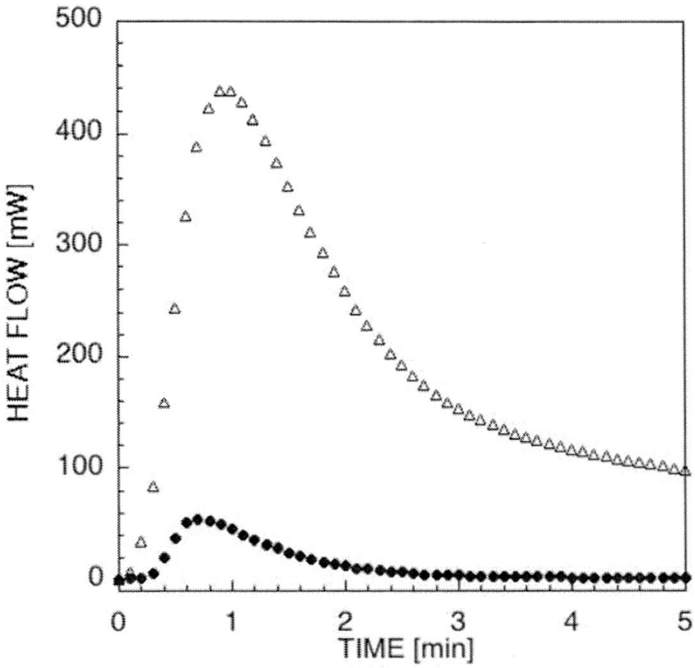

Figure 10. Heat release behavior in the reaction of HA (white triangle) and HACI (black circle) with Fe(Nh4)(SO4).

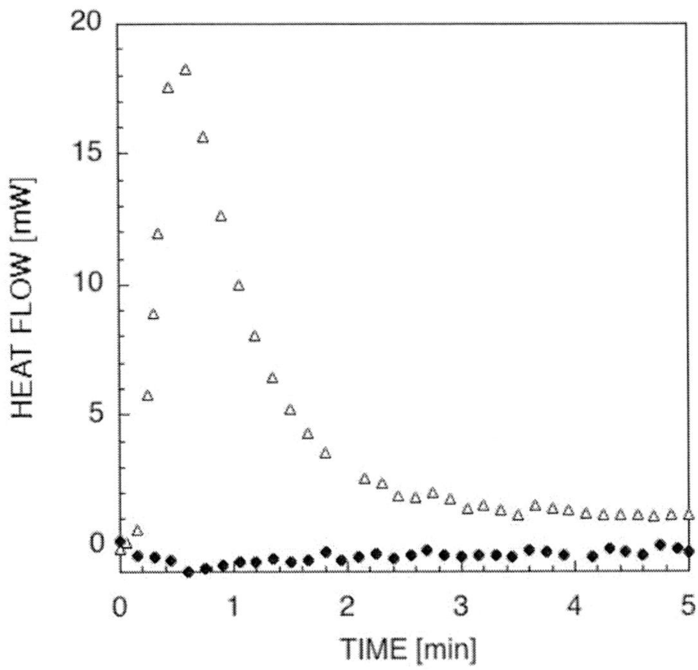

Figure 11. Heat release behavior in the reaction of HA (triangle) and HACI (black circle) with Cr3+.

3.3.2. Reactions with Cr^{3+} and Cr^{6+}

Cr^{6+} has the highest oxidation state of any chromium, and is a strong oxidizing reagent. On the other hand, Cr^{3+} has a lower oxidation state and is less oxidizing. The redox behavior of HA can be investigated by comparing these two oxidative states of chromium.

The heat flow profiles of the reactions with Cr^{3+} are shown in Figure 11. No heat release was observed for HACl in the reaction with Cr^{3+}. The UV - Vis spectra exhibited no absorption change, either. The mixture suggested that Cr^{3+} existed as $[Cr(H_2O)_6]^{3+}$ and no redox reaction occurred by the reductive HACl. Meanwhile HA exothermically reacted with Cr^{3+}. The reaction produced a sage-green precipitate, and the precipitation changed color to violet in a few hours. The green sage precipitation was $Cr(OH)_3$, which is generated from Cr^{3+} in an alkaline solution. However, the generation mechanism of the violet precipitate has not been currently understood. The precipitation suggested Cr^{3+} is enough to stable to HA species, in other words, HA worked as neither oxidizing nor reducing agent for Cr^{3+} despite acidity of solution.

Figure 12 shows the exothermic behavior caused by Cr^{6+}. In contact with CrO_4^{2-}, HACl bubbles and the yellow CrO_4^{2-} changed to a champagne color; the reactions of HACl completed within 30 minutes. The UV-Vis absorptions of CrO_4^{2-} disappeared (at 257 and 351nm), and new peaks suggesting the presence of $[Cr(H_2O)_6]^{3+}$ appeared at 562nm.

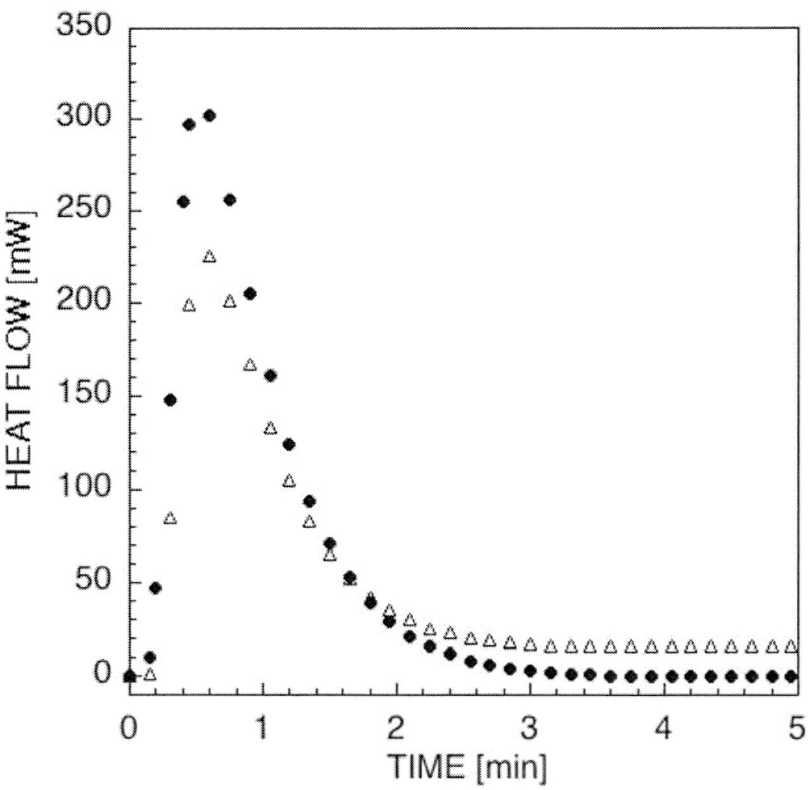

Figure 12. Heat release behavior in the reaction of HA (triangle) and HACI (black circle) with Mn7+.

As far as the reaction with Cr^{6+}, the acidic HACl exhibited a higher reactivity than HA. These results can be explained by the well-known fact that the oxidative ability of Cr^{6+} is higher in an acidic solution than it is in a basic solution. The HA continuously generated heat for more than the maximum test duration, and a violet precipitate was obtained in HA. The origin of the continuous heat generation was unclear, however, transition among several oxidation states of chromium along with HA redox reaction or catalytic ability of surface of the precipitates would be plausible as what was behind.

3.3.3. Reactions with Mn^{7+}

Figure 13 shows the heat flow curves of HA and HACl caused by MnO_4^-. During the mixing, Mn^{7+} exhibited a strongly oxidizing character as well as Cr^{6+}. When the MnO_4^- solutions were injected into the substrates, bubble evolved, and the deep violet color of MnO_4^- immediately disappeared. HA gradually produced a white precipitate. The precipitate was considered to be $Mn(OH)_2$.

HACl completed the reaction within 30 minutes, whereas HA continuously generated heat for longer than the maximum test durations. The UV-Vis absorptions of manganese in HACl were too weak to identify any products. However, $[Mn(H_2O)_6]^{2+}$ was plausible because of its weak UV-Vis absorption and its lowest redox potential among the manganese ions.

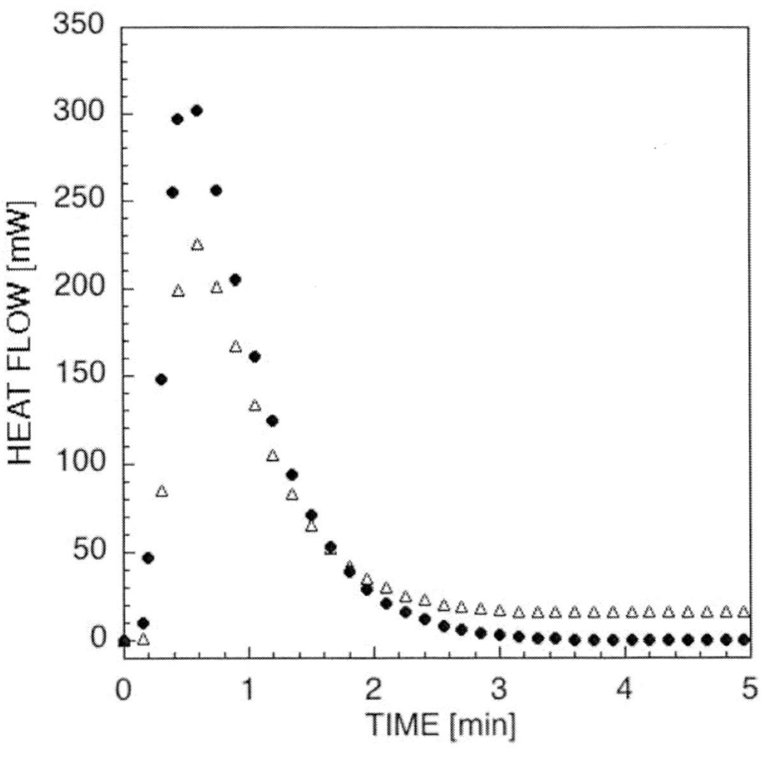

Figure 13. Heat release behavior in the reaction of HA (triangle) and HACI (black circle) with MN7+.

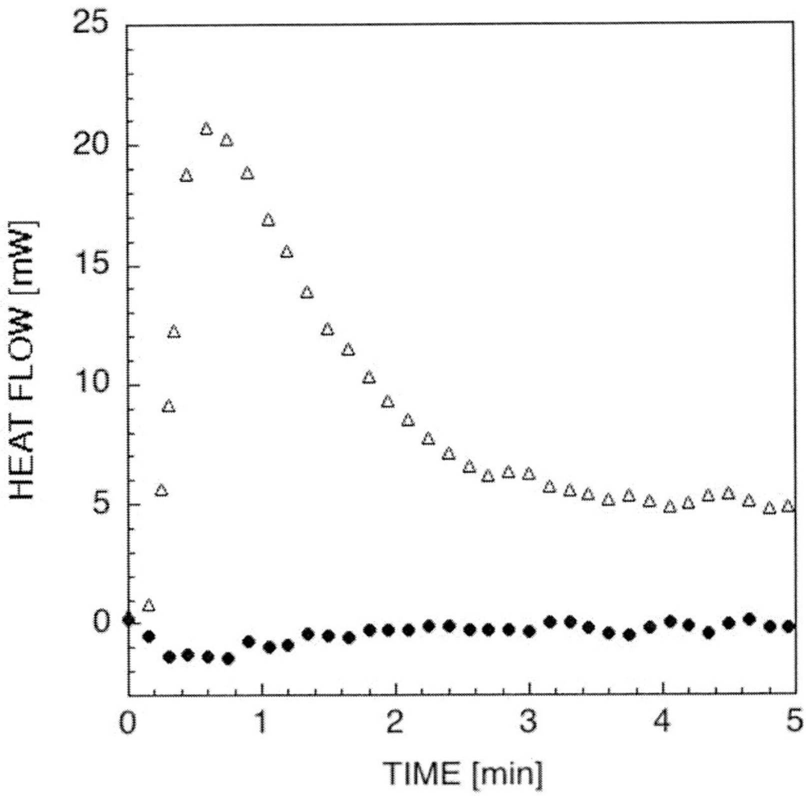

Figure 14. Heat release behavior in the reaction of HA (triangle) and HACI (black circle) with Co2+.

3.3.4. Reactions with Co^{3+} and Co^{2+}

Both of Co^{3+} and Co^{2+} are intermediate oxidation states of cobalt. They can act as either an oxidizer or a reducing agent, depending on the circumstances. The heat flow curves of the substrates caused by Co^{2+} are shown in Figure 14. The UV-Vis spectra showed no absorption change in HACl. No heat release was observed in the mixing, either. From these results, it can be deduced that no redox reaction occurred in the substance. However, Co^{2+} continuously generated heat with HA and produced a brown precipitate.

In the reaction of HACl and Co^{3+}, neither peak in the heat flow curve nor evidence of a redox reaction was observed. The orange color of $[Co(NH_3)_6]^{3+}$ remained in HACl and UV-Vis spectra indicate the existence of the species. Upon mixing with HA, no reaction seemed to occur at the beginning of the monitoring. However, the heat flow curve started to gradually rise following the injection, and it exhibited a peak about 10 hours later (Figure 15). After the induction period, the mixture provided a brown precipitate although it exhibited no change in color during the early hours.

3.3.5. Reactions with Cu^{2+}

Copper has three major oxidation states: Cu^{2+}, Cu^{+}, and Cu^{0}. Since Cu^{2+} is the highest oxidation state, Cu^{2+} is considered liable to undergo reduction despite its low oxidation number.

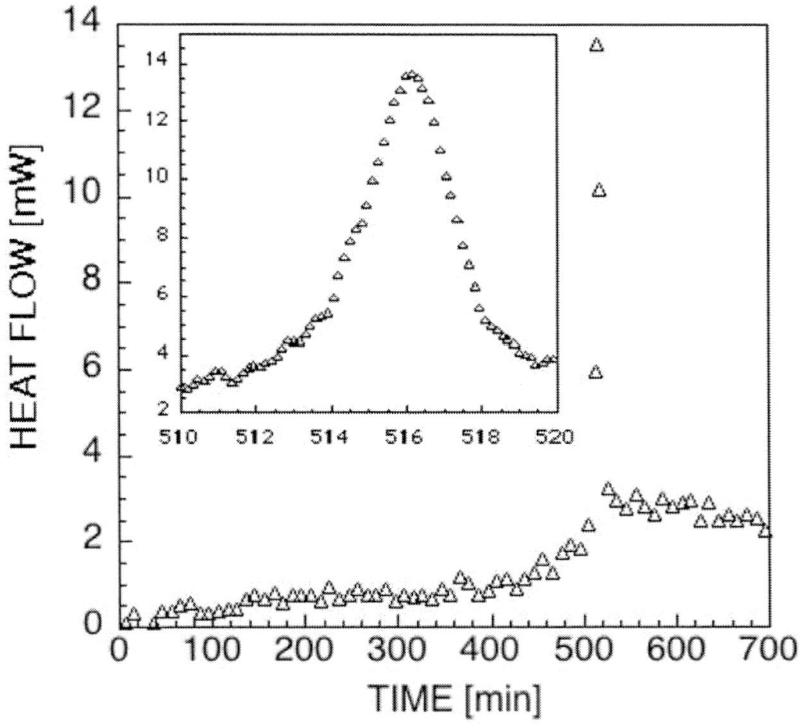

Figure 15. Heat release behavior in the reaction of HA (triangle) with Co3+.

The heat flow curve caused by Cu^{2+} is shown in Figure 16. The blue color of $[Cu(H_2O)_6]^{2+}$, which originated from $Cu(NH_4)_2(SO_4)$ was too visually faint to be determine whether the reactions occurred in HACl. UV-Vis spectroscopy doesn't cover the range in which the absorption peaks of $[Cu(H_2O)_6]^{2+}$ appear [38]. As another plausible product, $[Cu(NH_3)_4]^{2+}$, which has an absorption peak is at 590nm, was not detected, whereas the APTAC experiments showed the complex from 50%HA and copper [37]. In order to generate $[Cu(NH_3)_4]^{2+}$, ammonia-rich conditions are required and the mild condition in this experiment did not allow this complex to be generated. In HA, a yellow green film was observed on the liquid surface, which changed to brown. The appearance of a brown precipitate suggested that Cu_2O formed.

Apart from the spectroscopic results, calorimetric experiments showed the exothermic behavior of HA and HACl. Both substrates continuously generated a heat flow. The maximum heat flow of HA was not high; however, a continuous release of heat resulted in a high overall heat of the reaction.

3.3.6. Comparing Reactivity among the Three Chemicals

H_2O_2, NH_2NH_2, and NH_2OH consist of two electronegative atoms and hydrogen atoms. Despite the similar chemical structure, they exhibited different behavior in contact with Fe^{3+}. H_2O_2 and NH_2OH exhibited high reactivity compared to NH_2NH_2. Fe^{3+} reacted as a catalyst in H_2O_2 and NH_2OH while NH_2NH_2 released small heat along with precipitation. As for precipitation, NH_2OH also generate precipitation. NH_2OH, however, continuously released heat by the reaction on surface of precipitated gel, considered as $Fe(OH)_3$.

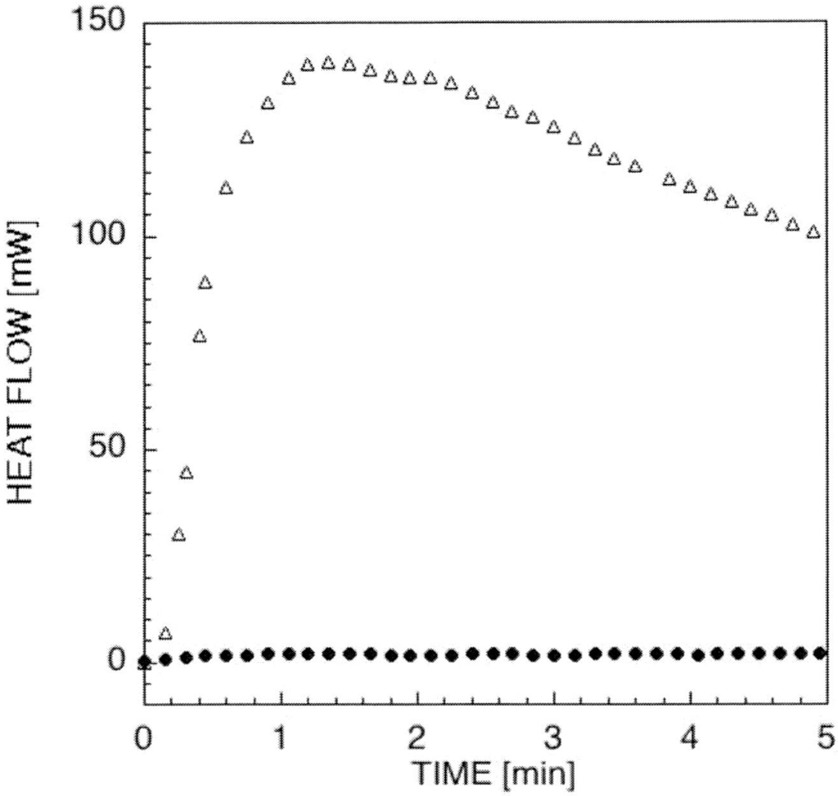

Figure 16. Heat release behavior in the reaction of HA (triangle) and HACI (black circle) with Cu2+.

Table 5 summarize the result of the reaction of NH_2OH and transition metals. All mixture of NH_2OH generate precipitate because of its basicity while its acidic chloride salt prevented precipitation and showed moderate redox reactions in contact with high oxidative metals. NH_2OH seems to maximize the reactivity under the basic condition while hydrogen peroxide works in acidic condition.

The chemicals structures and reactivity implies that oxygen serves important roll in terms of heat releasing reactions. In the reactions, transition metals works catalytic. On the other hand, NH_2NH_2 has a tendency to work as ligand to remain in complex while the others decompose immediately after the contact with metals.

Table 5. Summary of the reactivity of HA and HACl

Metal	HA	HACl
Fe^{3+}	Precipitation	$Fe^{3+} \rightarrow Fe^{2+}$
Cr^{3+}	Precipitation	No reaction
Cr^{6+}	Precipitation	$Cr^{6+} \rightarrow Cr^{3+}$
Mn^{7+}	Precipitation	$Mn^{7+} \rightarrow Mn^{2+}$
Co^{2+}	Precipitation	No reaction
Co^{3+}	Precipitation	No reaction
Cu^{2+}	Precipitation	Not distinguishable

4. The Interaction Features of Oxygen and Nitrogen with Fe^{3+}: The Reaction between Fe^{3+} and Substituted NH_2OH

This section relates some investigation to deepen understanding of the nature of reactivity of NH_2OH by using of structural change. The investigation can be divided into two parts: steric hindrance on Fe^{3+}, and steric hindrance on NH_2OH. Steric hindrance on Fe^{3+} also contributes to hydrophilicity. The previous section related the manner that precipitation decelerated the heat release of NH_2OH. Aqueous solution of NH_2OH is basic and easy to precipitate metal hydroxide. By using suitable ligands for Fe^{3+}, precipitation can be prevented and observed an interaction. Two complexes of iron are focused in this article: $[Fe(CN)_6]^{3-}$ and $Fe(EDTA)^-$.

Additionally, intended steric crowding for NH_2OH enables to restrict reaction center. For example, NH_2OH is expected to reaction with metal on oxygen if methyl group attaches on nitrogen atom. By using of steric hindrance, the affinity of oxygen or nitrogen to transition metal was able to be investigated. At the same time, the acidic condition was controlled by using NaOH since the acidity has influence on the reactivity of NH_2OH shown in the previous section.

4.1. Exothermic Behavior Caused by Coordinated Fe^{3+}

CN- is known as a ligand which strongly coordinates to metals. In the complex, six CN- species occupy coordination sites around Fe^{3+}. The donating atom is C, and Fe^{3+}, C, and N atoms form a line in the six directions [39] (Figure 17). EDTA is the common chelating reagent that exists in the complex as $EDTA^{4-}$ and is associated with the liberation of four H^+. The complex with Fe^{3+} has the highest stability of all EDTA complexes, excluding that with V^{3+}. The previous X-ray study of Fe(EDTA)- revealed that Fe was surrounded by six donating atoms of EDTA and an O atom of H_2O which could easily substitute with substrate [40]; the geometry about Fe^{3+} is pentagonal bipyramidal (Figure 18).

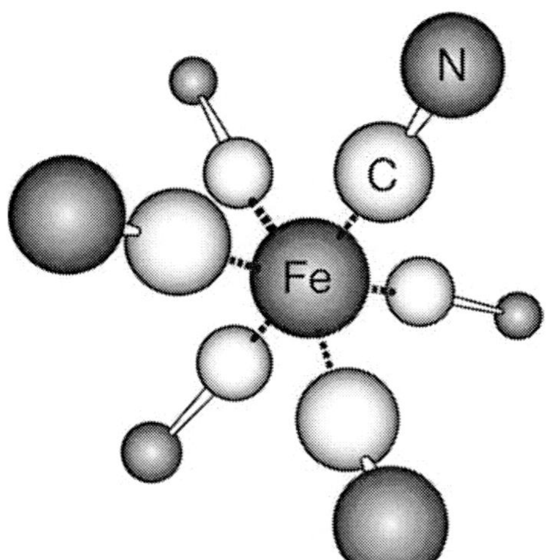

Figure 17. The Structure of [Fe(CN)6]3-.

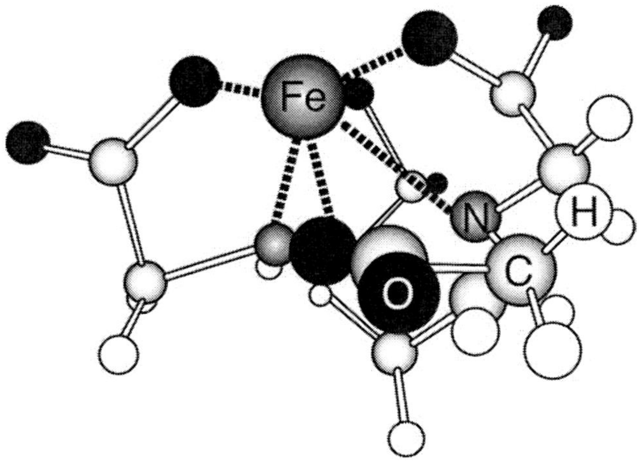

Figure 18.

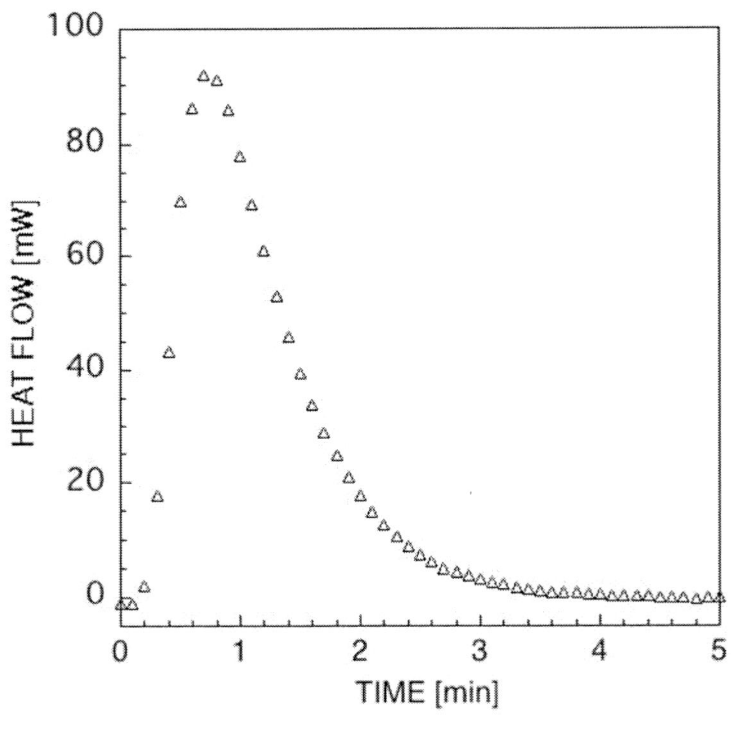

Figure 19. Heat release behavior in the reaction of HA and $Fe(CN)_6$3-.

Figures 19 and 20 show the expanded views of the exothermic behavior caused by $Fe(CN)_6^{3-}$ and $Fe(EDTA)^-$. NH_2OH notably reacted with both Fe^{3+} species. The heat flow of NH_2OH provided by $Fe(EDTA)^-$ rose slowly, compared with that of free Fe^{3+} from $Fe(NH_4)(SO_4)$. However, the maximum height of the curve was 489mW, which was larger than that of free Fe^{3+}. Overall heat of reaction was 421.93J. The gas, including NH_3, also evolved. The heat release caused by $Fe(EDTA)^-$ ended within 100 minutes. The difference to free Fe^{3+} in the heat of the reaction was due to the $EDTA^{4-}$ that dissolved Fe^{3+}/Fe^{2+} species in basic NH_2OH and leads free contact with NH_2OH homogeneously in solution, whereas Fe^{3+}

precipitated as the reaction proceeded and decreased the effective area for the reaction with NH_2OH. $Fe(CN)_6{}^{3-}$ also triggered an exothermic reaction of NH_2OH, however, the maximum heat of reaction(99.11mW) and overall heat of reaction(6.26J) were low compared with that of the other Fe^{3+} species. In the reaction mixture, $Fe(CN)_6{}^{4-}$ was supposed to exist because the yellow color of $Fe(CN)_6{}^{3-}$ disappeared and $[Fe(CN)_6]^{3-}$/ $[Fe(CN)_6]^{4-}$ remain stable under high pH condition. CN- ligand dissolve the Fe^{3+}/Fe^{2+} species even in basic conditions, and also prevent precipitation as well as $EDTA^{4-}$. In case of CN, no catalytic function appeared.

Considering the stability, the redox reaction of $[Fe(CN)_6]^{3-}$ can be described with outer sphere mechanism. The mechanism is an electron transfer in outer complex contact without ligand substitution, or with faster electron transfer than ligand substitution. The reaction manner implies $[Fe(CN)_6]^{3-}$ has no site to contact directly with NH_2OH. Six CN- occupy coordinate sites of Fe and prevent further intrusion of ligands onto inner sphere of Fe. Meanwhile, Fe(EDTA)- keep a coordination site open so as to interact with HA as shown in Figure 20. Not only electric effect but also steric effect has influence on the reaction with NH_2OH

4.2. Exothermic Behavior of Substituted Hydroxylamine

NH_2OH includes oxygen and nitrogen atoms both of which can interact with transition metals. The chemical structure attracts the researchers' interests and raises questions on how it interacts with metals and how the initiation process starts.

Using steric hindrance is one of the ways to gain information about the reactivity. The steric crowding of ligands is expected to decrease associative process with metals. In case of blocking of position on nitrogen atom in NH_2OH, a metal ion should prefer the interaction on oxygen atom, and vice versa. The difference in reactivity can imply the affinity of oxygen and nitrogen atoms to a metal atom. This section also reports reactivity depending on the acidic condition since acidity of aqueous solution appeared to have an effect on the reactivity in previous section.

For the experiment, chemicals were chosen so as to include one or two adequate space-filling substituent groups: *N*-Methylhydroxylamine hydrochloride, $NH(CH_3)OH$ HCl (abbreviated as NHACl indicating methyl group attaching to a nitrogen atom), *O*-Methylhydroxylamine hydrochloride, NH_2OCH_3 HCl (OHACl), *N,N*-Dimethylhydroxylamine hydrochloride, $N(CH_3)_2OH$ HCl (NNHACl), *N,O* - Dimethylhydroxylamine hydrochloride $NH(CH_3)OCH_3$ HCl (NOHACl). Corresponding basic substances were prepared by adding equivalent NaOH aqueous solution. They were $NH(CH_3)OH$ (NHA), NH_2OCH_3 (OHA), $N(CH_3)_2OH$ (NNHA), $NH(CH_3)OCH_3$ (NOHA). Aqueous solutions of substituted NH_2OH were carefully prepared to give 2mol/l.

The heat release behavior at mixing with iron species is obtained with SuperCRC, and summarized in Table 6. $2x10^{-3}$mol of substituted NH_2OH and $2x10^{-5}$mol of Fe^{3+} species were mixed. The result of NH_2OH is added to the summary for reference. All the substituted chemicals exhibited lower reactivity than NH_2OH. As a whole, acidic condition and steric hindrance on oxygen decelerate the interaction with metal. This phenomena is notable in the interaction with Fe(EDTA)-.

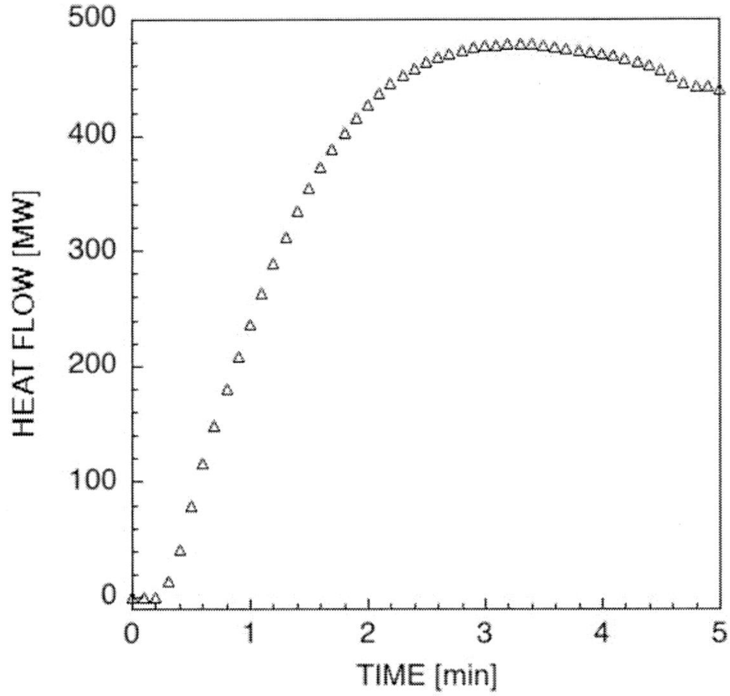

Figure 20. Heat release behavior in the reaction of HA and Fe (EDTA)-.

Ab inito calculation provides some insight from the viewpoint of electronic structure on this matter. Figures 21 and 22 show the optimized structure of NHA and NNHA by ab initio molecular orbital calculations using density functional theory (DFT) in conjunction with B3LYP approach with Gaussian03 [41]. The 6-31G+(d,p) basis set was used. For acidic condition, the calculation indicated molecules protonated on nitrogen are more stable rather than on oxyegn.

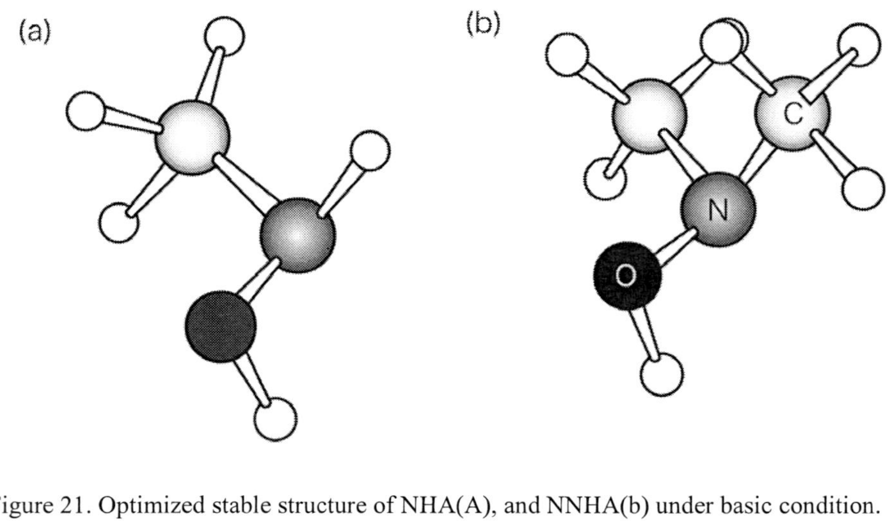

Figure 21. Optimized stable structure of NHA(A), and NNHA(b) under basic condition.

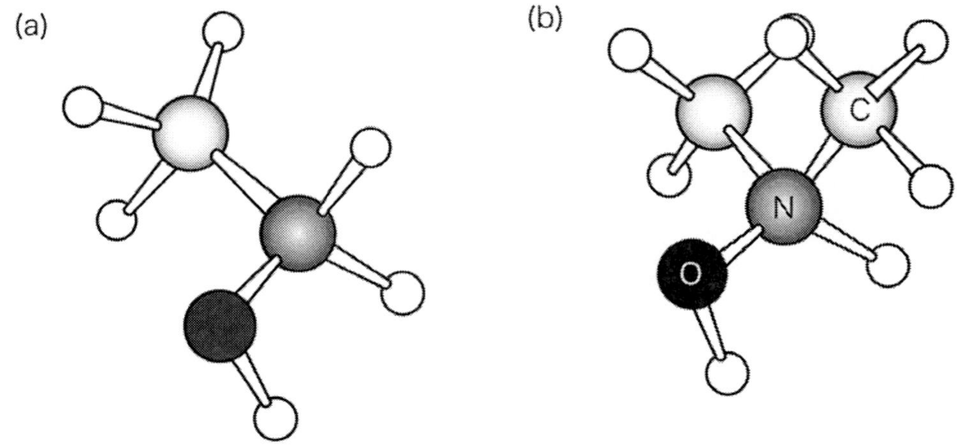

Figure 22. Optimized stable structure of protonated NHA(A), and NNHA(b) under acidic condition.

Mulliken charge analysis showed the difference on charge distribution in basic and acidic condition (Figure 23). Mulliken population analysis allocates charge onto atoms, and provides information about charge distribution in a molecule. The diameter of circles is proportional to amount of calculated charge. In basic condition, more charge lie on oxygen rather than nitrogen. On the contrary, optimized structures in basic condition shows more charge on nitrogen rather than oxygen. This result is consistent with the experimental result that substituted HA reacts more in basic condition with Fe^{3+} since Fe^{3+} prefers electrons in order to form coordination linkage.

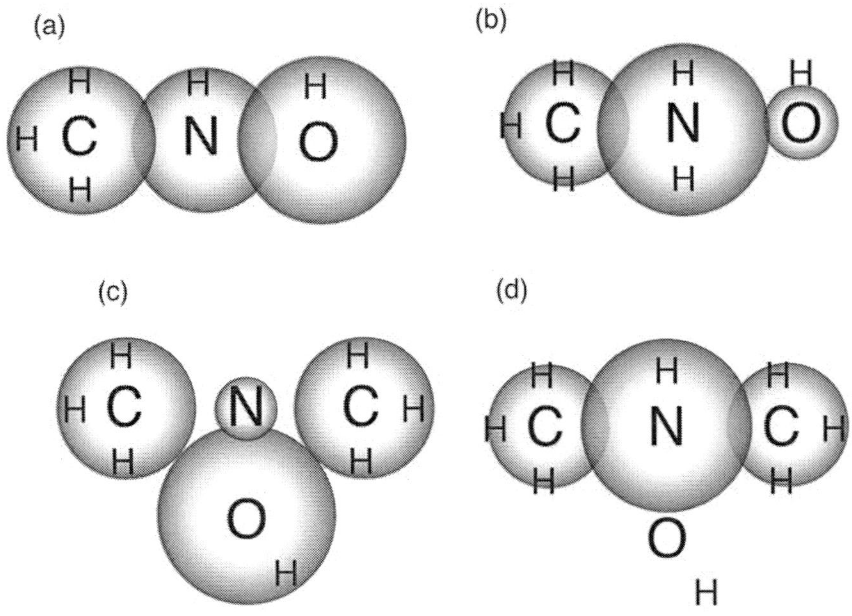

Figure 23. Charge distribution calculated by Muliken population analysis; (a)NHA, (b) protonated NHA, (c) NNHA, and (d) protonated NNHA.

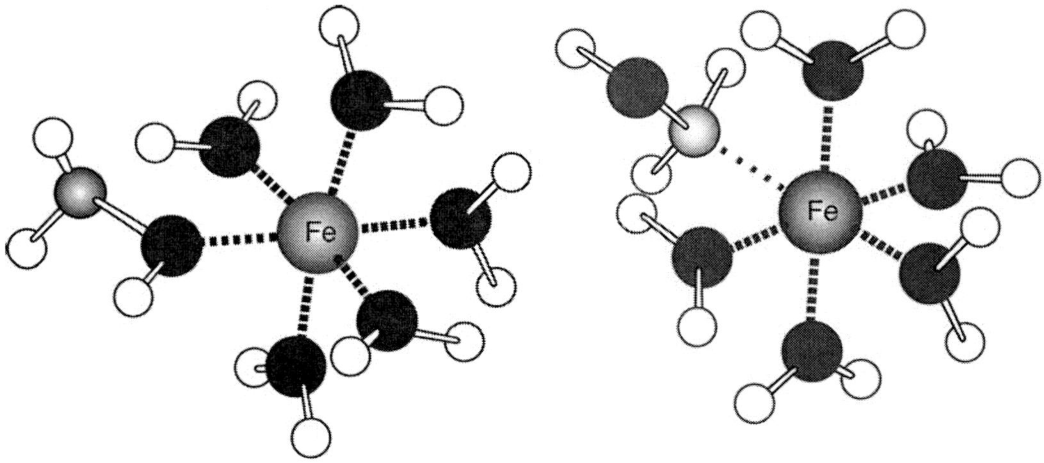

Figure 24. Optimized structure of Fe(NH2OH): (a) interacting through 0, (b) interacting through N.

The molecular orbital calculation also provided some insight of the interaction manner between Fe^{3+} and NH_2OH. Using Hartree Fock approach and 3-21G basis set, the optimized structures were calculated simulating the interaction (Figure 24). For Fe^{3+}, only high-spin states were considered. Frequency calculations were performed to classify each stationary point. The structure was calculated assuming hydrated Fe^{3+} with six H_2O molecles. One of the H_2O surrounding Fe^{3+} is substituted with NH_2OH. Based on the calculation results, interaction through the oxygen atom caused extend the N-O bond in NH_2OH; the length of N-O is 1.4692A, and 1.4899A in the complex. The coordination bond between Fe^{3+} and the oxygen atom is 2.0733A. In the optimized structure simulating interaction through the nitrogen, however, 2.2185A is the most stable distance between Fe^{3+} and the nitrogen atom. Based on the N-O distance in the complex (1.4704A), N-Fe^{3+} interaction may be difficult to represent the linkage formation.

Table 6. Summary of overall heat of reaction

		Under acidic condition			
	HACl	**NHACl**	**OHACl**	**NNHACl**	**NOHACl**
Fe^{3+}	3.39	4.53	0.26	0.28	0.20
$[Fe(CN)_6]^{3-}$	2.96	0.36	2.50	1.15	0.03
Fe(EDTA)-	2.02	0.74	1.01	0.07	0.18
		Under basic condition			
	HA	**NHAOH**	**OHAOH**	**NNHAOH**	**NIHAOH**
Fe^{3+}	**>238**	20.6	0.71	1.70	2.45
$[Fe(CN)_6]^{3-}$	**99.1**	5.29	0.05	2.29	4.30
Fe(EDTA)-	**489**	**165**	0.22	**133**	3.47

5. Conclusion

The nature of hazardous materials is their unstability, in other words, high reactivity. A molecule with high reactivity can interact with neighboring molecules and change their structure and itself. Hence, some hazardous materials have been utilized widely in chemical reaction in chemical and other industries. Some highly reactive materials, however, decompose by themselves if trigged by a slight stimulus such as heat, friction, impact, and so on. The decomposition generates sometimes stable gas molecules; the enthalpy difference between an unstable reactant and stable products turns into heat. Stable gas products expand by heat, and causes destruction. Alternatively, the heat ignites flammable gas product and causes fire. In order to prevent disasters caused by hazardous materials, knowledge of the reactivity is inevitably required so as to manage them safely and maximize their useful properties.

This article overviewed current research on hazardous materials interacting with transition metals. The first part focused on sensitivity of a complex synthesized from copper and azole. The complex was stable enough to evaluate the change in sensitivity by complex formation. The interaction with transition metal has an influence on the sensitivity of azoles, especially to heat. Following this part, reaction behaviors of three simple unstable chemical, H_2O_2, NH_2NH_2, NH_2OH were observed by way of mixing experiments. Despite their similar chemical structures, the three chemicals exhibited different properties respectively. Acidic H_2O_2 shows high reactivity in contact with Fe^{3+}, while basic NH_2NH_2 released heat moderately, along with precipitation. Intermediate NH_2OH demonstrated changed behavior depending on acidity. The acidic solution showed easy-to-understand redox reaction with high oxidated metals, while the basic solution generated precipitation and released heat continuously with some metals. The results proved the idea that oxygen plays a key role in the reaction. Based on the idea, the last section related the reaction manner of Fe^{3+} and NH_2OH by using of steric hindrance. The experimental result and ab initio molecular orbital calculation supported the idea.

In the field of hazardous materials, most of the studies are case studies, or derivatives from accidents. In this way, researchers tend to focus on a quantitative study on released power, or resulting damage. There have been quite a few studies with this inductive approach. The research related in this article is trials to overview the properties of hazardous materials by way of chemical structure, and require further development. There are various kinds of questions remaining regarding the principle of hazardous materials' properties. Further intensive research is expected to clarify the subject and contribute to safe handling of hazardous materials.

References

[1] Urben, P.G.; Pitt, M.J.; Battle, L.A.; Ed; *Bretherick's Handbook of Reactive Chemical Hazards*, fifth edition; Butterworth Heinemann: Oxford, 1995.

[2] *Recommendations of the Transportation of Dangerous Goods*, Twelfth revised edition, United Nations, 2001.

[3] Shriver, D.F.; Atkins,P.W.; Langford,C.H. *Inorganic Chemistry*, second edition; Oxford University Press: Oxford, 1994; Chapter 6.

[4] Martell,A.E.; Calvin, M.; *Chemistry of Metal Chelate Compounds*; Prentice-Hall: NewYork, 1952; p380.

[5] Kumasaki, M.; Miyasaka, R.; Kiuchi, H.; Wada, Y.; Arai, M.; Tamura, M., J. Japan *Explosives Soc.* 2001, Vol.62, 109-116.

[6] Krishnan, S.; Swami,R.D. *J. Propul. Power* 1998, Vol. 14, 295-300.

[7] Kishore, K.; Sunitha,M.R. *AIAAJ* 1979, Vol. 17, 1119-1123.

[8] Moore, D.S.; Robinson, S.D. *Adv. Inorg. Chem.* 1998, Vol.32, 171-239.

[9] Haasnoot, J.G.*Coord. Chem. Rev.* 2000, Vol.200-202, 131-185.

[10] Kowhakul,W.; Miyasaka,R.; Kumasaki,M.; Wada,Y.; Arai,M.; Tamura,M. J. *Japan Explosives Soc.* 2002, Vol.63, 362-366.

[11] Kowhakul, W.; Kumasaki,M.; Wada,Y.; .Arai,M; Tamura,M. *Sci. Tech. Energetic Materials* 2005, Vol.66, 229-232.

[12] Kowhakul,W.; Kumasaki, M.; Arai, M. *Sci. Tech. Energetic Materials* 2005, Vol.66 425-430.

[13] Kowhakul, W.; Kumasaki, M.; Wada, Y.; Arai, M.; Tamura, M. *Sci. Tech. Energetic Materials* 2006, Vol.67, 87-90.

[14] Mihina, J.S.; Herbst, R.M. *J. Org. Chem.* 1950, Vol.15, 1082.

[15] Yoshida, T; Safety of Reactive Chemicals, *Industrial Safety Series* volume 1; Elsevier, 1987,250-265.

[16] Dixon, W.J.; Masseg, F.J. *Introduction to Statistical Analysis*, second Edition, 1957, McGraw Hill , 318.

[17] Japan Explosives Society, *Japan Explosives Society Standard* (IV) Sensitivity test method, 1996, 65-68.

[18] Japan Explosives Society, *Japan Explosives Society Standard* (IV) Sensitivity test method, 1996, 76-80.

[19] Yoshida,T.; Ding, D. *Safety technique of chemical substances* 1996, Tokyo Progress System , 108.

[20] Aochi, T.; Miyake, A.; Ogawa, T.; Matsunaga, T.; Nakagawa, Y.; Iida, M. J. *Japan Explosives. Soc.* 1997, Vol.58, 202-210.

[21] Sakata, K.; *Materials of SG society for the study, third meeting at Tokyo*, 1992.

[22] Williams,J.K.; Palopoli,S.F.; Brill, T.B. *Combust. Flame* 1994, Vol.98, 197.

[23] Ohno, Y.; Akutsu, Y.; Arai, M.; Tamura, M.; Matsunaga, M. J. *Japan Explosives Soc.* 1999, Vol. 60, 110-117.

[24] Frish, M.J.; Trucks, G.W. Schlegel, H.B.; Gill, P.M.W.; Johnson, B.G; Robb, M.A.; Cheeseman, J.R.; Keith, T.; Petersson, G.A.; Montgomery, J.A.; Taghavachari, K.; Al-Laham, M.A.; Zakrzewski, V.G.; Ortiz, J.V.; Foresman, J.B.; Cislowski, J., Stefanov, B.B.; Nanayakkara, A.; Challacombe, M.; Peng, C.Y.; Ayala, P.Y.; Chen, W.; Wong, M.W.; Andres, J.L.; Replogle, E.S.; Gomperts, R.; Martin; R.L.; Fox, D.J.; Binkley, J.S.; Defrees, D.J.; Baker, J.; Stewart, J.P., Head-Gordon, M.; Gonzalez, C.; Pople, J.A. *Gaussian* 94, Revision D.3. Gaussian, Inc., Pittsburgh, 1995.

[25] Kawaguchi, S.; Kumasaki, M.; Wada, Y.; Akutsu, Y.; Arai, M.; Tamura, M. J. *Japan Explosives Soc.* 2001, Vol.62, 16-22.

[26] Kumasaki,M.; Fujimoto, Y.; Ando, T. J. *Loss Prev. Ind.* 2003, Vol.16, 507-512.

[27] Kumasaki, M.; *J. Haz. Mat.* 2004, Vol.115, 57-62.

[28] Kowhakul, W.; Kumasaki, M.; Arai, M.; Tamura, M.; *J. Loss Prev. Ind.* 2006, Vol.19, 452-458.

[29] Kumasaki, M.; *J. Loss Prev. Ind.* 2006, Vol. 19, 307-311.
[30] Bailar, JR J.C.; Emeleus, H.J.; Nyholm, R; Trotman-Dickenson, A.F. *Comprehensive Inorganic Chemistry*, Volume 2, Pergamon Press, 1973, 272-273.
[31] Schmidt, E. W. *Hydrazine and Its Derivatives, Preparation, Properties, Applications*, second edition, Volume 1, John Wiley & Sons, 2001, 498.
[32] Aliev, R. Y.; Kuliev, A.D.; Klyuchnikov, N.G.;Ingibitory Korroz. Met., Moscow 1974, 86, 135-138.
[33] Bailar, JR J.C.; Emeleus, H.J.; Nyholm, R; Trotman-Dickenson, A.F. *Comprehensive Inorganic Chemistry*, Volume 3, Pergamon Press, 1973, 1006.
[34] Fenton,H.J.H. *J.Chem.Soc.* 1894, 65, 899-910.
[35] Wardman, P.; Candeias, L.P. *Radiaton Research* 1996, 145, 523-531.
[36] Kremer, M.L. *Trans. Faraday Soc.*1963, 59, 2535-2542.
[37] Cisneros, L.O.; Rogers, W. J.; Mannan, M. S. *J. Hazard. Mat.* 2002, A95, 13-25.
[38] The Chemical Society of Japan, *Handbook of Chemistry: Pure Chemistry II.* Third edition, 1993, Maruzen: Tokyo.
[39] Wilkinson, Sir G. FRS, Comprehensive Coordination Chemistry, *The Synthesis, Reactions, Properties and Applications of Coordination Compounds*, Volume4 Middle Transition Elements, Pergamon Books, 1987, 220-221.
[40] Hoard,J. L.; Lind,M.; Silverton, J. V. *J. Am. Chem. Soc.* 1961, 83, 2770-2771.
[41] Frisch, M. J.; Trucks, G. W.; Schlegel,H. B.; Scuseria,G. E.; Robb, M. A.; Cheeseman,J. R;. Montgomery, Jr.,; Vreven,J. A. T.; Kudin,K. N.; Burant,J. C.; Millam, J. M;. Iyengar,S. S.; Tomasi,J.; Barone,V.; Mennucci,B.; Cossi,M.; Scalmani,G..; Rega, N.; Petersson, G. A.; Nakatsuji, H.; Hada, M.; Ehara, M.; Toyota, K.; Fukuda, R.; Hasegawa, J.; Ishida,M.; Nakajima, T.; Honda, Y.; Kitao,O.; Nakai,H.; Klene,M.; Li,X.; Knox, J. E.; Hratchian, H. P.; Cross, J. B.; Bakken, V.; Adamo, C.; Jaramillo, J.; Gomperts, R.; Stratmann, R. E.; Yazyev, O.; Austin, A. J.; Cammi, R.; Pomelli, C.; Ochterski, J. W.; Ayala, P. Y.; Morokuma, K.; Voth, G. A.; Salvador, P.; Dannenberg, J. J.; Zakrzewski, V. G..; Dapprich, S.; Daniels, A. D.; Strain, M. C.; Farkas, O.; Malick, D. K.; Rabuck, A. D.; Raghavachari, K.; Foresman, J. B.; Ortiz, J. V.; Cui, Q. A.; Baboul, G.; Clifford, S.; Cioslowski, J.; Stefanov, B. B.; Liu, G.; Liashenko, A.; Piskorz, P.; Komaromi, I.; Martin,R. L.; Fox, D. J.; Keith, T.; Al-Laham, M. A.; Peng, C. Y.; Nanayakkara, A.; Challacombe,M.; Gill,P. M. W.; Johnson, B.; Chen, W.; Wong, M.W.; Gonzalez, C.; Pople, J.A. *Gaussian* 03, Revision D.01, Gaussian, Inc., Wallingford CT, 2004.

In: Research Progress in Materials Science
Editors: W. Olsson and F. Lindberg, pp. 55-101

ISBN: 978-1-60456-629-1

Chapter 3

LITHOAUTOTROPHIC NITROGEN REMOVAL IN UPFLOW ANAEROBIC SLUDGE BED REACTOR AND ITS MICROBIAL CHARACTERIZATION

Young-Ho Ahn
School of Civil and Environmental Engineering
Yeungnam University, Gyungsan, 712-749 South Korea

Abstract

Recently, several novel autotrophic biological nitrogen removal processes have been developed, including partial nitritation, denitrification by nitrifier, anaerobic ammonium oxidation and its combination. As compared with the classical nitrogen removal process via nitrification and denitrification, these novel processes have various advantages in economical and environmental aspects, such as no need for external carbon addition, negligible sludge production and less energy and oxygen requirement.

This chapter deals with a brief review and recent laboratory scale experimental developments in lithoautotrophic nitrogen removal under an anaerobic environment: anaerobic ammonium oxidation by *Planctomycetes*-like bacteria and *Nitrosomonas*-like bacteria. The review addresses the specifics of process, microbial diversity, performance characteristics and the future challenges for application. In laboratory experiments, sustainable autotrophic nitrogen elimination from ammonium-rich wastewaters such as synthetic wastewater (100~200 mg/L NH_4-N) and high strength slurry-type piggery waste (56 g COD/L and 4.3 g/L NH_4-N) was investigated by using an upflow anaerobic granular sludge bed reactor. In the research, two types of anaerobic autotrophic nitrogen elimination, namely by *Planctomycetes* and *Nitrosomonas* species, were observed according to the type of the substrate. In the experiment using synthetic wastewater, the microbial community, based on cloned 16S rDNA analysis, indicated that an anoxic ammonium oxidizer (autotrophic denitrifying nitrifier) such as *Nitrosomonas eutropha* and *Nitrosomonas europaea* and other autotrophic denitrifiers mainly contributed to nitrogen elimination. The reactor showed quite low nitrogen conversion (0.02~0.03 kg N/m^3-d). FISH (Fluorescence *in situ* Hybridization) analysis of the biomass treating piggery waste demonstrated that *Planctomycetes*-like bacteria (especially, Candidatus "*Kuenenia stuttgartiensis*") played a significant role in nitrogen removal, contrary to the foregoing experiments for ammonium-rich synthetic wastewater. Average nitrogen conversion was 0.59~0.66 kg N/m^3-d (0.3~0.32 kg NH_4-N/m^3-d). The specific microbial activity in the batch experiment was 0.1 g NH_4-N/g VSS/day (0.77 g T-N/g

VSS/day), indicating that potential denitritation is also quite high. Based on mass balance, the result revealed that the Anammox reaction was influenced less by the degree of partial denitritation, and the Anammox bacteria did not compete with denitritation bacteria. The autotrophic nitrogen-removing granular sludge showed quite favorable physical characteristics in biomass settleability and size.

Keywords: Anammox, denitrification, Fluorescence *in situ* Hybridization (FISH), hydroxylamine, hydrazine, lithoautotrophic nitrogen removal, granular sludge, microbial community, piggery waste

Introduction

Since wastewater discharges containing nitrogen can be toxic to aquatic life, cause oxygen depletion and eutrophication in receiving water, and affect chlorine disinfection efficiency, reducing the nitrogen level from the discharges is necessary (Metcalf and Eddy, 2003). From various wastewaters, nitrogen compounds can be removed by a variety of physicochemical and biological processes. Because biological nitrogen removal is more effective and relatively inexpensive, it has been adopted widely in favor of the physicochemical processes (US EPA, 1993).

The classical biological nitrogen removal proceeds slowly due to low microbial activity and yield. The process has generally not performed with wastewater containing a high concentration of nitrogen. In the past few years, several novel and cost-effective biological nitrogen removal processes have been developed, including partial nitritation, nitrifier denitrification, anaerobic ammonium oxidation (the Anammox process), and its combined system (completely autotrophic nitrogen removal over nitrite, Canon). In Figure 1, the nitrogen cycle is recently re-classified according to a potential biochemical pathway (van Loosdrecht, 2004). A combined system for nitrogen removal based on partial nitritation with Anammox has various advantages such as no need for external carbon addition, negligible sludge production and less energy and oxygen requirement than conventional process (Jetten *et al.*, 2002).

This chapter first reviews the novel processes for biological nitrogen removal in its performance and microbial characteristics briefly, followed by recent laboratory experimental developments in lithoautotrophic nitrogen removal under an anaerobic environment. The review addresses the specifics of process, microbial diversity, performance characteristics and the future challenges for application. The classical nitrogen removal process (thus, nitrification and denitrification), aerobic denitrification and other lithoautotrophic denitrification are also discussed briefly for its comparison together. The experimental research focused on the development of an Anammox granular sludge bed process and its microbial characterization using molecular biological technique such as fluorescence in situ hybridization (FISH) and 16S rDNA sequence analysis. The effect of hydroxylamine and hydrazine as the Anammox intermediates was also estimated.

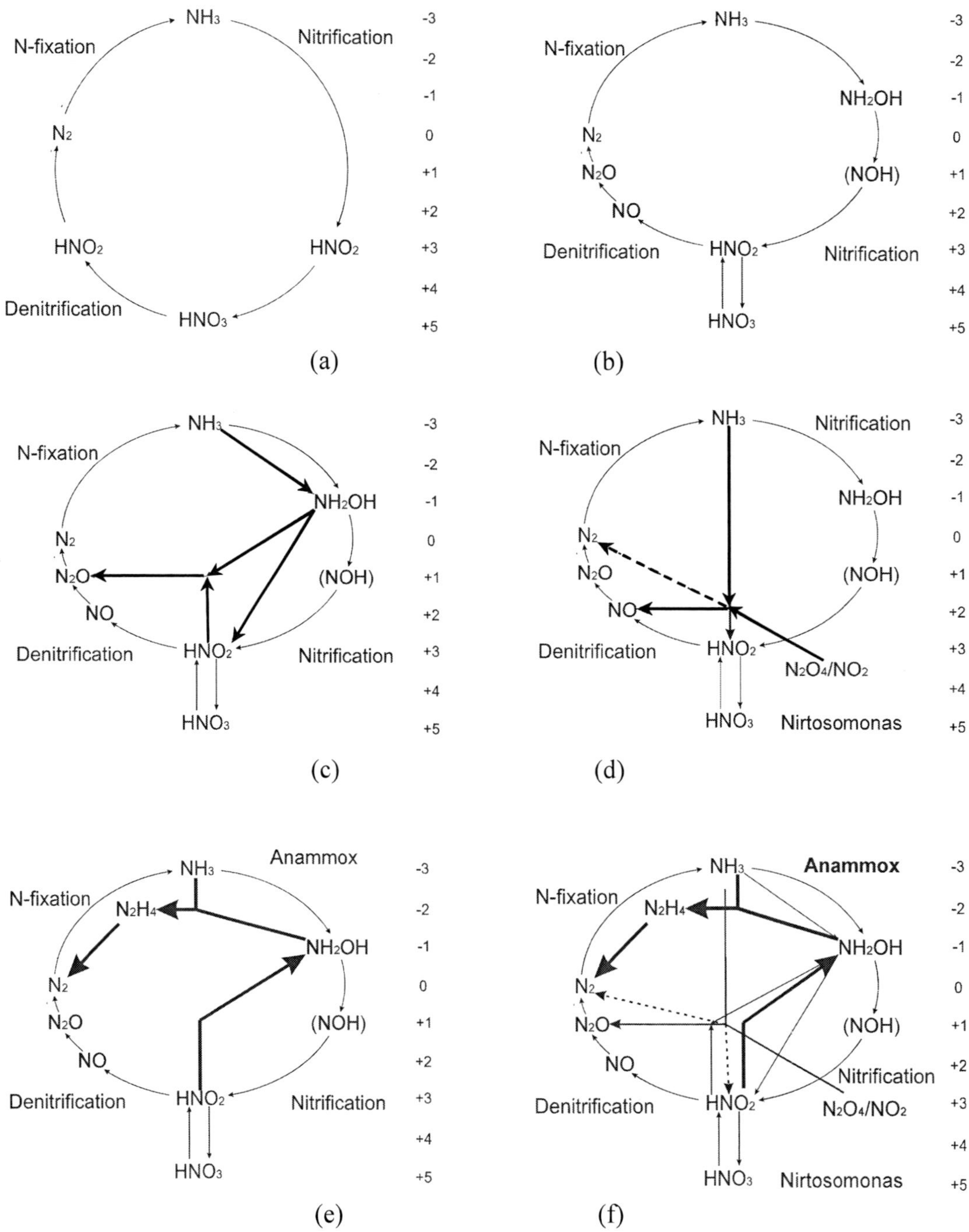

Figure 1. Nitrogen cycle (van Loosdrecht, 2004). (a)classical N-cycle; (b) Sharon process; (c) Nitrosomonas aerobic denitrification or aerobic deammonification; (d) Nitrosomonas denitrification; (e) Anaerobic ammonium oxidation; (f) Overall nitrogen web

Biological Nutrient Removal: A Review

Conventional Nitrification/Denitrification

In general, the biological elimination of nitrogen from wastewater requires a two–step process, nitrification and denitrification, as shown in Figure 1 (a). Nitrification which is the first step is the chemolithoautotrophic oxidation of ammonia to nitrate and a strict aerobic biochemical process conducted in two sequential oxidative stages: ammonia to nitrite (ammonia oxidation) and nitrite to nitrate (nitrite oxidation). Each stage is performed by different bacterial genera of either ammonia or nitrite oxidation. The most commonly recognized genus of bacteria that carries out the ammonia oxidation step is *Nitrosomonas*; however, *Nitrosococcus*, *Nitrosopira*, *Nitrosovibrio*, and *Nitrosolobus* are also able to oxidize ammonium to nitrite. These ammonium oxidizers are genetically diverse, but related to each other in the beta subdivision of the *Proteobacteria*. In the second stage, *Nitrospira*, *Nitrospina*, *Nitrococcus*, and *Nitrocystis* are known to sustain. However, the most famous genus of the nitrite oxidizer is *Nitrobacter*, which is closely related genetically within the alpha subdivision of the *Proteobacteria* (Rittmann and McCarty, 2001; Teske *et al.*, 1994). These microorganisms use ammonia and nitrite as an energy source and molecular oxygen as an electron acceptor. The carbon source for the nitrifying bacteria is carbon dioxide. Based on type of main process, comparative performance in aerobic ammonium oxidation is summarized in Table 1. It represents a wide range of ammonia uptake rate (AUR) according to reactor configuration, substrate type and influent ammonium concentration, etc.

$$55\ NH_4^+ + 76\ O_2 + 109\ HCO_3^- \rightarrow C_5H_7O_2N + 54\ NO_2^- + 57\ H_2O + 104\ H_2CO_3 \quad \text{(Eq.1)}$$

$$400\ NO_2^- + NH_4^+ + 4\ H_2CO_3 + HCO_3^- + 195\ O_2 \rightarrow C_5H_7O_2N + 3\ H_2O + 400\ NO_3^- \quad \text{(Eq.2)}$$

$$NH_4^+ + 1.83\ O_2 + 1.98\ HCO_3^- \rightarrow 0.021\ C_5H_7O_2N + 0.98\ NO_3^- + 1.041\ H_2O + 1.88\ H_2CO_3 \quad \text{(Eq.3)}$$

As the second step, denitrification is generally performed by a heterotrophic bioconversion process under anaerobic (anoxic, precisely) condition. The oxidized nitrogen compounds (NO_2^- and NO_3^-) is reduced to gaseous nitrogen compounds by heterotrophic microorganisms that use nitrite and/or nitrate as alternative electron acceptors (instead of oxygen) for energy production and organic matters as carbon source and energy source. Denitrifiers are common among the gram-negative alpha and beta classes of the *Proteobacteria*, such as *Pseudomons*, *Alcaligenes*, *Paracoccus*, and *Thiobacillus*. Some gram-positive bacteria (such as *Bacillus*) and a few halophilic Archaea (such as *Halobacterium*) are able to denitrify (Zumft, 1997).

The process in environmental biotechnology is accomplished with a variety of electron donors and carbon sources such as: methanol, acetate, glucose, ethanol, and a few others (Table 2). Because methanol (CH_3OH) was relatively inexpensive, it gained widespread use (Rittmann and McCarty, 2001).

$$NO_2^- + 0.53\ CH_3OH + 0.67\ H_2CO_3 \rightarrow 0.04\ C_5H_7O_2N + 0.48\ N_2 + 1.23\ H_2O + HCO_3^- \quad \text{(Eq.4)}$$

$$NO_3^- + 1.08\ CH_3OH + 0.24\ H_2CO_3 \rightarrow 0.056\ C_5H_7O_2N + 0.47\ N_2 + 1.68\ H_2O + HCO_3^- \quad \text{(Eq.5)}$$

Lithoautotrophic Denitrification

Denitrification process can be performed by various chemoorganotrophic, lithoautotrophic, and phototrophic bacteria and some fungi. Most denitrifiers are anaerobic heterotrophic organisms that transfer redox equivalents from the oxidation of a carbon source to a nitrogen oxide under anaerobic conditions. However, autotrophic denitrifiers utilize inorganic sulfur compound, hydrogen, ammonia, or nitrite (Table 3) (Zumft, 1997). It is also found that the oxidation of Fe(II) can be also coupled to a complete denitrification process (Straub *et al.*, 1996):

$$10\ FeCO_3 + 2\ NO_3 + 24\ H_2O \rightarrow 10\ Fe(OH)_3 + N_2 + 10\ HCO_3^- + 8\ H^+ \quad \text{(Eq.6)}$$

Aerobic Denitrification

Generally denitrification is nearly exclusively a facultatively anaerobic or microaerophilic trait. With some exceptions, there are no examples where denitrification occurs in an obligate anaerobic bacterium (Zumft, 1997). The complete denitrification at high dissolved oxygen (DO) concentrations was first observed in *Paracoccus* species. In batch culture experiments, *Thiosphaera pantotropha* which is an autotrophic bacterium produced dinitrogen gas from ammonium and/or nitrite and nitrate in well-mixed aerobic cultures (Robertson and Kuenen, 1984). On the basis of identical 16S rRNA and a high phenotypic similarity, it was transferred to *Paracoccus denitrificans* (Ludwig *et al.*, 1993). It was also found that *Pseudomonas stutzeri* SU2 which was isolated from activated sludge of a sequencing batch reactor treating piggery wastewater, rapidly reduced nitrate to nitrogen gas without nitrite accumulation under aerobic conditions (Su *et al.*, 2001). Oguz (2004) stated that in batch and complete mixed reactor with activated sludge 0.07 kg N/m^3/d of nitrogen conversion and 56.7 % removal rate were observed, producing most equivalent amounts of N_2 and N_2O. Most aerobic denitrifiers are heterotrophic organisms. Bacteria which co-respire nitrate and oxygen are widespread in the environment. The system may offer several potential advantages because it occurs directly in aerated bioreactors in the presence of readily biodegradable organics. However, the respiratory pathway little-studied yet.

Table 1. Comparative performance in aerobic ammonium oxidation

Process	Reactor	Substrate	Influent NH_4-N (mg/L)	AUR (ammonia uptake rate) mg NH_4-N_{ox} /L/hr	mg NH_4-N_{ox} /g VSS/hr	AOR (%)	SRT (day)	Reference
Nitrification -Denitrification	BNR	Sewage	18.8~26.5[a]	-	1.8~2.0	100	15	Randall *et al.*(1992)
	BNR	Sewage	8.1~23.2[a]	-	1.5~5.7	79~100	5	Randall *et al.* (1992)
	BNR	Sewage	21.6[a]	-	7.2	93	2.7	Randall *et al.*(1992)
	BNR	Sewage	12.3[a]	-	6.9	44	1.5	Randall *et al.*(1992)
Nitritation	SBR	Synthetic	400	-	4	-	40	Pollice *et al.*(2002)
	SBR	Synthetic	350~540	-	3.2	-	24	Pollice *et al.*(2002)
	SBR	Synthetic	400~500	-	4.8~6.1	-	14	Pollice *et al.*(2002)
	SBR	Synthetic	500	-	25.8	-	10	Pollice *et al.*(2002)
	SBR	Synthetic	500	-	10.0	-	5	Pollice *et al.*(2002)
	SBR	Synthetic	500	-	22.9	-	3	Pollice *et al.*(2002)
Nitritation -Denitrification[b]	CSTR	Digester supernatant	1,000	3.8~15.8	-	-	4[c]	Gil *et al.*(2002)
	Biofilm	Digester supernatant	1,000	6.5~8.8	-	-	4[c]	Gil *et al.*(2002)
Aerobic denitrification	SBR	Rural wastewater	100-110	-	1.1~2.0	56~79	16~32 (0.5~1)[c]	Bernat *et al.*,(2003)
Partial nitritation (PN)	CSTR	Sludge liquor	1,176	24.2	-	53	1	Van Dongen *et al.*(2001)
	CSTR	Urine	7,300	32.5	-	47	4.8	Udert *et al.*(2003)
	SBR	Urine	2,240	11.7	-	51	>30	Udert *et al.*(2003)
	CSTR	Sludge liquor	417±14	12.7~15.9	64~90	45~64	0.63~0.83	Choi (2004)
	Sharon	Synthetic[d]	500~1,000	16.6~20.8	100~150	42~60	1.1[c]	Mosquera-Corral *et al* (2005)
	Sharon	Digester effluent	1,000	12.5~20	-	30~40	1[c]	Mosquera-Corral *et al* (2005)
	Biofilm	Synthetic	800~1,000	10.2~57.2	-	49	-	Fux *et al.*(2004)
	Biofilm	Sludge liquor	660±50	30.9	-	53	1.7	Fux *et al.*(2004)
	Biofilm	Sludge liquor	610±20	26.0	-	48	-	Fux *et al.*(2004)
PN-Anammox	SBR	Digester supernatant	619~657	14.6	-	58	0.85	Fux *et al.*(2002)

* AOR, Ammonia oxidation rate (NH_4-N_{in} - NH_4-N_{out}/NH_4-N_{in}) x 100; a, Total NH_4-N oxidized; b, 20~35°C; c, hydraulic retention time (day); d, ratio of TOC/N =0~0.2.

Table 2. Organic requirement in heterotrophic denitrification

	Carbon source	Organic requirement (g COD/g N)	Reference
NO_2-N	Acetic acid	1.56	Abeling and Seyfried (1992)
	Acetic acid	2.0	Akunna *et al*. (1993)
	Lactic acid	2.8	Akunna *et al*. (1993)
	Methanol	2.3	US EPA (1993)
	Methanol	2.1~2.6	Ho and Choi (2000)
NO_3-N	Acetic acid	2.08	Abeling and Seyfried (1992)
	Acetic acid	3.7	Akunna *et al*. (1993)
	Lactic acid	4.1	Akunna *et al*. (1993)
	Methanol	3.75~4.5	US EPA (1993)
	Methanol	7.35	Nyberg *et al*. (1992)
	Raw sewage	5.2	Rogalla *et al*. (1992)
	Piggery waste	8.44	Bae *et al.* (2001)
NO_x-N	Acetate	2.07	Narkis *et al*. (1979)
	Methanol	4.2	Narkis *et al*. (1979)
	Piggery waste	6.42	Bae *et al.* (2001)

Table 3. Microbial diversity of autotrophic denitrifying and Anammox bacterial species

Type	Species	
Aerobic denitrification	*Magnetospirillum magnetotacticum** (Bazylinski *et al.* 1983)	
	Paracoccus denitrificans (Ludwig *et al.*, 1993)	
	*Pseudomonas carboxydoflava** (Moir *et al.,* 1996)	
	*Pseudomonas stutzeri** (Su *et al.*, 2001; Körner and Zumft, 1989)	
	Thiosphaera pantotropha (Robertson and Kuenen, 1984)	
S oxidizing denitrification	*Beggiatoa*	(Zumft, 1992)
	Thiobacillus	(Zumft, 1992; Batcheler and Lawrence,1978)
	Thioploca	(Zumft, 1992)
		Paracoccus denitrificans (Ludwig *et al.*, 1993)
		Thiosphaera pantotropha (Robertson and Kuenen, 1984)
H_2 oxidizing denitrification	*Paracoccus*	(Zumft, 1992)
		P. denitrificans (Castignetti *et al.*1984, Robertson *et al.* 1988)
		P. stutzeri (Körner *et al.* 1989)
	Pseudomonas	(Zumft, 1992)
		P. carboxydohydrogena (Frunzke and Meyer, 1990)
	Ralstonia	(Zumft, 1992)
NO_2^- or NH_4^+ oxidation	*Nitrosomonas*	*N. europaea* (Poth and Focht, *1985)*
		N. eutropha (Bock *et al.* 1995, Schmidt and Bock, 1997)
	Nitrobacter	*N. vulgaris* (Ahlers *et al.* 1990)
Anaerobic ammonium oxidation	*Candidatus Brocadia anammoxidans* (Jetten *et al.* 2001)	
	Candidatus Kuenenia stuttgartiensis (Schmid *et al.* 2000)	
	Candidatus Scalindua brodae (Schmid *et al.* 2003)	
	Candidatus Scalindua wagneri (Schmid *et al.* 2003)	
	Candidatus Scalindua sorokinii (Schmid *et al.* 2003)	

* heterotrophic.

Table 4. Physiological parameters for anaerobic and aerobic ammonium oxidation (Jetten et al., 2001; Strous et al., 1999b)

Parameters	Anammox $NH_4^+ + NO_2^- \rightarrow N_2$	Nitrification $NH_4^+ + O_2 \rightarrow NO_2^-$	Unit
pH range	6.7-8.3	variable	
Temperature range	20-43	≤ 42	°C
Free energy	-367	-275	kJ/mol
Biomass yield	0.07	0.08	mol/mol C
	0.07	0.1	g protein/g NH_4-N
Aerobic rate	0	200-600	nmol/min/mg protein
Anaerobic rate	60	2	nmol/min/mg protein
Growth rate	0.003	0.04	/h
Doubling time	10.6	0.73	days
K_S NH_4^+	5	5-2,600	μM
K_S NO_2^-	<5	n.a.	μM
K_S O_2	n.a.	10-50	μM
NO_2 inhibition of			
NH_4^+ consumption	Ki =0.8, α=0.8	Usually	g NO_2-N/L
NO2- consumption	Ki=1, α=0.7	n.a.	g NO_2-N/L
Protein content of biomass	0.6	variable	g protein/g SS
Protein density	50	variable	g protein/L biomass

n.a. - not applicable; *KS*, affinity constant.

Table 5. Comparison of various nitrogen removal processes

Reaction	First phase		Second phase		Remark
	Oxygen	Alkalinity	Alkalinity	Organic*	
Nitrification	4.57 [4.18]	7.14 [7.07]	(3.57)	3.7	1), 2)
Nitritation	3.43 [3.16]	7.14 [7.07]	(3.57)	2.3	1), 2)
Partial nitritation+	1.71~2.06	3.57	0.24	-	3)
CANON	1.94	3.68	-	-	4)

* Based on methanol.

+ 50~60% partial nitritation.

[ ] Calculated by combined dissimilation-synthesis equations.

( ) Alkalinity production in heterotrophic denitrification/denitritation.

1) McCarty, *et al.* (1969); 2) Ritmann and McCarty (2001); 3) Van Dongen *et al.* (2001); 4) Sliekers *et al.* (2003).

Table 6. Comparative performance in nitrogen removal

Process Type	Reactor	Substrate	Seed Biomass	Influent NH_4-N/NO_2-N (mg/L)	SDNR mgNO_2-N/g VSS/hr	SDNR mgNO_3-N /g VSS/hr	NUR kg N/m^3/d	NUR mg N/g VSS/hr	N removal (%)	Reference
Nitrification -Denitrification	BNR	Municipal w/w	AS	-	-	3~30 [a]	-	-	-	Ekama *et al.* (1984)
	BNR	Municipal w/w	AS	-	-	1.6~1.92 [a]	-	-	-	Randall *et al.*(1992)
	Batch	D. supernatant	AS	1,000 / 0	-	6~13	0.31~0.68 [b]	-	-	Gil *et al.*(2002)
Nitritation -Denitrification	Batch	D. supernatant	AS	1,000 / 0	13~23	-	0.64~1.12 [b]	-	-	Gil *et al.*(2002)
Aerobic Denitrification	SBR	Rural w/w	AS	100 / 0	-	0.51~1.56	-	-	43~52	Bernat *et al.* (2003)
	CSTR	Municipal w/w	AS	29.3 /0	-	-	0.07	-	56.7	Oguz (2004)
Anammox	FBR	Synthetic	Denitrifying	90~130/ n.a.	-	-	0.7~1.5	-	-	Mulder *et al.*(1995)
	FBR	Synthetic	Denitrifying	420/550	-	-	4.8	18.2~21.0	-	Van de Graaf *et al.*(1996)
	FBR	Synthetic	Denitrifying	70~840/70~840	-	-	1.8	7.5	83~85	Strous *et al.*(1997b)
	Fixed bed	Synthetic	Denitrifying	70~840/70~840	-	-	1.1	-	92~99	Strous *et al.*(1997b)
	SBR	Synthetic	Denitrifying	300/400	-	-	-	54	-	Strous *et al.*(1998)
	SBR	Synthetic	Anammox	450/460	-	-	0.75±0.2	33.3	-	Van Dongen *et al.* (2001)
	SBR	Synthetic	Anammox	375/375	-	-	-	18.3	-	Datena-Mora *et al.*(2004)
	CSTR	Synthetic	Anammox	7~750/8.4~1,150	-	-	0.003~0.58	-	90~94	Guven *et al.*(2003)
	Gas-lift	Synthetic	Anammox (80%)	257/202	-	-	0.31	8.8	-	Sliekers *et al.*(2002)
	Gas-lift	Synthetic	Anammox (80%)	1,360/1,370	-	-	8.9	-	83	Sliekers *et al.*(2003)
	Gas-lift	Synthetic	Anammox	900/1,100	-	-	-	37.5	88	Dapena-Mora *et al.*(2004)
	UASB	Synthetic	Anammox	709~1,027/ 27~1,409	-	-	2.3~6.4	-	86~92	Imajo *et al.*(2003)
	FBR	D. effluent	Denitrifying	1,100~2,100/70~840	-	-	1.5	6.3	81~99	Strous *et al.*(1997)
	Batch	Urine	Anammox	4,100/3,400	-	-	1~1.3	-	-	Udert *et al.* (2003)
	UASB	Piggery waste	UASB Granule	2,160 / 2,500	-	-	0.6~0.7	2.7~3.3	65~71	Ahn *et al.* (2004)
	Batch	Piggery waste	UASB Granule	1,280 / 2,070	-	-	1.15	32.1	70	Ahn and Kim (2004)

Table 6. Continued

Process Type	Reactor	Substrate	Seed Biomass	Influent NH_4-N/NO_2-N (mg/L)	SDNR		NUR		N removal (%)	Reference
					mgNO_2-N/g VSS/hr	mgNO_3-N/g VSS/hr	kg N/m^3/d	mg N/g VSS/hr		
PN (Sharon) - Anammox	CSTR-FBR	D. supernatant	Anammox	584 / 0	-	-	0.71	-	83	Jetten *et al.* (1997)
	CSTR-SBR	D. liquor	Anammox	1,176 / 0	-	-	0.75±0.2	8.0 [a]	-	Van Dongen *et al.* (2001)
	CSTR-SBR	D. supernatant	Anammox	619~657 / 0	-	-	0.6~2.4	12.5 [a]	90	Fux *et al.*(2002)
CANON	SBR	Synthetic	Anammox (80%) +Nitrifier (20%)	-	-	-	0.04~0.11	2.2~4.9	36~92	Third *et al.* (2001)
	Gas-lift	Synthetic	Anammox (80%) +Nitrifier (20%)	131 / 0	-	-	0.064	1.8	-	Slieker *et al.*(2002)
	Gas-lift	Synthetic	Anammox (80%) +Nitrifier (20%)	1,545 / 0	-	-	1.5±0.2	-	42±4.7	Slieker *et al.*(2003)
	UASB	D. liquor	Granule (90%) + AS (10%)	420 / 0	-	-	0.06~0.08	-	80~95	Ahn and Choi (2005)
OLAND	SBR	Synthetic	Nitrifying	1,000 / 0	-	-	0.05	1.2	40	Kuai *et al.*(1998)
	Batch	Synthetic	Nitrifying	-	-	-	0.001~0.007	-	-	Philips et al. (2002)
	RBC	Synthetic	RBC biomass	840 / 0	-	-	1.05±0.06	-	89±5	Pynaert *et al.* (2003)
Deammonification	RBC	Synthetic	RBC biomass	-	-	-	0.3	-	-	Seyfired *et al.*(2001)
	Biofilm	Synthetic	-	150 / 0	-	-	1.49	-	98.9	Helmer *et al.*(2001)
	UASB	Synthetic	UASB Granule	100~200 / 50	-	-	0.02~0.03	-	30~50	Ahn and Kim (2004)

FBR, Fluidized bed reactor; SBR, sequencing batch reactor; UASB, upflow anaerobic sludge bed reactor; RBC, rotating biological contactor; PN, partial nitritation (Sharon); D, Anaerobic digester; AS, activated sludge; SDNR, specific denitrification rate; NUR, nitrogen uptake rate; a, mg N/mg SS/hr; b, based on reactor volume of anoxic zone.

Sulfur- and Hydrogen-oxidizing Autotrophic Denitrification

In an autotrophic denitrification, the energy sources is derived from inorganic oxidation-reduction reactions with elements such as hydrogen or various reduced-sulfur compounds (HS^-, H_2S, S, $S_2O_3^{2-}$, $S_4O_6^{2-}$ or SO_3^{2-} etc.) as electron donor, while the carbon sources are inorganic carbon compounds (e.g. CO_2 or HCO_3^-). The early research on physiological and biochemical studies of sulfur-oxidizing bacteria was performed by Baalsruud and Baalsruud (1954). *Thiosphaera pantotropha* is a facultatively anaerobic, autotrophic sulfur bacterium. (Robertson and Kuenen, 1984). *Thiobacillus denitrificans* is a widely distributed and well-characterized obligate chemolithoautotrophic bacterium. It is well known for its ability to couple the oxidation of inorganic sulfur compounds (such as hydrogen sulfide and thiosulfate) to denitrification, as shown in Eq.7. Because of difficulty and economics in hydrogen utilization, much more attention has been concentrated recently on sulfur-based autotrophic denitrification. Limestone used often a source of alkalinity and inorganic carbon (Batchelor and Lawrence 1978). It has been reported that the process is a reliable and simple, however, process by-products (e.g. sulfate, hardness increase, and H_2S gas produced in the upper part of the reactor) should be adequately controlled in following step.

$$1.14\ S^o + NO_3^- + 0.669\ H_2O + 0.337\ CO_2 + 0.0842\ HCO_3^- + 0.0842\ NH_4^+ \rightarrow 1.114\ SO_4^{2-} + 0.5\ N_2 + 0.0842\ C_5H_7O_2N + 1.228\ H^+ \quad \text{(Eq.7)}$$

Denitrifying Nitrifiers

Various proteobacterial oxidizing bacteria which are generally considered as aerobic chemolithoautotrophs have been found to dominate different wastewater plant and natural ecosystem. *Nitrosomonas*-like microorganisms nitrify and denitrify simultaneously even under fully oxic or anoxic conditions with N_2 as main product (Bock *et al.*, 1995; Jetten *et al.*, 1999; Poth, 1986; Schmidt and Bock, 1997; Schmidt *et al.*, 2003).

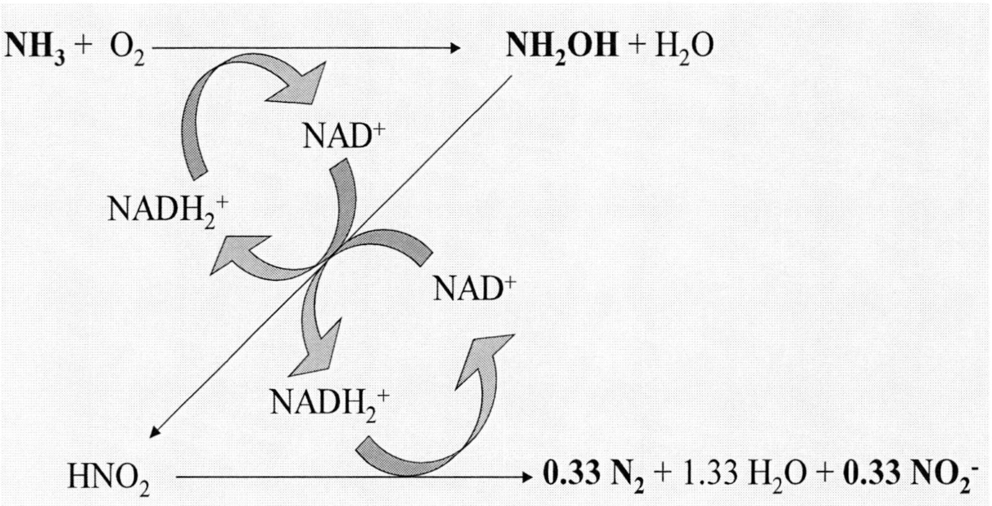

Figure 2. Possible biochemical pathway of ammonia to dinitrogen and nitrite (Binswanger *et al.*, 1997).

Nitrosomonas eutropha is an obligate lithoautotrophic nitrifying bacterium and also a denitrifying organism that uses hydrogen as electron donor and nitrite as electron acceptor. Schmidt and Bock (1997) described that denitrification activity of *N. eutropha* could be stimulated by adding gaseous nitrogen oxide under anaerobic condition. *N. europaea* also produce N_2O in denitrification process (Poth and Focht, 1985). These microorganisms are mainly detected in bioreactor which has a combination concept of aerobic and anaerobic ammonia oxidation, thus Sharon (single reactor system for high-rate ammonium removal over nitrite), NO_x, Canon (completely autotrophic nitrogen removal over nitrite), OLAND (oxygen-limited nitrification and denitrification), and aerobic deammonification (Jetten *et al.*, 1999; Schmidt *et al.*, 2003).

Figure 1 (b) to (d) represents the potential nitrogen cycle by this concept (van Loosdrecht *et al.*, 2004). Possible biochemical pathway of the process may be described as shown in Figure 2. Hydroxylamine is a reaction intermediate in the oxidation of ammonia to nitrite. *N. europaea* are capable of growing mixotropically on ammonia and hydroxylamine. In the absence of oxygen, hydroxylamine was oxidized with nitrite as electron acceptor, while nitrous oxide was produced (De Bruijn *et al.*, 1995). Under anoxic or oxygen-limited conditions, the oxygen source is unknown. However, nitrite, NO_2, and NO could theoretically have been used as the oxygen source in a heretofore unknown type reaction. Although the trace gas NO and NO_2 (NO_x) are toxic to microorganisms, NO_2 concentration up to 50 ppm have no toxic effects on the anaerobic ammonium oxidation of *N. eutropha* (Schmidt and Bock 1997; Zart and Bock, 1998).

Anaerobic Ammonium Oxidation

The anaerobic ammonium oxidation (anammox) is a lithoautotrophic biological conversion process, mediated by a group of *Planctomycete* bacteria, as shown in Table 3 (Strous *et al*, 1999a). The remarkable advance in the molecular biological technique has disclosed a great variety of information on biodiversity of the Anammox bacteria (Schmid *et al.*, 2003). These bacteria have a highly unusual physiology, in that they live by consuming ammonia in the absence of oxygen. Anammox is highly exergonic and linked to energy metabolism of the organisms involved. This process involves the oxidation of ammonia with nitrite as the electron acceptor to yield gaseous nitrogen (Eq. 8). Also, the anammox organisms grow with CO_2 as sole carbon sources and uses nitrite as electron donor here to produce cell material (Eq. 9) (Madigan *et al.*, 2003; van de Graaf *et al.*, 1995; Jetten *et al.*, 2001). The overall process, referred as "Anammox," is the central energy generating pathway for these microbes. Even though the *Nitrosomonas* species can also oxidize ammonium anaerobically as well as aerobically (Fux and Siegrist, 2004; Lee *et al.*, 2004; Schmidt and Bock, 1997; Zart and Bock, 1998), the "Anammox" are used exclusively for nitrogen removal by these *Planctomycetes*-like bacteria (Jetten *et al.*, 2001).

$$NH4+ + NO2- \rightarrow N2 + 2\ H2O \qquad \text{(Eq.8)}$$

$$CO2 + 2\ NO2- + H2O \rightarrow CH2O + 2\ NO3- \qquad \text{(Eq.9)}$$

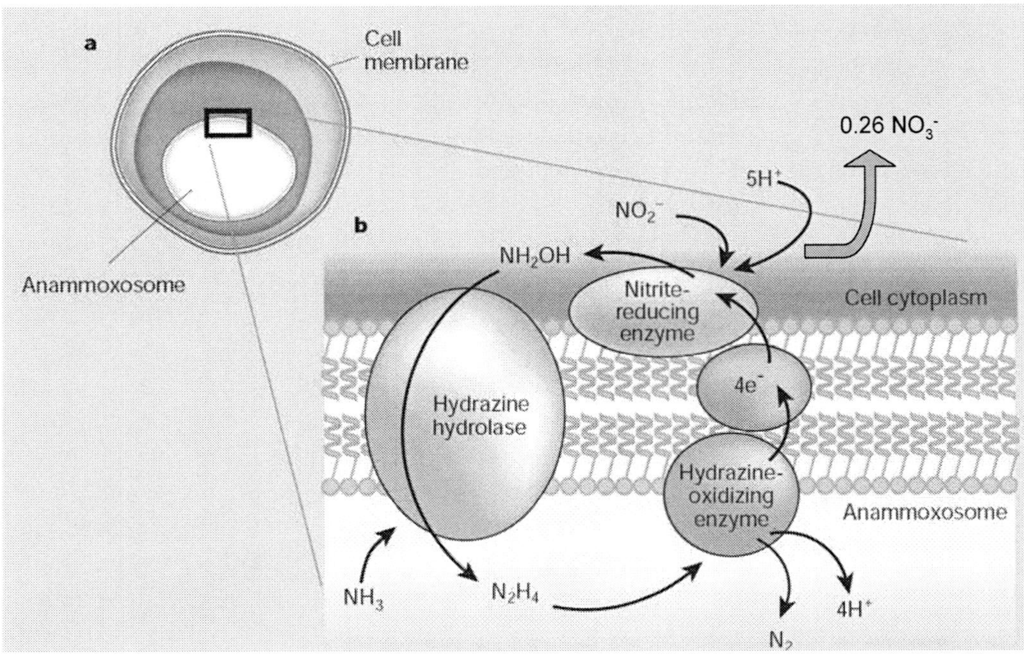

Figure 3. Biochemical pathway of Anammox reaction.

The Anammox reaction was observed in a denitrifying-fluidized bed reactor treating the effluent from a methanogenic reactor (Mulder et al., 1995) but Broda (1977) had postulated two thermodynamically possible stoichiometries for ammonium oxidation in the absence of oxygen, in which the molar ratios of ammonium to nitrite were 1:1 and 1: 1.67. From several basic studies, the stoichiometry of the Anammox reaction based on mass balance over Anammox enrichment cultures was represented by (Eq.10) (Strous et al., 1998). The main product of Anammox process is dinitrogen gas, but about 10% of the nitrogen feed is converted to nitrite. The overall nitrogen balance gave a ratio of ammonium to nitrite conversion of 1: 1.31±0.06 and 1: 0.22±0.02 for the ratio of nitrite conversion to nitrate production (van de Graaf et al., 1996). However, the NO2-N/NH4-N removal in the various Anammox reactors is in the range of 0.5~4 depending on substrate, operating conditions and reactor configuration (Ahn et al., 2004; Strous et al, 1999b).

$$NH4+ + 1.32\ NO2- + 0.066\ HCO3- + 0.13\ H+ \rightarrow 0.066\ CH2O0.5N0.15 + 1.02\ N2 + 0.26\ NO3- + 2.03\ H2O \quad (Eq.10)$$

Figure 1 (e) represents the potential nitrogen cycle by this concept (van Loosdrecht et al., 2004). The specifics of the Anammox pathway are quite remarkable, and especially starting, as shown in Figure 3, are the highly toxic and reactive intermediates that are formed-hydroxylamine and hydrazine, the second of which finds a very different uses as rocket fuel (DeLong, 2002). Hydrazine is a rarely found as an intermediate in microbial nitrogen conversions. In batch experiments on effect of hydrazine by the Anammox culture, Schalk et al. (1998) observed the conversion of hydrazine with nitrite as an electron acceptor. However the Anammox was not possible to grow with hydrazine. Particularly, it was noteworthy that in the Anammox, hydrazine plays a role as an important electron donor in conversion of nitrite to hydroxylamine, meaning that the Anammox is a distinct process as compared to other nitrogen bioconversion processes. Several important physiological parameters using an SBR system have been determined, such as biomass yield, growth rate, and affinity constants, etc, as shown in Table 4 (Strous et al., 1998, 1999b).

Combined Autotrophic Nitrogen Removal Process

Partial Nitritation-Anammox Process

Nitrogen Compounds can be eliminated from ammonium-rich wastewater by anaerobic ammonium oxidation (Anammox). However, ammonium in substrate must partly be pre-oxidized to nitrite (55~60% of ammonium) before feeding into the Anammox process. Thus, the Anammox process needs to be applied by series operation with partial nitritation process such as Sharon (single reactor high activity ammonium removal over nitrite) process. The Sharon process is ideally suited to remove nitrogen from waste streams with a high ammonium concentration (> 0.5 g/L). When coupled with the Anammox process only 50% of the ammonium needs to be converted to nitrite, allowing adequate influent condition for the Anammox reaction. Generally, the partial nitritation process is performed in a single, stirred tank reactor with unique operating conditions: no sludge retention, about 1 day of HRT, 30~40oC of Temperature and 6.6~7.0 of pH. This results in a stable nitrification with nitrite as end-product (Jetten et al., 1997). For full-scale operation, a CSTR or SBR with suspended biomass is recommended by Fux et al. (2004). Based on long-term laboratory and pilot scale experiments using a moving-bed biofilm reactor, they described that none of the selection criteria applied such as a high ammonium loading rate, high free ammonia or low oxygen concentration led to selective suppression of nitrite oxidation. Long-term nitrite production without nitrate accumulation can be unreliable because control of sludge age is difficult in biofilm system.

In treatment of anaerobic digester effluent, no extra addition of base is necessary since the digester effluent generally contain enough alkalinity (in the form of bicarbonate). Jetten et al. (1997) reported that in the partial Sharon-Anammox treating digester effluent, the overall nitrogen ammonium removal was 83% under a total nitrogen load of about 0.8 kg N/m3/d.

Canon Process

The concept of Canon process is also the combination of partial nitritation and Anammox. However, this process performs two sequential reactions (Eq. 11 and 12) in a single and aerated reactor, implying that the two groups of bacteria cooperate in the whole process. The nitrifiers oxidize ammonia to nitrite, consume oxygen and so create anoxic conditions the Anammox process need.

$$NH_4^+ + 0.75\ O_2 + HCO_3^- \rightarrow 0.5\ NH_4^+ + 0.5\ NO_2^- + CO_2 + 1.5\ H_2O \quad \text{(Eq.11)}$$

$$NH_4^+ + 1.32\ NO_2^- + 0.066\ HCO_3^- + 0.13\ H^+ \rightarrow 0.066\ CH_2O_{0.5}N_{0.15} + 1.02\ N_2 + 0.26\ NO_3^- + 2.03\ H_2O \quad \text{(Eq.12)}$$

$$NH4+ + 0.85\ O2 \rightarrow 0.44\ N2 + 0.11\ NO3- + 1.43\ H2O + 0.14\ H+ \quad \text{(Eq.13)}$$

Strous (2000) described that a co-culture of aerobic and anaerobic ammonium oxidizing bacteria could be established under oxygen limited condition. The Anammox bacteria are reversibility inhibited by low concentration of oxygen (0.5% air saturation) (Strous et al., 1997a). The oxygen limited conditions below 0.5% air saturate provide an adequate environment on a stable interaction between Nitrosomonas-like aerobic microorganisms and Planctomycete-like anaerobic bacteria. Schmidt et al. (2002) described that addition of nitrogen oxides (NO2) to a mixed population of Brocadia anammoxidans and Nitrosomonas induced simultaneously specific anaerobic ammonia oxidation activities of up to 5.5 mmol NH4+/g protein/h by B. anammoxidans and 1.5 mmol NH4+/g protein/h by Nitrosomonas. The ammonia oxidizing activity of B. anammoxidans was not inhibited by NO2 concentration of 100 ppm. Based on the experimental results, a model on the competition and cooperation between these two bacteria was introduced (Figure 4).

Two other oxygen limited processes for one-step ammonium removal, thus OLAND and deammonification, have been described by Kuai and Verstreate (1998) and Helmer et al. (2001), respectively. The main difference between these and Canon is that OLAND and deammonification make use of the denitrification activity of conventional aerobic nitrifiers, whereas Canon incorporates the Anammox process (Jetten et al., 2001).

Comparison of Nitrogen Removal Process

Compared with the conventional nitrogen removal process, the process using anaerobic ammonium oxidation appears to be a sustainable biotechnology to removal ammonia in fully autotrophic bioprocess. The process coupled with partial nitritation process remains a challenge for future application in the removal of ammonium from wastewater with high ammonium concentrations. The combined process offers various advantages, such as less oxygen and alkalinity demand, no need for organic carbon (Table 5), less nitrite and nitrate production, no production of undesirable by-products like N2O, and negligible sludge production, etc.

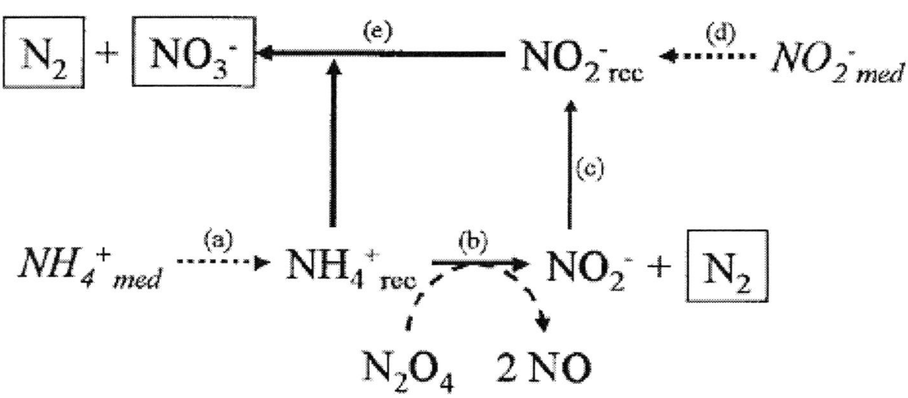

Figure 4. Schematic model describing the competition and cooperation between B. Anammoxidans and Nitrosomonas sp. under anoxic conditions (Schmidt et al., 2002).

Table 6 represents the comparative performance of various novel processes in microbial nitrogen removal. These novel processes were mainly investigated in laboratory and pilot scale. Most Anammox reactors used the enriched Anammox biomass as seed biomass, originated from an Anammox-SBR reactor in which 80% of the biomass consisted of Anammox bacteria (Strous et al., 1998). The processes showed better nitrogen conversion rate. The combination of the partial nitritation and Anammox processes and its applications are currently underway. Since the Canon process has a compact reactor configuration with excellent biomass retention, it may be more economic and efficient options for wastewater treatment. In spite of its high potential for successful application, only a few researchers have studied the Canon process, particularly by using synthetic wastewater and laboratory scale (Sliekers et al., 2002, 2003; Third et al., 2001). The nitrogen loading and removal rate was quite lower than for Anammox. As shown in Table 6, the nitrogen conversion represents quite different trend according to type of seed biomass, substrate, type of process, reactor configuration and concentration of influent ammonium, etc. The nitrogen removal process using denitrification nitrifier (thus, Nitrosomonas-like microorganisms) represents lower nitrogen conversion, relatively.

The optimization and application of Anammox and its combination process on pilot- or full-scale plant remain many challenges yet. One of the main challenges in Anammox process is to decrease the long start up time because the Anammox bacteria have very low biomass yielding rate. Although several types of bioreactors including sequencing batch reactor (SBR) have been successfully applied for the cultivation of the Anammox microorganisms in different laboratories, but many others conducting studies on this process agree that the detection and mass enrichment of the Anammox bacteria are still remaining as a significant obstacle. The accumulation of enough Anammox biomass is required for fast process start up. Although effect of hydrazine as a drive force of Anammox process has been described (Ahn and Kim, 2004; DeLong, 2002; Schalk et al., 1988), control methods to stimulate the activities of Anammox and new strategies on process control have to be developed and evaluated.

Second is the competition between autotrophic Anammox bacteria and heterotrophic denitritation bacteria (or other autotrophic denitrifiers) during startup procedure. Because the Anammox process requires nitrite as an electron donor, a pre-partial nitritation process should be initiated. Also, denitrification is a faster process than the Anammox process (Jetten et al., 2001). This situation could result in competition between the Anammox bacteria and the combined microbial mixture. Actually, this will be a common problem in treatment of high strength wastewater containing carbon compounds as well as nitrogen. Toh and Ashbolt (2002) reported the Anammox activity was greatly affected by the addition of phenol but recovered by using chloramphenicol, which inhibited denitrification. Furthermore, there are some descriptions that the Anammox was less competitive with denitrification (Ahn et al., 2004; Dong and Tollner, 2003).

Under favourable conditions, various autotrophic nitrogen removal bacteria involving Anammox bacteria could be co-cultivated in the anaerobic or oxygen-limited bioreactor such as the Canon reactor, giving a promise of complete nitrogen removal by elimination of undesirable by-products (NO3-N and N2O, etc.). In a co-culture of denitrifying nitrifier and anaerobic ammonium oxidizing bacteria under these conditions, denitrifying nitrifier can be predominant species due to difference of both bacteria in physiological characteristics (Table 4), unless the process has enough Anammox biomass. Under the point of view, two-in series operation of partial nitritation and Anammox process appears to be more attractive. The phase separation in physiological characteristics may provide better environment to maintain process stability and to reduce start up time of the process.

Even though there are still many unknown reactions in novel nitrogen removal process, it appears that the autotrophic nitrogen removal microorganisms can contribute to sustainable nitrogen removal under anaerobic (or anoxic) conditions, depending on the type of electron donor in the substrate. Because most lithoautotrophic organisms, including Anammox bacteria, are characterized by a low maximum growth rate, a reactor with high biomass retention, such as immobilization or granulation process (biofilm or UASB) is required. The lithoautotrophic granular sludge reactor will provide more stable and higher nitrogen conversion rate although a longer start up time is required. The development and application of microbial Anammox granulation remain another challenge towards implementation of the sustainable nitrogen removal process in a future wastewater treatment plant.

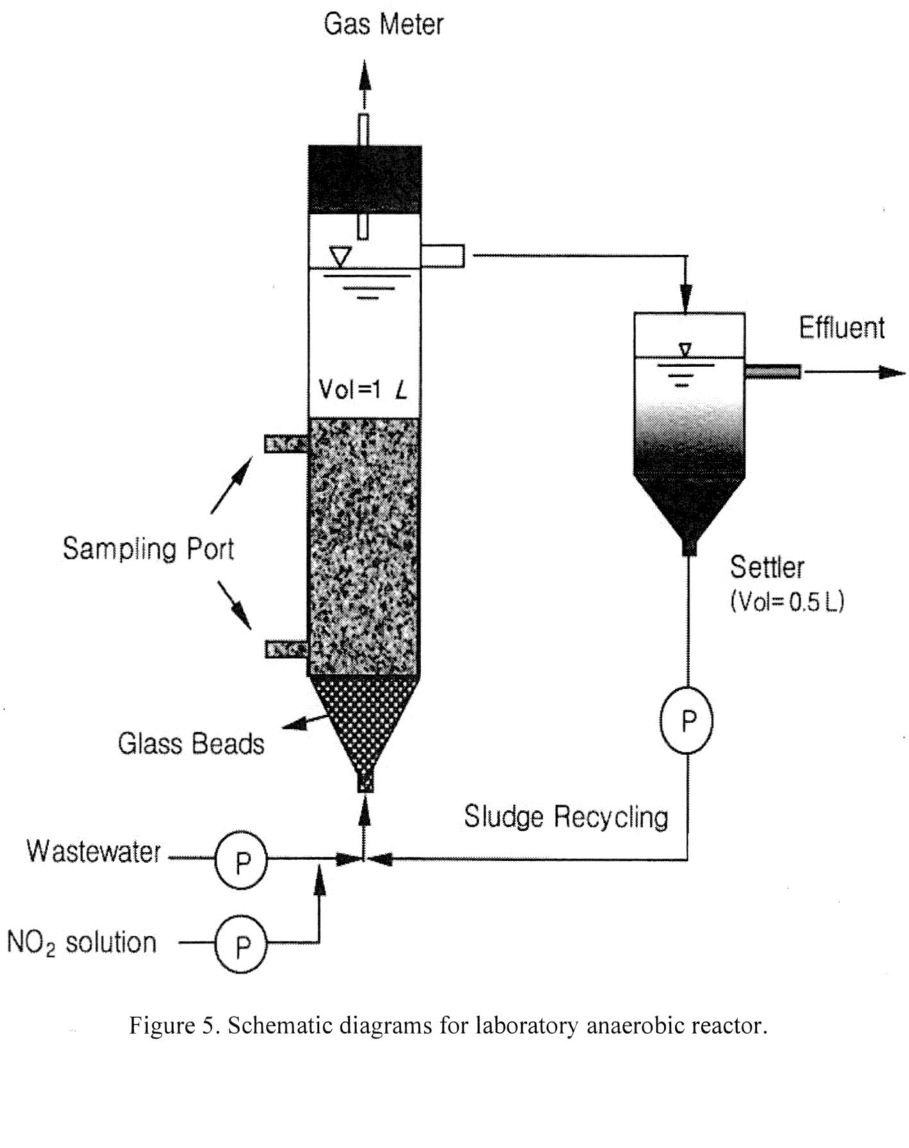

Figure 5. Schematic diagrams for laboratory anaerobic reactor.

Summary

The conventional nitrogen removal process comprised of autotrophic nitrification and heterotrophic denitrification are often used for treating nitrogen in wastewater. Since the process requires significant energy and carbon source, more research should be directed toward development and application of more economical process.

The discovery of Anammox microorganisms has greatly improved the understanding of the nitrogen cycle. In environmental biotechnology, the Anammox provide great promise for removal of nitrogen from wastewater, containing high concentration of ammonium. The discovery of the versatility of aerobic ammonium oxidizer in denitrification also led to the development of new processes such as Canon, Sharon, OLAND and NOX process. The combination of the different groups of nitrogen elimination microorganisms and the process optimization will provide better performance in nitrogen removal. Even though novel technologies described in the review meet a few challenges for the introduction and application in full-scale plant, it might offer better solution for treatment of high strength wastewater. In physiological characteristics of bacteria, two-in series operation of partial nitritation and Anammox process appears to be more attractive. The phase separation may provide better environment to maintain process stability and to reduce start up time of the process. Research to stimulate the activity of Anammox is required with accumulation of Anammox biomass as predominant species. In the future, full-scale introduction of lithoautotrophic nitrogen elimination biotechnology will lead to substantial saving in energy and resources and better management of water environment.

Experimental

Laboratory Reactor Setup

Two laboratory scale 1.5 L upflow anaerobic sludge bed (UASB) glass reactors (internal diameter 6cm) with a 0.5 L settling tank were operated at mesophilic (35°C) condition (Figure 5). The reactor was operated using a semi continuous feeding system. The substrate was fed four times daily in a fill-and-draw mode. During the overall operating period, the reactor constantly operated with a hydraulic retention time (HRT) of 5 days. The settled sludge in the settler was recycled into the bottom of the UASB reactor, resulting in the upflow rate of about 1Q. 0.7 L of granular sludge (18.6 g VS/L and 65% VS/TS) was inoculated as seed biomass. Originally, the granular sludge was collected from a full-scale UASB reactor treating brewery wastewater, and seeded into the reactors after microbial acclimation (more than about 4 months) in a laboratory batch reactor with slurry-type piggery waste.

Substrate

The experiment was performed by using ammonium-rich inorganic synthetic wastewater (Reactor A) and slurry-type piggery waste (Reactor B) as a real waste. The Reactor A was operated with nitrogen concentration (100~200 mg NH_4-N/L and 50 mg NO_2-N /L), in order to decrease the nitrite inhibition in Anammox reaction. To induce the Anammox reaction, hydroxylamine and hydrazine, which are well known as intermediates in Anammox reaction, were also added to the substrate and monitored from reactor effluent. To avoid potential deficiencies in trace element supply, a mineral medium solution (20 mL per L influent) was added to the synthetic substrate. The medium solution contained KH2PO4 13.61 g/L, NH4Cl 49.20 g/L, CaCl4 4.44 g/L, MgCl2·6H2O 8.13 g/L and 10 mL/L of a trace element solution. The trace salt stock solution was prepared according to Ahn et al. (2000).

A slurry-type piggery waste with high strength (56 g COD/L and 5 g T-N/L) was used as the substrate for Reactor B. The substrate had a high ammonium fraction (about 90% of NH4-N/T-N), and ammonium nitrogen loading was about 0.43 kg NH4-N/m3-day. To induce the Anammox reaction, nitrite stock solution (NaNO2) with the same flowrate was fed to the separated feed line with the substrate. Nitrite nitrogen loading was 0.36 kg NO2-N/m3-day for Phase 1 and 0.5 kg NO2-N/m3-day for Phase 2, corresponding to that the total nitrogen loading and influent NO2-N/ NH4-N ratio were 0.83 kg T-N/m3-day and 0.8: 1 for Phase 1, and 1.02 kg T-N/m3-day and 1.2: 1 for Phase 2, respectively. The pH of the influent was 8.2~8.5 and not controlled throughout the reactor operation. The characteristics of the piggery waste and the influent mixture are represented in Table 7.

Analysis

The nitrogen (ammonium, nitrite and nitrate) concentration was measured colorimetrically by Standard Methods (APHA et al., 1998). In the operation of the reactor, the pH (Orion 720, USA), bicarbonate alkalinity (Buchauer, 1998) and gas production (Wet-test gas meter, Sinagawa Model W-NK-0.5A, Japan) were monitored daily. Hydroxylamine and hydrazine were measured as described by Frear and Burrell (1955) and Watt and Chrisp (1952), respectively.

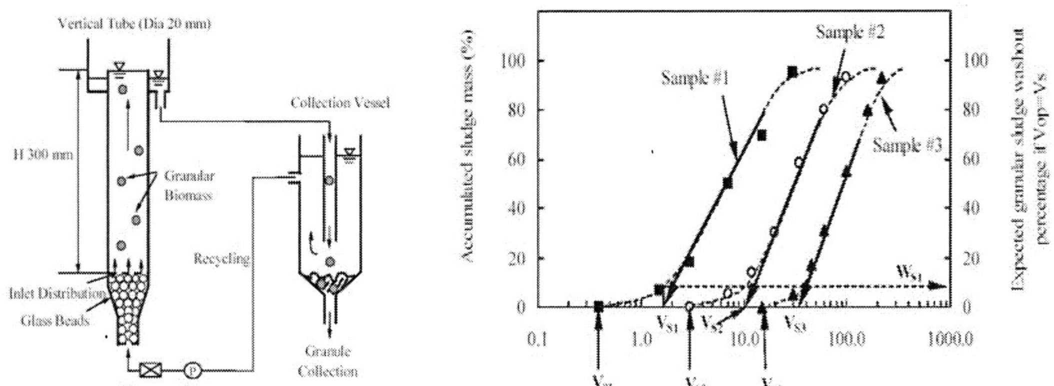

(a) (b)

Figure 6. Settleability assessment protocol (Ahn and Speece, 2003); (a) Schematic diagram for settling device; (b) Development of settleability profiles.

Table 7. Characteristics of slurry-type piggery waste

Parameters	Phase 1		Phase 2	
	Raw waste	Influent mixture*	Raw waste	Influent mixture*
pH	8.4 ± 0.1	8.4 ± 0.1	8.6 ± 0.1	8.6 ± 0.1
TCOD	56,200 ± 1,370	30,100 ± 690	55,500 ± 3,520	30,600 ± 1,760
SCOD	47,000 ± 1,760	25,600 ± 880	30,500 ± 3,960	18,100 ± 1,980
T-N	4,760 ± 400	4,150 ± 200	5,230 ± 440	5,110 ± 220
TKN	4,710 ± 400	2,350 ± 200	5,180 ± 440	2,590 ± 220
NH_4–N	4,300 ± 330	2,150 ± 170	4,320 ± 170	2,160 ± 80
NO_2–N	0	1,800	0	2,500
NO_3–N	50	25	50	25
SO_4^{2-}	420	210	420	210
T-P	479 ± 10	249 ± 5	1,260 ± 110	630 ± 60
S-P	105 ± 6	53 ± 3	629 ± 145	314 ± 73
TS	28,600 ± 840	14,300 ± 420	37,900 ± 7,660	19,000 ± 3,830
VS	21,700 ± 1,250	10,800 ± 630	23,900 ± 5,230	12,000 ± 2,610
TSS	17,800 ± 390	8,890 ± 190	15,500 ± 5,470	7,770 ± 2,740
VSS	16,300 ± 290	8,150 ± 140	14,200 ± 5,090	7,080 ± 2,540
TA	8,370 ± 990	4,180 ± 500	15,800 ± 1,040	7,910 ± 520
BA	5,150 ± 770	2,580 ± 380	12,400 ± 1,900	6,180 ± 960
VA(as HAc)	229 ± 44	114 ± 22	337 ± 102	168 ± 51

* after adding nitrite stock solution (3.6 g NO_2-N/L for Phase 1, 5 g NO_2-N/L for Phase 2); Unit: mg/L, except pH.

A gas chromatograph (Tremetrics Model 9000, USA) with a TCD detector and a Hayesap Q (80/100) column was utilized to measure gas composition (N2O, N2, CO2, NH3 and CH4). The temperature for the column was kept at 35°C, 120°C for the injector, and 120°C for the detector. The helium carrier gas had a flow rate of 30 mL/min. Data integration was accomplished using a Varian 4270 Integrator.

The sludge granulation during the operation was examined by microscopic analysis (Olympus BX60F5, Japan) and by settleability assessment protocol (Ahn, 2000; Ahn and Speece, 2003). The settleability assessment protocol is an indirect method to quantify the physical characteristics of the anaerobic granular sludge. The detail analysis method and an example for the application were introduced in the literature (Ahn, 2000; Ahn and Speece, 2003; Ahn et al., 2001). The settleability assessment protocol is to figure out critical upflow velocity, which granules are washed out of the reactor, and fractions of washout sludge at that upflow velocity. From the settleability profile as shown in Figure 6, several important points can be determined by graphical approximation method as follows:

- V0: zero washout upflow velocity of sample (the crossing point between the settleability profile and the x-axis)
- VS: critical upflow velocity of sample to allow some amount of granular sludge washout (=WS) (the crossing point between the extension of the linear section of the settleability profile and the x-axis)
- WS: the expected granular sludge washout percent of sample at Vop = VS

- V60: the velocity means that the accumulated washout sludge mass were 60% of the net sludge mass.

Anammox Activity Test of Granular Biomass

Anammox biomass activity from granular sludge treating piggery waste (Reactor B) was measured by using a 0.5 L batch reactor. At last phase of operation, granular sludge (100 mL) was collected from the bottom of reactor and was inoculated into the batch reactor. Initial ammonium and nitrite concentrations in the reactor were set to 1,280 mg/L and 2,070 mg/L using the influent mixture of the UASB reactor, resulting that the initial NO2-N/NH4-N ratio in the batch reactor was 1.6 for allowing unlimited condition in nitrite concentration.

Microbial Community Analysis

At the start-up and end phase of each reactor operation in the experiment, microbial community was analyzed by using phylogenic analysis of 16S rDNA and Fluorescence In situ Hybridization (FISH) technique.

Phylogenic Analysis of 16S rDNA

Genomic DNA of the isolates was purified by the procedure (Figure 7) of Hallin and Lindgren (1999). Genomic DNA extractions from the sludge samples were performed using a modified method of Lee et al. (1996). The quality of extracted DNA was analyzed by applying the standard agarose gel electrophoresis. DNA concentrations were measured by absorbance level at 260 nm. PCR and amplification of 16S rDNA from chromosomal DNA were carried out in a DNA thermal cycle (Model 480; Perkin-Eelmer, Norwalk, Conn. USA) with universal bacterial primers, 27F (AGAGTTTGATCMTGCTCAG) and 1492R (GGTTACCTTTGTTACGACTT). After an initial heating to 94oC for 4 min, 30 cycles consisting of 94oC (1 min), 57oC (1 min), and 72oC (2 min) with a final 10 min extension period at 72oC were performed. The PCR product was purified with a QIAquick PCR purification kit (Qiagen, Germany), cloned into pGEM-T vector (Promega, Madison, WI, USA) according to the manufacturer's instruction (Promega, Madison, WI, USA). Clones containing appropriate-sized inserts were identified by agarose gel electrophoresis of PCR products obtained from host lysates by PCR with primers complementary to the vector at sites flanking the insertion site. Unique clones were identified by restriction fragment length polymorphism (RFLP) analysis of the insert (Manz et. al., 1996). The clones representing the dominant restriction fragment groups were selected from a reactor, respectively. These clones were sequenced by PCR cycling sequencing of quick-prepped recombinant plasmids by using cycling was performed with ABI PRISM Big Dye Terminator cycle sequencing kit according to the manufacturer's instructions (Applied Biosystems, Foster City, CA, USA). The 16S rRNA partial sequences of sludge samples obtained in Reactor A are available from the EMBL nucleotide sequence database under accession no.AY609339 to AY609349.

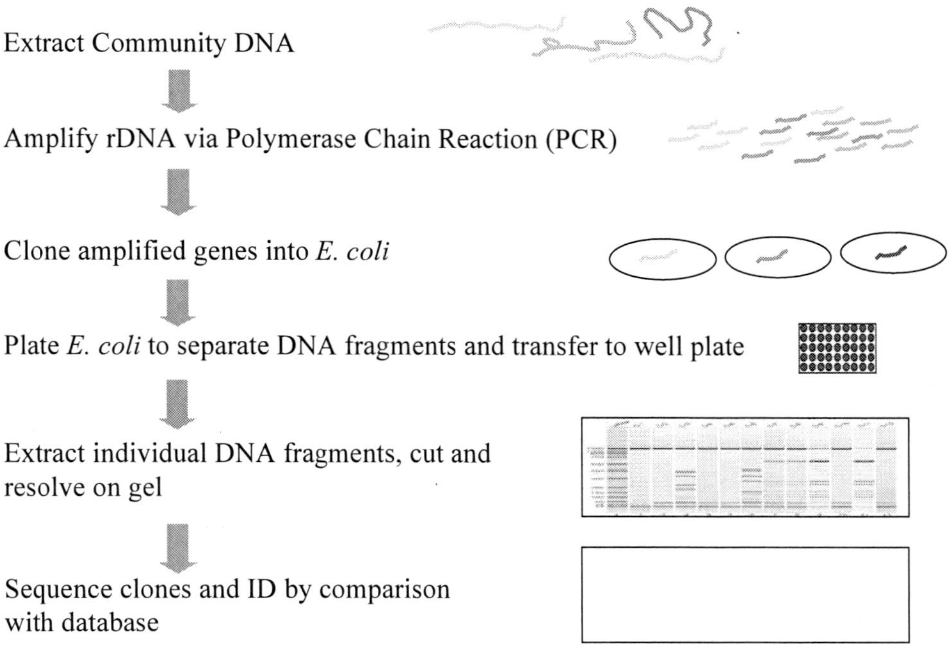

Figure 7. Procedure of microbial community analysis.

Fluorescence In situ Hybridization

At the end of the experiment using piggery waste, FISH was performed to analysis the microbial dynamics of Anammox. The granular sludge sample was fixed with 4% paraformaldehyde for at least 1hour and then exposed to 50% ethanol in PBS (phosphate-buffered saline) for 12 hrs at 4oC. after fixation, it was centrifuged at 14,000 rpm for 5 min. Using 1 x PBS the washed cell-pellets were resuspended and spotted on Teflon-coated slides (Cel-Line), dried at 46oC for 30 min, and dehydrated in 50%, 80% and 90% (v/v) ethanol (3 min each), 50:50(v/v) ethanol-xylene and 100% xylene. Dehydrated granular sludge were embedded in melted paraffin wax and sectioned in thickness of 10 μm. The 16S rRNA-targeted oligonucleotide probes used in the experiment are summarized in Table 8.

Table 8. Oligonucleotide probe sequences, target organisms, formamide concentrations required for in situ hybridization buffer and references

Probes	Specificity	Sequences (5' to 3')	% FA*/ mM[NaCl]	Reference
Pla46	Planctomycetales	GACTTGCATGCCT AATCC	25/159	Neef et al. (1998)
Kst1275	Candidatus Kuenenia stuttgartiensis	TCGGCTTTATAGG TTTCGCA	25/159	Schmid et al. (2000)
NSO190	Ammonia-oxidizing β-roteobacteria	CGATCCCCTGCTT TTCTCC	55/20	Mobbarry et al. (1996)
NIT3	Nitrobacter spp.	CCTGTGCTCCATG CTCCG	35/80	Wagner et al. (1996)

* % formamide in the hybridization buffer.

Air dried, fixed samples on the Teflon-coated slide were hybridized with 20 μL hybridization solution and each about 100 ng probe for Anammox (Pla46, Kst1275) and nitrifier(NSO190, NIT3) in an isotronically equilibrated humidity chamber at 46oC for 10 hrs. After incubation and washing, the cells were dual-strained for total bacteria with DAPI (4, 6-diamidine-2-phenylindole, 0.33 μg/mL) for 2 hrs in the dark. The mean values for each order, genus and group-specific bacteria and total cell counts were calculated from the counts of 15 randomly chosen fields using epifluorescence microscope (Zeiss Axioplan, Germany) and CLSM (Zeiss Pascal, Germany). The results were expressed in ratio (%) of the number of individual group-specific bacteria to the number of total bacteria.

Results and Discussion

Nitrogen Removal from Ammonium-rich Synthetic Wastewater

Reactor A was first operated with nitrogen concentrations of 100 mg NH4-N/L and 50 mg NO2-N /L because nitrite (> 0.1 g NO2-N/L) could inhibit Anammox reaction (Strous et al., 1999b). To examine impact of reaction intermediate of Anammox, hydroxylamine (0.42~14.7 mg N/L) and hydrazine (0~15.8 mg N/L) were also added to the influent constantly. Nitrogen loading during start-up period was 0.03±0.005 kg T-N/m3-day. As shown in Figure 8, the elimination efficiency of ammonium was over 90% during the overall operating period. At start-up time, the concentration of nitrite in effluent showed higher than that of the influent. However, the reactor reveals that the nitrite reduction occurred gradually, meaning that microbial acclimation occurred as well. Particularly, the nitrite and nitrate decreased faster, when more Anammox intermediates were added. As the behavior of reactor at start-up time, autotrophic ammonium oxidation to nitrite/nitrate is generally unusual under anoxic condition. However, the addition of hydroxylamine could result in lithoautotrophic ammonium oxidation to nitrite/nitrate (Böttcher and Koops, 1994). Even though the oxygen source under an anoxic condition is unknown, nitrite, NO2, and NO could theoretically have been used as the oxygen source (Bock et al., 1995; Schmidt and Bock, 1997). When the addition rate of the intermediates increased, the nitrite and nitrate temporarily increased, but was removed faster. The concentrations of the Anammox intermediates in effluent were very low compared with those of the influent. The concentration of hydrazine in effluent gradually decreased according to the operation time. Particularly the nitrite and hydrazine was dramatically disappeared during the batch operation. It was also observed that both nitrogen compounds were eliminated simultaneously. These results mean that proper content of Anammox intermediates (particularly, hydrazine) in substrate are an important parameter for success of the anaerobic nitrogen removal process. The nitrogen conversion rate and nitrogen removal were 0.02±0.002 kg N/m3-day and 50~64 %, respectively.

Table 9 represents the results of the secondary operation for Reactor A. The results clearly represents that ammonium oxidation and partial nitrite accumulation occurred between Day 130 to 175, resulting in a total N and ammonium removal of 48% and 92%, respectively. Due to higher nitrite concentration in the effluent, it was decided to increase ammonium loading. Double ammonium loading resulted in a rapid nitrite reduction for the next 10 days. After then, a little increase of NH4 and NO2 in the effluent was observed, but the reactor showed stable performance after another 5 days, resulting in 60% of NH4 removal. The nitrogen conversion rate was similar throughout the entire period (0.02~0.023 kg N/m3-day).

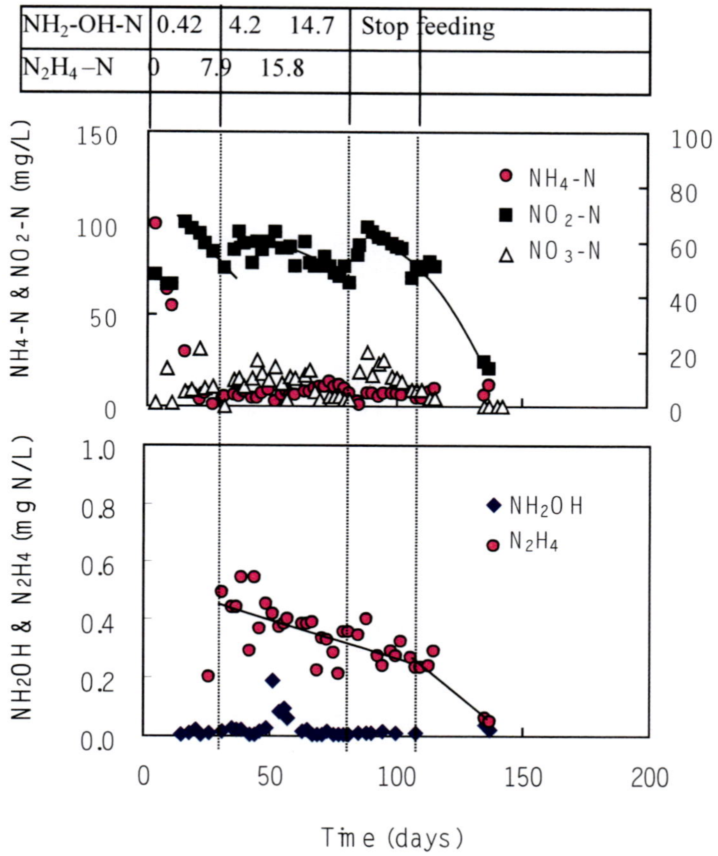

Figure 8. Nitrogen conversion from synthetic wastewater (Reactor A, Influent concentration: 0.1 g NH4-N/L, 0.05 g NO2-N/L).

Table 9. Operating result during second operation of Reactor A

Items	Test period (days)	
	150~175	190~200
Influent NH_4-N (mg/L)	100	200
Influent NO_2-N (mg/L)	50	50
Effluent NH_4-N (mg/L)	8±4	86±11
Effluent NO_2-N (mg/L)	75±11	44±9
Effluent NO_3-N (mg/L)	5±6	47±8
T-N loading (kg/m^3-d)	0.03	0.05
T-N conversion (kg/m^3-d)	0.02±0.002	0.02±0.003
NH_4-N conversion (kg/m^3-d)	0.018	0.023
T-N removal (%)	48±6	30±5
NH_4-N removal (%)	92±5	57±7

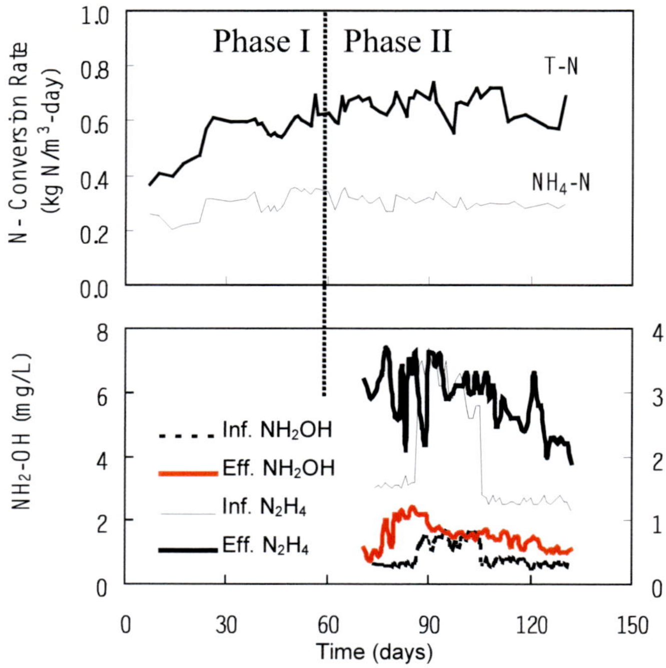

Figure 9. Nitrogen conversion from piggery waste.

According to recent research, there are some results that the ammonium removal at anaerobic (or anoxic) process treating inorganic wastewater, can be performed by anoxic ammonium oxidizer such as Nitrosomonas eutropha (Schmidt and Bock, 1997), lithoautotrophic denitrifying nitrifier such as Nitrosomonas europaea (Poth and Focht, 1985), Anammox organism with sulfate reduction bacteria (Fdz Polanco et al., 2001), as well as Anammox organism such as Brocadia anammoxidans (Jetten et al., 2001). The performance of the reactor in this experiment showed as that of the Canon process (Sliekers et al., 2002). It could be presumed that the ammonium removal in the experiment was performed by some microorganisms, such as an anoxic denitrifying autotrophic nitrifier, autotrophic denitrifier or Anammox bacteria, based on the result of this experiment and above literature.

Nitrogen Removal from Slurry-type Piggery Waste

Contrary to the foregoing experiments for synthetic wastewater, the anaerobic ammonium removal from the real strong nitrogenous waste, piggery waste, was performed successfully as shown in Figure 9. As compared to that of synthetic wastewater, the reaction quickly started in spite of the high ammonium concentration. After day 26 from start-up, the reactor showed a stable condition. Under stable operation, the average T-N and NH4 –N removal in Phase I were 71% and 74%, and 65% and 70% for Phase II, respectively. Most nitrite was removed after start-up and a little by-product of nitrate nitrogen was produced (i.e. < 5 mg/L for Phase 1 and <15 mg/L for Phase 2) during the overall operation period. The average nitrogen loading in the experiment was about 1.0 kg N/m3-d. Average nitrogen conversion was 0.59 kg T-N/m3-day (0.32 kg NH4-N/m3-day) for Phase I, and 0.66 kg T-N/m3-d (0.30 kg NH4-

N/m3-d) for Phase II. Also, the average specific nitrogen conversion rate was 0.06 kg T-N/kg VSS/day for Phase I and 0.08 kg T-N/kg VSS/day for Phase II. In spite of the higher nitrogen loading in Phase II, the specific nitrogen conversion rate was 33% higher than that in Phase II. Gas production in the reactor was 0.45 L/d and 0.57 L/d (@STP) for Phase I and Phase II, respectively. The gas produced from the reactor respectively contained 95% of dinitrogen gas and 5% of CO2 for Phase I and 100% of dinitrogen gas for Phase II, which means that the large amount of ammonium removal in the reactor was not due to ammonia stripping at the operating condition of a high pH. Generally 100~180 days of the operating time was required to reach approximately 1 kg N/m3 reactor/day of loading rate even though the enriched Anammox culture was used as seed biomass (Van Dongen et al., 2001). The result of this research demonstrates that, in the UASB reactor, the anaerobic ammonium removal from the strong nitrogenous waste such as piggery waste can be performed successfully. Furthermore, it appears that by using granular sludge as the seed biomass, the Anammox reaction can start more quickly, implying that an UASB-type reactor have many advantages; such as an easy start-up and operation.

As shown in Figure 9, the piggery waste used as a substrate contained 0.7~1.5 mg/L of hydroxylamine and 2.5~7 mg/L of hydrazine, depending on the sampling time at the piggery slurry storage pit. However, the concentrations of hydroxylamine and hydrazine in the effluent were quite higher (about double) than those of the influent. Also, those concentrations in the effluent gradually showed a tendency to decrease according to the operating time, indicating enhancement of the process stability. The results also indicate that hydroxylamine and hydrazine contents in the substrate would be an important parameter in anaerobic ammonium removal.

The slurry-type piggery waste which was used in the experiment contained a large amount of organics. Somewhat of COD removal in the experiment was observed and slightly increased with the ammonium removal. During the stable period, the COD loading rate was about 6 kg COD/m3-day and about 41~55% of SCOD removal was observed (4.7~5.2 kg COD/kg NH4-N removal). In Anammox reaction, which is a lithoautotrophic bioprocess, organic removal cannot be expected. Therefore, the COD removal could be explained by heterotrophic microbial activation in the reactor. Three types of heterotrophic organisms such as nitrite/nitrate denitrifying bacteria, acidogens and sulfate reducing bacteria could be considered due to the characteristics of the substrate and process. However, contribution of methanogens in the reactor can not be expected because the methane gas was not produced. The production of by-product nitrate nitrogen in the Anammox process can also result in the heterotrophic denitrification.

The organic acid analysis on the effluent revealed that the main organic acids were butyric (45% of COD) and acetic acid (6.3% of COD), while only a little acetic acid is contained in the substrate. This result demonstrates that acid fermentation (particularly butyric acid) would occur in the reactor. The possible butyrate fermentation could be expressed as (Eq.14) which is analogous with ethanol fermentation (Rittmann and McCarty, 2001). From this equation, COD consumption will be 11.4 g/g NH4-N removal due to the biomass synthesis in acid fermentation. Consequently, it could be conjectured acid fermentation followed by denitrification and sulfate reduction (even quite small) using the butyric acid as a carbon source for the following (Eq.14 to 17):

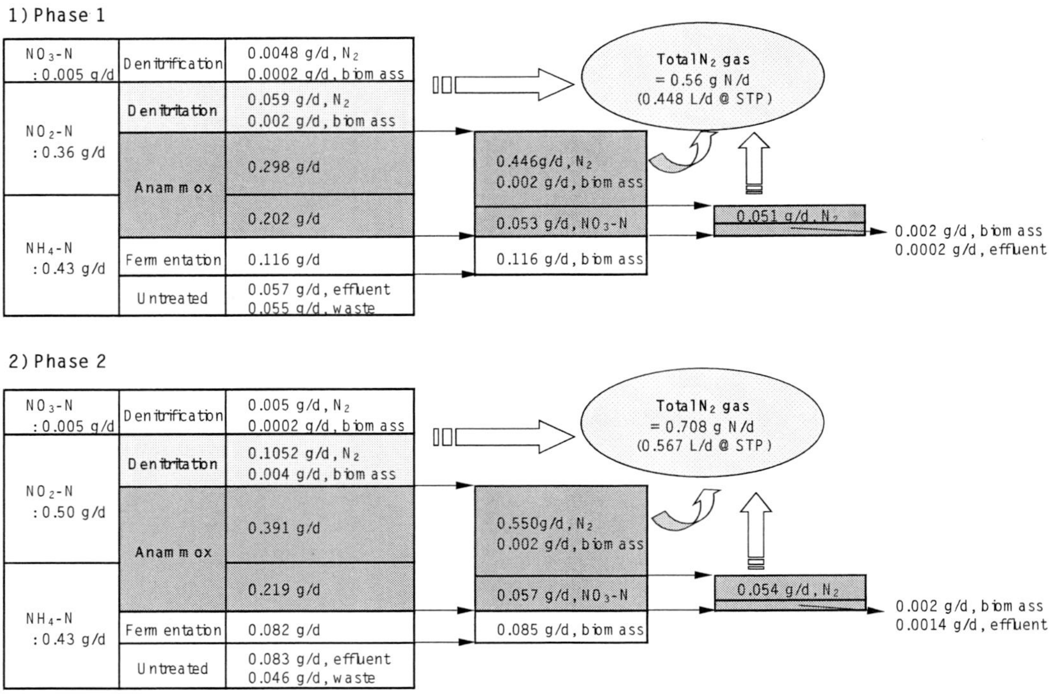

Figure 10. Nitrogen mass balance.

C6H12O6 + 0.2 NH4+ + 0.2 HCO3- → 0.2 C5H7O2N + CH3CH2CH2COOH + 1.2 CO2 + 1.8 H2O (Eq.14)

NO2- + 0.19 CH3CH2CH2COOH + H2CO3 → 0.037 C5H7O2N + HCO3- + 1.14 H2O + 0.585 CO2 + 0.481 N2 (Eq.15)

NO3- + 0.29 CH3CH2CH2COOH + H2CO3 → 0.034 C5H7O2N + HCO3- + 1.54 H2O + 0.986 CO2 + 0.483 N2 (Eq.16)

SO42- + 0.4 CH3CH2CH2COOH + 1.5 H+ → 1.6 CO2 + 1.6 H2O + 0.5 H2S + 0.5 HS- (Eq.17)

1) Phase 1

$COD_{nitrite}$: 0.41 g/d	Nitrite reduction	0.41 g/d
COD_{org} : 4.70 g/d	Sulfate reduction	0.016 g/d
	NO_2-N denitritation	0.134 g/d
	NO_3-N denitrification	0.190 g/d
	Biomass synthesis by fermentation	1.327 g/d
	Untreated	1.607 g/d (effluent)
		1.430 g/d (waste)

2) Phase 2

$COD_{nitrite}$: 0.57 g/d	Nitrite reduction	0.57 g/d
COD_{org} : 3.05 g/d	Sulfate reduction	0.016 g/d
	NO_2-N denitritation	0.240 g/d
	NO_3-N denitrification	0.200 g/d
	Biomass synthesis by fermentation	0.970 g/d
	Untreated	1.072 g/d (effluent)
		0.549 g/d (waste)

Figure 11. COD mass balance.

Based on the possible stoichiometries (Eq.14) to (Eq.17) and the experimental results for both Phases, mass balance of the nitrogen excepted organic nitrogen fraction was summarized in Figure 10. In the constant ammonium nitrogen loading (0.43 kg N/m3-day for both Phases), the ammonium removal by the Anammox was higher for Phase II (0.219 g/d, 51%) than that of Phase I (0.202 g/d, 46%). The nitrite removal for the Anammox reaction as well as the partial denitritation also increased at Phase II, while the acid fermentation decreased at Phase II. This result demonstrates that the Anammox activity at Phase II was higher than that for Phase I. Although, it is not clear whether the higher nitrite uptake for the Anammox reaction was due to the higher Anammox activity or the higher nitrite contents in the substrate, it appears that the higher nitrite content may enhance the potential of partial denitritation. By the Anammox and partial denitritation in the reactor, total nitrogen conversion to dinitrogen gas was 0.56 g/d (67% of T-N) for Phase 1 and 0.708 g/d (69% of T-N) for Phase 2, respectively. Also, total biomass production in Phase 1 (16.8%) was a little higher than that (10.5%) for Phase II, due to higher fermentation activity at Phase I.

Table 10. Summary of operating results of Reactor B

Parameters		Phase 1	Phase 2
Steady state period (day)		26~45	80~130
Influent (NO_2-N/NH_4-N) ratio		0.8: 1	1.2: 1
T-N loading (kg N_{tot}/m^3-d)		0.84	1.02
NH_4-N loading (kg NH_4-N/m^3-d)		0.43	0.43
NO_2-N loading (kg NO_2-N/m^3-d)		0.36	0.50
Effluent NO_3-N (mg/L)		2	14
$SCOD_{org}$ removal (%)		35.4 (31)	46.8 (30)
by cell synthesis in acid fermentation		28.2	31.8
by denitritation/denitrification		6.9	14.5
by sulfate reduction		0.3	0.5
BA production*,	observed	(0.42)	(1.44)
		(0.79)	(1.35)
NH_4-N removal (%)		74 (73.5)	70 (61.7)
by Anammox		47	51
by fermentation		27	19
NO_2-N removal (%)		99.8	100
by Anammox (%)		82.8	78
by denitritation (%)		17.0	22
Total nitrogen removal	(kg N_{tot}/m^3-c	0.59	0.66
		71 (82)	65 (80)
Average specific N removal (kg N_{tot}/g VSS-d)		0.064	0.08
Total. ($SCOD_{org}$ /NO_2-N) removal		4.64	2.85
Total. (NO_2-N/NH_4-N) removal		1.13	1.65
(NO_2-N/NH_4-N)$_{rem}$ ratio by ANAMMOX		1.48	1.79

( ) Based on effluent quality.

* g $CaCO_3$/g $SCOD_{org}$ removal.

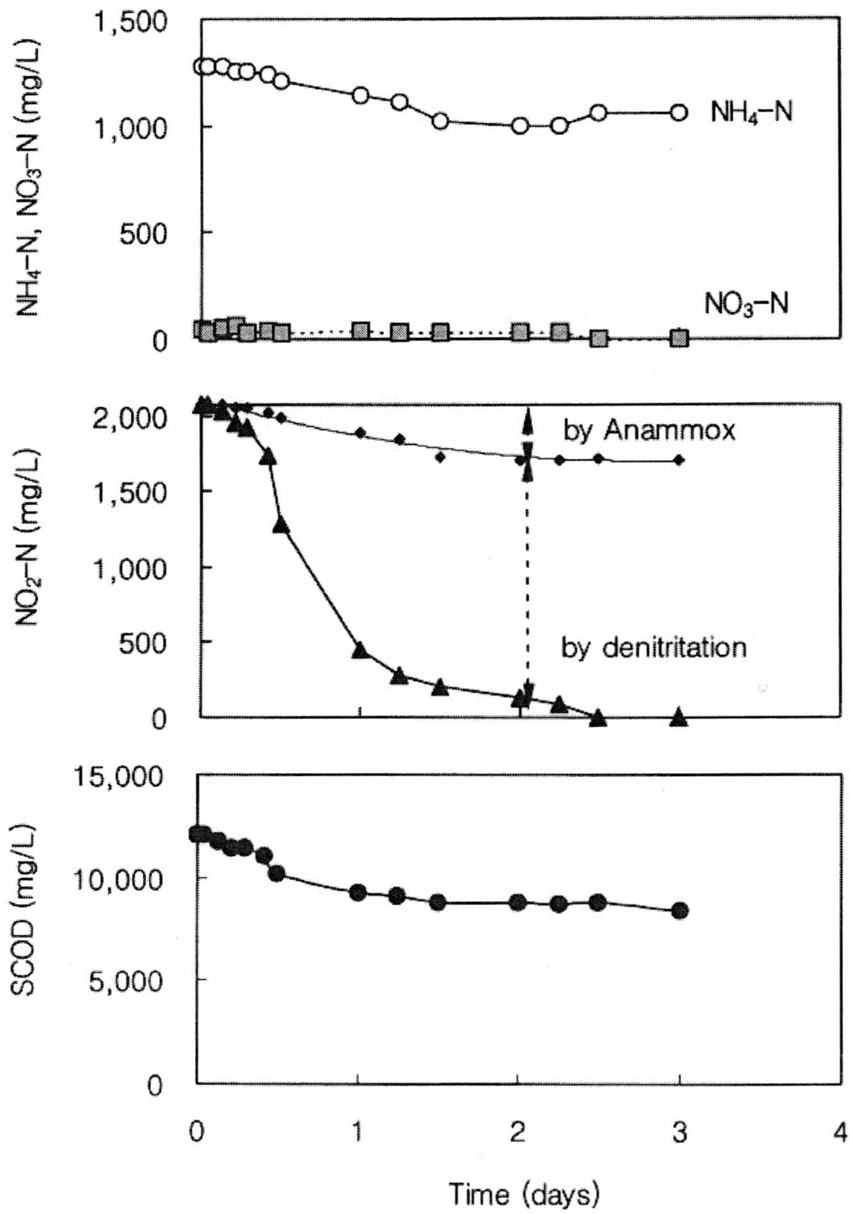

Figure 12. Microbial activity of Anammox sludge treating piggery waste.

Figure 11 represents the soluble COD balance for both Phases. With the organic loading of about 6 kg COD/m3-d (0.65 kg COD/kg VSS/day), the removal of SCOD, which the COD equivalent of nitrite excepted, was 35.4% (1.66 g/day) for Phase I and 46.8%(1.43 g/day) for Phase II, respectively. COD removal by sulfate reduction (Eq.17) was about 0.016 g/d for both Phases. Carbon consumptions for partial denitritation (Eq.15) and nitrate denitrification (Eq.16) were 2.9% and 4% of COD removal for Phase I and 7.9% and 6.6% of COD removal for Phase II, respectively.

As shown in Table 10, which summarized the comparative results for both Phases, a large fraction of COD removal was mainly due to the biomass synthesis in the acid fermentation.

The ammonium nitrogen removal by the Anammox reaction for Phase I and Phase II were 47% and 51%, respectively and 27% and 19% by fermentation. The higher nitrite content (about 50%) resulted in a higher nitrite nitrogen removal by the partial denitritation (Phase II), which led to higher bicarbonate alkalinity production as shown in Table 2. Contrary to that of Phase II, the production of bicarbonate alkalinity for Phase I was lower than the calculated value. This may be due to CO2 gas solubility and HCO3-/H2CO3 fraction in the reactor (CO2 fraction was 5% for Phase I but 0% for Phase II). Also, in spite of the higher nitrite denitritation, the NO2-N/NH4-N removal ratio by the Anammox was higher for Phase II, implying a higher Anammox activity. This result demonstrates that the Anammox reaction was less influenced by the degree of partial denitritation, and the Anammox bacteria did not compete with denitritation bacteria. By the Anammox reaction, the NO2-N/NH4-N removal ratio was 1.48 and 1.79 for Phase I and Phase II, while the total NO2-N/NH4-N removal ratio was 1.13 and 1.65 for Phase I and Phase II, respectively.

The result of the Anammox activity test using a batch reactor is shown in Figure 12. This figure shows that the Anammox reaction must be almost fully established within 1.5 days. According to these results, the maximum specific nitrogen conversion rate was 0.1 kg NH4-N/kg VSS/day (0.77 kg T-N/kg VSS/day). Compared to the ammonium conversion rate, the maximum specific nitrite conversion rate was quite higher (0.67 kg NO2-N/kg VSS/day), indicating that nitrite denitritation mainly occurred due to the unlimited condition of nitrite (initial NO2-N/NH4-N ratio = 1.6). In the batch test, 3,280 mg/L of COD was removed within 1.5 days, and the methane gas production was not observed. Literature was reported that the organic requirement for unit nitrite removal (g COD/g NO2-N) is 2.0 g COD for acetic acid and 2.8 g COD for lactic acid (Akunna et al., 1993). Also, Ho and Choi (2000) reported that 1.8~2.1 g COD/g NO2-N and 2.1~2.6 g COD (as methanol)/g NO2-N in the nitrite nitrification-denitrification process is necessary for piggery waste. If the organic requirement is assumed to be 2.1 g COD/g NO2-N, then the NO2-N/NH4-N removal ratio by the Anammox is 1.17.

Microbial Granulation and Settleability

In the early stage of reactor set-up, dark gray-colored granular sludge as seed biomass was inoculated in both anaerobic reactors. After the operation of the reactors, the color of sludge bed biomass was gradually changed to reddish brown as shown in Figure 13 (granular sludge treating ammonium-rich synthetic wastewater) and Figure 14 (granular sludge treating slurry-type piggery waste), which was accompanied by an increase in the content of cytochrome (van de Graaf et al., 1996). This peculiar biomass was grown from at the bottom of the reactor and attached on the wall of the influent feed line.

From the microscopic analysis at the end phase of the experiment, the granular sludge with a diameter of 1~2mm was observed at the lower part of sludge bed, as shown in Figure 15 and Figure 16. The size of granule decreased gradually according to operating time whereas the colour of biomass was clearly changed, as compared to that of seed sludge. Reddish granules can be observed from the micrographs of sludge treating piggery waste (Figure 16), but brownish granules for synthetic wastewater (Figure 15).

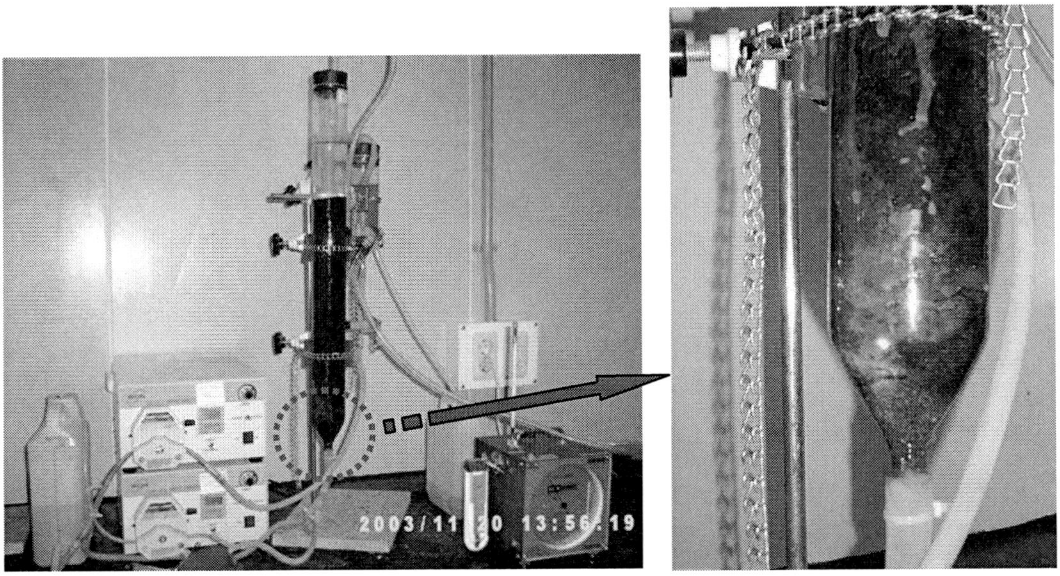

Figure 13. Picture of anaerobic reactor treating synthetic wastewater.

Figure 14. Picture of anaerobic reactor treating piggery waste.

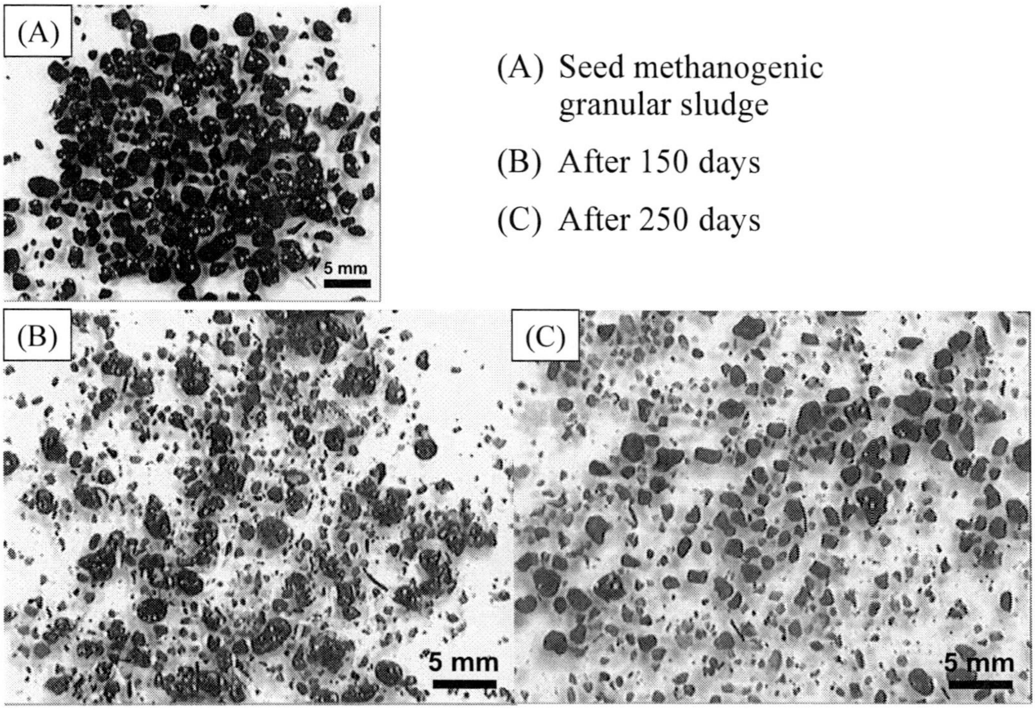

Figure 15. Micrographs of granular sludge treating synthetic wastewater.

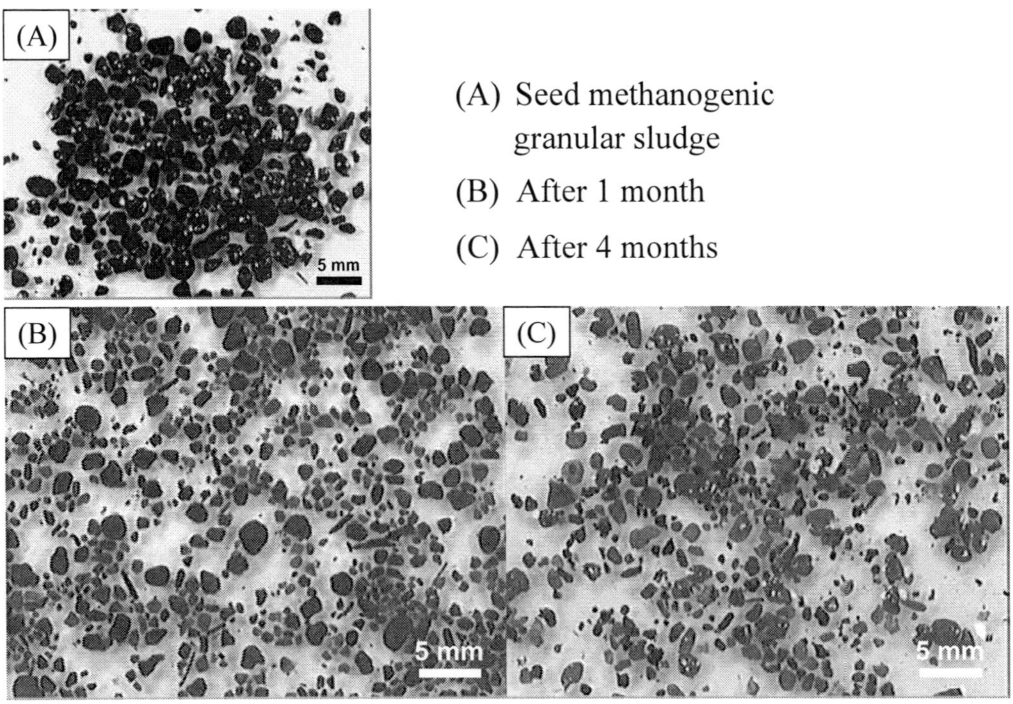

Figure 16. Micrographs of granular sludge treating piggery waste.

Table 11. Comparative granular biomass settleability

Sludge type	Substrate	Seed sludge		Cultivated biomass		Reference
		V_S	V_{60}	V_S	V_{60}	
C-removal						
	Sucrose	7	14	10~12	50	Ahn (2000)
	Dairy waste	7	14	8~10	35	Ahn (2000)
	VFAs (HAc: HPr=1:1)	10	45	2	4~5	Ahn (2000)
	Brewery waste	6	12	9~17	25~44	Ahn and Speece (2003)
N-removal						
	Synthetic inorganic	2.2	6	1.6	3~4	This research
	Piggery waste	10	15	4	5.5	This research
Slurry-type piggery waste		n.a.	n.a.	0.1	0.8	This research
Municipal sludge	Thickened	n.a.	n.a.	4	6	Ahn (2000)
	Digested	n.a.	n.a.	4	6	Ahn (2000)

Note: Unit, m/hr; V_S, critical upflow velocity; V_{60}, the velocity that the accumulated washout sludge mass were 60% of the net sludge mass.

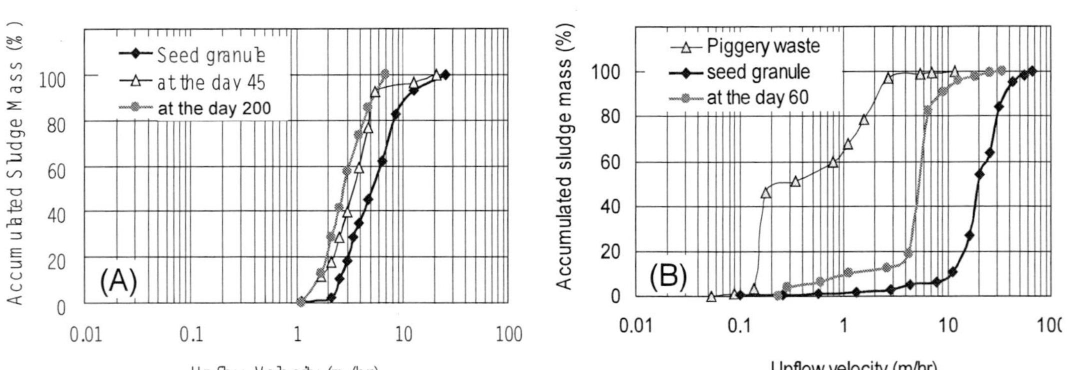

Figure 17. Settleability profile of granular sludge.

There is a simple and low labour technique to monitor stability of sludge bed by using an improved settleability assessment protocol of anaerobic granular biomass (Ahn, 2000; Ahn and Speece, 2003). Two important parameters from the protocol can be determined simply, such as critical upflow velocity (VS) which granules are washed out of the reactor, and potential fractions of washout sludge at that upflow velocity. The settleability profiles of granular sludges were shown in Figure 17. Based on this figure, Table 11 summarized the settleability characteristics of granular sludge, as compared to other granular sludge treating various organic wastes. The results revealed that the granular sludge had quite favourable settleability even though it was worse than that of the early seeded granular sludge. These results also reveal that nitrogen-removing granular sludges show poor settleability as compared to various organic-removing heterotrophic granules. Critical upflow velocity for granular sludge treating ammonium-rich synthetic wastewater was about 2 m/hr and 4 m/hr for granular sludge treating piggery waste. These physical characteristics could be used as basic parameters in operation and design of autotrophic nitrogen removal process with upflow-type reactor configuration.

Microbial Characterization

Seed Granular Sludge

The granular sludge used as seed biomass was originally collected from a full-scale UASB reactor treating brewery wastewater. Generally, UASB granules treating organic wastewater contain various types of anaerobic microorganisms, such as acidogens, acetogens and methanogenic bacteria, etc. However, It was expected that population dynamics might have been changed during about 4-months acclimation period with slurry-type piggery waste in parent batch reactor. Therefore, to compare the initial community structure and microbial diversity, microbial 16S rDNA clone library of seed granular sludge was set up and screened by RFLP. Table 12 shows the phylogenic diversity of the domain clones based on 16S rDNA sequences identified by BLAST databases. 10 unique RFLP patterns were found in the sludge sample, and most of them belong to well recognized phyla, Clostridium group which is a subclass of proteobacteria. Clostridium species are gram-positive, rod-shaped and spore-formers. These generally obligate anaerobes are ubiquitous saprophytes, and Clostridia employ butyric fermentation pathways to generate energy (Madigan et al., 2003). Two RFLP types, P2 and P5 is a close relative of Tissierella praeacuta, being that a member of the Clostridium subphylum of the gram-positive bacteria (Farrow et al., 1995). The other dominant clone, P3, is a close relative of an anaerobic thiosulfate-reducing bacterium, Fusibacter paucivorans (Ravot et al., 1999). The P4 group could be matched to a known sequence of an aerotolerant acetogen, Clostridium glycolicum (Kusel et al., 2001). The P7, P9 and P10 groups were most closely matched with gastrointestinal bacteria (Leser et al., 2002). The results reveal that the microbial community of seed sludge that acclimated in piggery waste during several months may be consisted of fermentation bacteria, acetogens, and gastrointestinal bacteria (from pig), but few methanogenic bacteria.

N-removing Granular Sludge Treating Ammonium-rich Synthetic Wastewater

To more fully characterize the bacterial community of N-removing granular sludge treating ammonium-rich synthetic wastewater, 16S rDNA directly extracted from the reactors was cloned and analyzed. Cloned 16S rDNA was analyzed by RFLP analysis. Total 129 clones were analyzed and RFLP patterns which appeared more than twice classified into eleven dominants (Table 13 and Figure 18). K1 group is the most dominant clones from the clone library. The K1 and K6 group could be matched to a known sequence of uncultured eubacterium, which is known as denitrifying bacteria isolated from the anoxic reactor of a combined system treating landfill leachate, belonging to the genera Thaura, Acidovorax, and Alcaligenes (Etchebehere et al., 2001). Two dominant RFLP types, K2 and K4 groups, are most closely matched with Nitrosomonas eutropha and Nitrosomonas europaea, respectively, which are well known denitrifying nitrifiers capable of ammonium-oxidizing in aerobic and anaerobic conditions (Bock et al., 1995; Jetten, et al., 1999; Poth, 1986; Schmidt et al., 2003). The other dominant groups of K3 and K5 are related on nitrogen removing bacterium, isolated in nitrifying-denitrifying activated sludge from an industrial sewage treatment plant (Juretschko et al., 2002). The RFLP types K8 is a close relative of Curacaobacter baltica OS 140 (namely Rheinheimera baltica in gamma subclass of proteobacteria), which is a blue-coloured bacterium isolated from the central Baltic Sea (Brettar et al., 2002). The K9 and K11 groups are a close relative of those of anaerobic dechlorinating consortia, phylogenically (Schlotelburg et al., 2000; von Wintzinggerode et al., 1999). The K10 was characterized of beta subclass of proteobacteria capable of growing in mesophilic and thermophilic environment (Lapara et al, 2000). Phylogenic analysis of the predominant RFLP types led to the identification of species in the microbial community. As shown in Figure 18, we found that the results showed quite different microbial community, as compared to those of seed sludge. Most sequence types are closely related to known anoxic denitrifying autotrophic nitrifiers and other denitrifiers, meaning that the anaerobic ammonium removal in this reactor was performed by Nitrosomonas-like bacteria and autotrophic denitrifier, but not Planctomycetales-like Anammox bacteria.

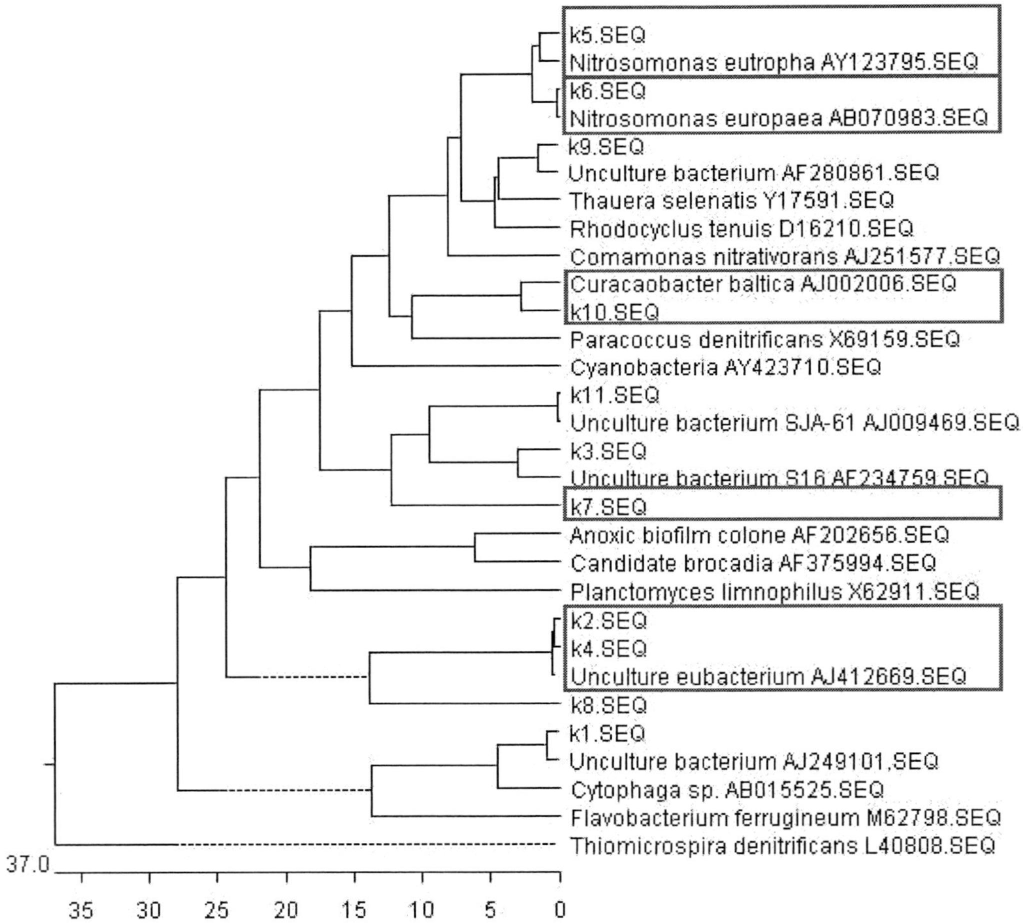

Figure 18. Phylogenic tree of the domain clones based 16S rDNA sequences (synthetic wastewater).

Table 12. Phyla from the dominant bacteria represented in the anammox clone libraries determined by BLAST comparisons of partial clone insert sequences (seed sludge)

Clones	16S rRNA gene sequencing			Subclass of proteobacteria
	Closest match (Accession No.)	No. of nucleotide compared	% Similarity with closes match	
P1 (13)[a]	*Clostridium disporicum* (Y18176)	474	92%	Cl [b]
P2 (11)	*Tissierella praeacuta* (X80833)	589	993%	Cl
P3 (10)	*Fusibacter paucivorans* (AF050099)	719	87%	Cl
P4 (9)	*Clostridium glycolicum* (AJ291746)	809	97%	Cl
P5 (3)	*Clostridium hastiforme* (X80841)	794	73%	Cl
P6 (9)	*Clostridium tertium* (Y18174)	827	94%	Cl
P7 (4)	uncultured bacterium (AF371835)	800	95%	Cl
P8 (6)	*Clostridium* sp (X75909)	445	91%	Cl
P9 (6)	uncultured bacterium (AF371689)	862	95%	Cl
P10 (5)	uncultured bacterium (AF371837)	775	94%	Cl

[a] The parenthesis is the number of colonies involved in same group. [b] *Clostridium* group.

Table 13. Summary of phylogenic diversity of the domain clones based on 16S rDNA sequences identified by BLAST databases in the reactor (synthetic wastewater)

Clones of Reactor	16S rRNA gene sequencing		
	Closest match	No. of nucleotides compared	% Similarity with closest match
K1 (32)	*Uncultured eubacterium* partial 16S rRNA gene (AJ412669)	877	96
K2 (13)	*Nitrosomonas eutropha* (AY123795)	794	96
K3 (13)	*Uncultured bacterium* partial 16S rRNA gene (AJ504594)	467	99
K4 (10)	*Nitrosomonas europaea* (ATCC 19718, BX321856)	767	97
K5 (9)	*Uncultured sludge bacterium* S16 16S rRNA gene (AF234759)	580	95
K6 (8)	*Uncultured eubacterium* partial 16S rRNA gene (AJ412669)	855	97
K7 (8)	*Uncultured bacterium clone* 265ds10 16S rRNA gene (AY212715)	757	92
K8 (5)	*Curacaobacter baltica* OS 140 16S rRNA gene (AJ002006)	859	96
K9 (3)	*Uncultured bacterium* SHA-25 16S rRNA gene (AJ249101)	844	96
K10 (3)	*Uncultured bacterium* mle1-33 (AF280861)	811	93
K11 (3)	*Uncultured bacterium* SJA-61 (AJ009469)	909	96

* The parenthesis is the number of colonies involved in same group.

Figure 19. SEM micrographs of seed granule (A) and Anammox granular sludge (B) treating piggery waste.

N-removing Granular Sludge Treating Piggery Waste

Figure 19 represents scanning electron micrographs (SEM) of Anammox granule treating piggery waste, as compared to those of seed granule. Both granules shows similar properties in size (1~2 mm) and superficial configuration, comprising of cocci- and rod-type bacteria, and filamentous bacteria together. Also, it seems that the granule has similar structural characteristics in physical aspect, as compared to those of methanogenic UASB granules treating organic wastewater (Ahn, 2000).

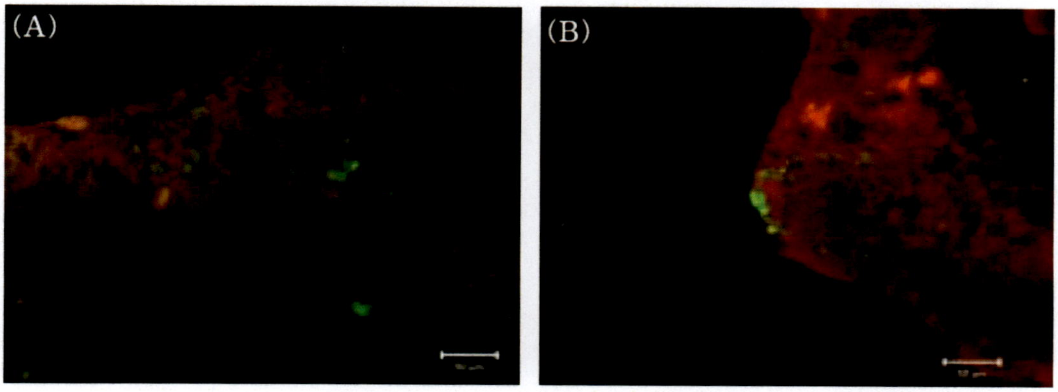

Figure 20. CLSM micrographs of Anammox granule. (A) In situ hybridization with FITC-labelled probe Pla46 (Cells of Planctomycetales are in green). (B) In situ hybridization with FITC-labelled probe Pla46 and CY3-labelled probe Kst1275 (Cells of Planctomycetales and Candidatus Kuenenia Stuttgartiensis are shown in green and red, respectively).

To characterize the microbial community of Anammox granular sludge treating piggery waste, the sludge was hybridized with probe Pla46 and Kst1275, which represent Planctomycetales and Candidatus "Kuenenia stuttgartiensis." Investigation of microbial community by means of confocal laser scanning microscope (CLSM), epifluorescence and phase contrast microscopy after nucleic acid staining showed, furthermore, that the Anammox biomass was mainly growing in the outer layer of the granule, as shown in Figure 20. The main reason might be affinity between substrate and microorganisms. From in-situ hybridization analysis, about 77% and 59% of cells were hybridized with Pla46 and Kst1275, respectively, meaning that Anammox bacteria were dominated the biomass in the reactor. However, very little ammonia-oxidizing (Nitrosomonas spp.) and nitrite-oxidizing bacteria (Nitrobacter spp.), which were hybridized with probes NSO190 and NIT3 were also detected. The results indicated that the anaerobic nitrogen removal from high strength slurry type piggery waste was mainly performed by the Planctomycetales-like Anammox bacteria, contrary to the foregoing experiments for ammonium-rich synthetic wastewater.

Conclusion

Based on recent research, at least two different possibilities for nitrogen elimination under an anaerobic environment have been reported: anaerobic ammonium oxidation by Planctomycetes-like bacteria and Nitrosomonas-like bacteria. The latter, also called denitrifying nitrifiers, was observed to reduce nitrite or nitrogen dioxide with ammonium or hydroxylamine under an anoxic condition. The novel anaerobic ammonium oxidation by Planctomycetes-like bacteria was first discovered in a denitrifying reactor. This reaction is much faster than in Nitrosomonas-like bacteria, but the peculiar lithoautotrophic microorganisms has a much lower growth rate, meaning that enrichment of the biomass as a dominant bacterium is not easy. These novel processes have various advantages in economical and environmental aspects, such as no need for external carbon addition, negligible sludge production and less energy and oxygen requirement. Even though these novel technologies meet a few challenges for application in a full-scale plant, they might offer a high potential solution for treatment of ammonium-rich wastewater.

In the experimental research, by using a laboratory scale upflow anaerobic granular sludge bed reactor, both types of anaerobic autotrophic nitrogen elimination were observed according to the type of the substrate. From treatment of ammonium-rich synthetic wastewater (100~200 mg/L NH4-N), the microbial community, based on cloned 16S rDNA analysis, indicated that an anoxic ammonium oxidizer such as Nitrosomonas eutropha and Nitrosomonas europaea and other autotrophic denitrifiers mainly contributed to nitrogen elimination, even low nitrogen conversion (0.02~0.03 kg N/m3-d). FISH (Fluorescence in situ Hybridization) analysis demonstrated that, contrary to the results from synthetic wastewater, Planctomycetes-like bacteria (especially, Candidatus "Kuenenia stuttgartiensis") played a significant role in nitrogen removal from slurry-type piggery waste (56 g COD/L and 4.3 g/L NH4-N), representing higher nitrogen conversion (0.59~0.66 kg T-N/m3-d, 0.3~0.32 kg NH4-N/m3-d). The specific microbial activity of the biomass at the end of operation was 0.1 g NH4-N/g VSS/day (0.77 g T-N/g VSS/day), indicating that potential denitritation is also still high. The results of mass balance also revealed that the Anammox reaction was influenced less by the degree of partial denitritation. From both experiments, autotrophic nitrogen-removing granular sludge showed quite favorable physical characteristics in settleability and size.

Acknowledgements

This research was partly supported by the Korean Science and Engineering Foundation (KOSEF) (Grant No.: R05-2002-000-00386-0, R05-2004-000-10012-0).

References

Abeling, U. and Seyfried, C.F. (1992). Anaerobic-aerobic treatment of high strength ammonium wastewater nitrogen removal via nitrite. *Wat. Sci. Tech.*, **26**,1007-1015.

Akunna, J.C., Bizeau, C. and Moletta, R. (1993). Nitrate and nitrite reductions with anaerobic sludge using carbon sources: glucose, glycerol, acetic acid, lactic acid and methanol. *Wat. Res.*, **27**(8), 1303-1312.

Ahlers, B., König, W. and Bock, E. (1990). Nitrite reductase activity in *Nitrobacter vulgaris*. *FEMS Microbiol. Lett.* **67**, 121~126.

Ahn, Y.H. (2000). Physicochemical and microbial aspects of anaerobic granular biopellets. *J. Environ. Sci. Health*, **A35**(9), 1617-1635.

Ahn, Y.H. and Choi, H.C. (2005). Nitrogen removal from sludge digester liquids in upflow sludge bed reactor with external aeration. *Proc. IWA Specialized Conf. on Nutrient Management in Wastewater Treatment Processes and Recycle Streams*. Krakow, Poland, 18-21 Sept.

Ahn, Y.H. and Kim, H.C. (2004). Nutrient removal and microbial granulation in anaerobic process treating inorganic and organic nitrogenous wastewater. *Wat. Sci. Tech.,* **50**(6), 207-216.

Ahn, Y.H. and Speece, R.E. (2003) Settleability assessment protocol for anaerobic granular sludge and its application. *Water SA*, **29**, 419-426.

Ahn, Y.H., Hwang I.S. and Min K.S., (2004). Anammox and partial denitritation in anaerobic nitrogen removal from piggery waste. *Wat. Sci. Tech.*, **49**(5), 145-153.

Ahn, Y.H., Min, K.S. and Speece, R.E. (2001). Pre-acidification in anaerobic sludge bed process treating brewery wastewater. *Water Res.,* **35**(18), 4267-4276.

Ahn, Y.H., Min, K. and Yun, Z. (2000). Enhancement of reactor performance and pelletization by reactor modification in UASB system. *J Environ Sci & Health*, **A35**, 1719-1733.

APHA, WEF and ASCE (1998). *Standard Methods for the Examination of Water and Wastewater*. 20th Eds. Washington DC. USA.

Baalsruud, K. and Baalsruud, K.S. (1954). Studies on *Thiobacillus denitrificans*. *Archiv Fur Mikrobiologie*, **20**, 34-62.

Bae, U.K., Ko, G.B., Lee, Y.W. and Lee, C.H. (2001). Piggery wastewater treatment using short-cut nitrogen removal technology. *Proc. Symposium of Current Issue in Korea: The Technologies and Policy for the Livestock Wastewater Management.* **23** Feb, Korea University, Seoul, Korea., 39.

Bazylinski, D.A. and Blakemore, R.P. (1983). Denitrification and assimilatory nitrate reduction in *Aquaspirillum magnetotacticum*. *Appl. Environ. Microbiol.* **46**, 1118~1124.

Batcheler, B. and Lawrence, A. (1978) Autotrophic denitrification using elemental sulfur. *J. WPCF*, **50**, 1986-2001.

Bernat, K., Wojnowska-Baryla, I. And Dobrzynska, A. (2003). Nitrogen oxidation and reduction in aerated single-stage activated sludge process. *Polish J. of Environ. Studies*, **12**(4), 387-394.

Binswanger, S., Siegrist, H. and Lais, P., (1997). Silmutane nitrifikation/denitrifikation von stark ammonium-belasteten Abwässern ohne orhanische kohlenstoffquellen. *Korrespondenz Abwasser*. **44**, 1573~1580.

Bock, E., Schmidt, I., Stuven, R. and Zart, D. (1995). Nitrogen loss caused by denitrifying *Nitrosomonas* cells using ammonium or hydrogen as electron donors and nitrite as electron acceptor. *Arch. Microbiol.* **163**, 16~20.

Böttcher, B. and Koops, H.P. (1994). Growth of lithotrophic ammonia-oxidizing bacteria on hydroxylamine. *FEMS Microbiol. Letter*, 122, 263-266.

Brettar, I., Christen, R. and Hofle, M.G. (2002). *Rheinheimera baltica* gen. Nov., sp. nov. a blue-coloured bacterium isolated from the central Baltic Sea. *Int. J. Sys. Evol. Microbiol.*, **52**(5), 1851-1857.

Broda, E. (1977). Two kinds of lithotrophs missing in nature. *Z. Allg. Mikrobiol.* **17**, 491~493.

Buchauer, K. (1998). A comparison of two simple titration procedures to determine volatile fatty acids in influents to wastewater and sludge treatment processes. *Water SA*, **24**(1), 49-56.

Castignetti, D. and Hollocher, T.C. (1984). Heterotrophic nitrification among denitrifiers. *Appl. Environ. Microbiol.* **47**, 620~623.

Choi, H.C. (2004). Performance of autotrophic nitrogen removal process with microbial granular sludge. MS Thesis, Yeungnam University (*In Korean*).

Dapena-Mora, A., Campos, J.L., Mosquera-Corral, A., Jetten, M.S.M. and Mendez, R.(2004). Stability of the AMAMMOX process in a gis-lift reactor and a SBR. *Biotechnology*. **110**, 159~170.

Damste J.S.S., Strous, M., Rijpstra, W.I.C., Hopmans, E.C., Geenevasen, J.A.J., Van Duin§, A.T., Van Niftrik, L.A. and Jetten, M.S.M. (2002). Linearly concatenated cyclobutane lipids form a dense bacterial membrane. *Nature*, **419**, 708-712.

De Bruijn, P., van de Graaf, A.A., Jetten, M.S.M., Robertson, L.A. and Kuenen, J.G. (1995). Growth of *Nitrosonomas europaea* on hydroxylamine. *FEMS Microbiol. Letter*, **125**, 179-184.

DeLong, E.F. (2002). All in the packaging. *Nature*, **419**, 676-677.

Dong, X. and Tollner, E.W. (2003). Evaluation of Anammox and denitrification during anaerobic digestion of poultry manure. *Bioresource Technol.* **86**, 139-145.

Ekama, G.A, Marais, G.v.R. and Siebritz, I.P. (1984). "Biological Excess Phosphorus Removal," *Design and Operation of Nutrient Removal Activated Sludge Processes*, Water Research Commission, Pretoria S.A.

Etchebehere C, Errazquin I, Brrandeguy E, Dabert P, Moletta R, and Muxi L (2001) Evaluation of the denitrifying microbiota of anoxic rectors. *FEMS Microbiol Ecol.*, **1224**, 1-7.

Farrow, J.A., Laqson, P.A., Hippe, H., Gauglithz, U. and Collins, M.D. (1995). Phylogenetic evidence that the gram-negative nonsporulating bacterium *Tissierella* (Bactereioides) *praeacuta* is a member of the *Clostridium* subphylum of the gram-positive bacteria and description of *Tissierella creatinini* sp. nov. *Int. J. Sys. Bacteriol.*, **45**(3), 436-440.

Fdz-Polanco F, Fdz-Polanco M, Fernandez N, Uruena MA, Garcia PA and Villaverde S (2001) New process for simultaneous removal of nitrogen and sulphur under anaerobic conditions. *Water Res.*, 35, 1111-1114.

Frear, D.S. and Burrell, R.C. (1955) Spectrophotometric method for determining hydroxylamine reductase activity in higher plants. *Analytical Chemistry*, **27**, 1664-1665.

Frunzke, K. and Meyer, O. (1990). Nitrate respiration, denitrification, and utilization of nitrogen sources by aerobic carbon monoxide-oxidizing bacteria. *Arch. Mocrobiol.*, **154**, 168-174.

Fux, C., Boehler, M., Huber, P., Brunner, I. and Siegrist, H., (2002). Biological treatment of ammonium-rich wastewater by partial nitritation and subsequent anaerobic ammonium oxidation (Anammox) in a pilot plant. *J. Biotechnol.* **99**, 295-306.

Fux, C. and Siegrist, H. (2004). Nitrogen removal from sludge digester liquids by nitrification /denitrification or partial nitritation/anammox: environmental and economical considerations. *Wat. Sci. Tech.*, **50**(10), 19-26.

Fux, C., Huang, D., Monti, A. and Siegrist, H. (2004). Difficulties in maintaining long-term partial nitritation of ammonium-rich sludge digester liquids in a moving-bed biofilm reactor (MBBR). *Wat. Sci. Tech.*, **49**(11-12), 53-60.

Gil, K.I., Lee, Y.H., Choi, E., Yun, Z. and Ha, J.S. (2002). Nitritation of anaerobic digester supernatant by syspended growth reactor and biofilm reactor. *J. KSEE*, **24**(8), 1415-1422 (*In Korean*).

Guven, D., Van de Pad-Schoonen, K., Schmid, M.C., Strous, M., Jetten, M.S.M., Sozen, S., Orhon, D. and Schmidt, I., (2003). Implementation of the anammox process for improved nitrogen removal. *IWA International Specialty Symposium on Strong Nitrogen and Agro-Wastewate*r, June 11~13, Seoul, Korea. 1, 387~394.

Helmer, C., Tromm, C., Hippen, A., Rosenwinkel, K-H., Seyfried, C.F. and Kunst, S. (2001). Single stage biological nitrogen removal by nitritation and anaerobic ammonium oxidation in biofilm systems. *Water Sci. Technol.*, **43**(1), 311~320.

Ho, J.H. and Choi, E. (2000). Nitrogen removal from piggery waste by nitrite nitrification. *Proc. of WEFTEC 2000 73rd Annual Conference & Exposition on Water Quality and Wastewater Treament.* Oct. 14-18, Anaheim, California, U.S.A., Session 30.

Imajo, U., Tokutomi, T., Furukawa, K. (2003). Granulation of Anammox microorganisms in up-flow reactors. *IWA International Specialty Symposium on Strong Nitrogen and Agro-Wastewater* June 11~13, Seoul, Korea, 1, 403~410.

Jetten, M.S.M., Horn, S.J., and Van Loosdrecht M.C.M. (1997). Towards a more sustainable minucipal wastewater treatment system. *Wat. Sci. Tech.*, **35**, 171-180.

Jetten, M.S.M., Schmid, M., Schmidt, I., Wubben, M., Van Dongen, U., Abma, W., Sliekers, O., Revsbech, N.P., Hubertus, J.E., Beaumont., Ottosen, L., Volcke, E., Laanbroek, H..J., Campos-Gomez, J.L., Cole, J., Van Loosdrecht, M., Mulder, J.W., Fuerst, J., Richardson, D., Van de Pas, K., Mendez-Pampin, R., Third, K., Cirpus, I., van Spanning, R., Bollmann, A., Nielsen, L.P., Op den Camp, H., Schultz, C., Gundersen, J., Vanrolleghem, P., Strous, M., Michael, W. and Kuenen, J.G.. (2002). Improved nitrogen removal by application of new nitrogen-cycle bacteria. *Re/Views in Environmental Science & Bio/Technology,* **1**, 51~63.

Jetten, M.S.M., Strous, M., Van de Pas-Schoonen, K.T., Schalk, J., Van Dongen, U.G.J.M., Van de Graaf, A.A., Logeman, S., Muyxer, G., Van Loosdrecht, M.C.M. and Kuenen, J.G., (1999). The anaerobic oxidation of ammonium. *FEMS Microbiol. Rev.* **22**, 421~437.

Jetten, M.S.M., Wagner, M., Fuerst, J., van Loosdrecht, M.C.M., Kuenen, J.G. and Strous, M. (2001). Microbiology and application of the anaerobic ammonium oxidation (anammox) process. *Curr. Opinion. Biotechnol,* **12**, 283~288

Juretschko, S., Loy, A., Lehner, A. and Wagner, M. (2002). The microbial community composition od a nitrifying-denitrifying activated sludge from an industrial sewage treatment plant analyzed by the full-cycle rRNA approach. *Sys. Appl. Microbiol.*, **25**(1), 84-99.

Körner, H., and Zumft, W.G. (1989). Expression of denitrification enzymes in response to the dissolved oxygen level and respiratory substrate in continuous culture of *Pseudomonas stutzeri. Appl. Environ. Microbiol.* **55**, 1670~1676.

Kuai L and Verstraete W (1998). Ammonium removal by the oxygen-limited autotrophic nitrification - denitrifcation system *Appl. Environ. Microbiol.* **64**, 4500~4506.

Kusel, K., Karnholz, A., Trinkwalter, T., Devereux, R., Acker, G. and Drake, H.L. (2001). Physiological ecology of *Clostridium glycolicum* RD-1, an aerotolerant acetogen isolated from sea grass roots. *Appl. Environ. Microbiol.*, **67**(10), 4734-4741.

Lapara, T.M., Nakatsu, C.H., Pantea, L. and Alleman, J.E. (2000). Phylogenetic analysis of bacterial communites in mesophillic and thermophilic bioreactors treating pharmaceutical wastewater, *Appl Environ Microbiol*, **66**, 3951-3959.

Lee, J.W., Kim H.C., Kwon, S.Y. and Ahn Y.H. (2004). Autotrophic nitrogen removal with anaerobic granular sludge and its microbial community. *Proc. IWA World Congress on Anaerobic Digestion (AD10)*, Montreal, Canada, 30 August- 3 Sept., 1183-1187.

Lee, S.Y., Bollinger, J.B., Bezdicek, D., and Ogram, A. (1996). Estimation of the abundance of an unculturable soil bacterial strain by a competitive quantitative PCR method. *Appl Environ Microbiol.*, **62**, 3787-3793.

Leser, T.D., Amenuvor, J.Z., Jensen, T.K., Lindecrona, R.H., Boye, M. and Moller, K. (2002). Culture-independent analysis of gut bacteria: the pig gastrointestinal tract microbiota revisited. *Appl. Environ. Microbiol.*, **68**(2), 673-6901.

Ludwig, W., Mittenhuber, G. and Friedrich, C.G. (1993). Transfer of *Thiosphaera pantotropha* to *Paracoccus denitrificans*. *Int. J. Syst. Bacteriol.*, **43**, 363-367.

Madigan, M.T., Martinko, J.M. and Parker, J. (2003). *Brock Biology of Microorganisms*. 10th eds., Pearson Educational International, NJ, U.S.A., 575.

Manz, W., Amann, R., Vancanney, M. and Scheifer, K.H. (1996). Application of a suite of 16S rRNA-specific oligonucleotide probes designed to investigate bacteria of the phylum cytophaga-flavobacter-bacteroides in natural environment. *Microbiology*, **142**, 1097-1106.

McCarty, P.L., Beck, L. and Amant, P. (1969). Biological denitrification of wastewater by addition of organic materials. *Proc. 24th Annual Purdue Indistrial Waste Conf.*, 1271.

Metcalf and Eddy (2003). *Wastewater Engineering-Treatment and Reuses*. 4th eds., McGraw-Hill, New York, U.S.A.

Mobarry, B.K., Wagner, M., Urbain, V., Rittmann, B.E. and Stahl, D.A. (1996). Phylogenic probes for analyzing abundance and spatial organization of nitrifying bacteria. *Appl. Environ. Microbiol.*, **62**, 2156-2162.

Moir, J.W.B., Crossman, L.C., Spiro, S. and Richardson, D.J. (1996). The purification of ammonia monooxygenase from *Paracoccus denitrificans*. *FEBS Lett.* **387**, 71~74.

Mosquera-Corral, A., González, F., Campos, J.L. and Méndez, R. (2005). Partial nitrification in a SHARON reactor in the presence of salts and organic carbon compounds. *Process Biochemistry*, **40**, 3109-3118.

Mulder, A., Van de Graaf, A.A., Robertson, L.A. and Kuenen, J.G. (1995). Anaerobic ammonium oxidation discovered in a denitrifying fluidized bed reactor. *FEMS Microbiol. Ecol.* **16**, 177~183.

Narkis, N., Rebhun, M. and Sheindorf, C.H. (1979). Denitrification at various carbon to nitrogen ratios. *Wat. Res.*, **13**, 93-98.

Neef, A., Amann, R.I., Schlesner, H., Schleifer, K.H. (1998). Monitoring a widespread bacterial group: *in situ* detection of Planctomycetes with 16S rRNA-targetted probes. *Microbiology (UK)*, **144**, 3257-3266.

Nyberg, U., Aspegren, H., Andersson, B., Jansen, J. la C. and Villadsen, I.S. (1992). Full-scale application of nitrogen removal with methanol as carbon source. *Wat. Sci. Tech.*, **26**, 1077-1086.

Oguz, M. (2004). *Volatile Fatty Acid Effects on Nitrite Removal and Nitrate Formation during Activated Sludge Treatment*. Ph. D. Dissertation, The University of Tennessee, Knoxville, U.S.A.

Philips, S., Wyffels, S., Sprengers, R. and Verstraete, W. (2002). Oxygen-limited autotrophic nitrification/denitrification by ammonia oxidizers enables upward motion towards more favorable conditions. *Appl. Microbiol. Biotechnol.* **59**, 557~566.

Pollice, A., Tandoi, V. and Lestingi, C. (2002). Influence of aeration and sludge retention time on ammonium oxidation to nitrite and nitrate. *Wat. Res.*, **36**, 2541-2546.

Poth, M., and Focht, D.D. (1985). ^{15}N kinetic analysis of N_2O production by *Nitrosomonas europaea*: an examination of nitrifier denitrification. *Appl. Environ. Microbiol.* **49**, 1134~1141.

Poth, M. (1986). Dinitrogen production from nitrite by a *Nitrosomonas* solate. *Appl. Environ. Microbiol.* **52**, 957~959.

Pynaert, K., Smets, B.F., Wyffels, S., Beheydt, D., Siciliano, S.D. and Verstraete, W. (2003). Characterization of an autotrophic nitrogen-removaing biofilm from a highly loaded lab-scale rotating biological contactor. *Appl. Environ. Microbiol.* **69**(6), 3626~3635.

Randall, C.W., Bernard, J.L. and Stensel, H.D. (1992). *Design and Retrofit of Wastewater Treatment Plants for biological Nutrient Removal*, Water Quality management Library 5, Technomic Publication, Lancaster, U.S.A.

Ravot, G., Magot, M., Fardeau, M.L., Patel, B.K., Thomas, P., Garcia, J.L. and Ollivier, B. (1999). *Fusibacter paucivorans* gen. nov., sp. nov., an anaerobic, thiosulfate-reducing bacterium from an oil-producing well. *Int. J. Sys. Bacteriol.*, **49**(3), 1141-1147.

Rittmann, B.E and McCarty, P.L. (2001). *Environmental Biotechnology: Principles and Applications*. McGraw-Hill, New York, NY 10020.

Robertson, L.A. and Kuenen, J.G. (1984). Aerobic denitrification: a controversy revived. *Arch. Microbiol.* **139**, 351-354.

Robertson, L.A., van Niel, E.W.J., Torremans, R.A.M. and Kuenen, J.G. (1988). Simultaneous nitrification and denitrification in aerobic chemostat cultures of *Thiosphaera pantotropha. Appl. Environ. Microbiol.* **54**, 2812~2818.

Rogalla, F., Badard, M., Hansen, F. and Dansholm, P. (1992). Upscaling a compact nitrogen removal process. *Wat. Sci. Tech.*, **26**, 1067-1076.

Schalk, J. Oustad, H., Kuenen, J.G. and Jetten, M.S.M. (1988). The anaerobic oxidation of hydrazine: a novel reaction in microbial nitrogen metabolism. *FEMs Microbiol. Letters*, **158**, 61-67.

Schmid, M., Twachtmann, U., Klein, M., Strous, M., Juretschko, S., Jetten, M., Metzger, J., Schleifer, K.H. and Wagner, M. (2000). Molecular evidence for genus level diversity of bacteria capable of catalyzing anaerobic ammonium oxidation. *Sys. Appl. Microbiol.* **23**, 93~106.

Schmid, M., Walsh, K., Webb, R., Rijpstra, W.I.C., van de Pas-Schoonen, K., Jan Verbruggen, M., Hill, T., Moffett, B., Fuerst, J., Schouten, S., Damste, J.S.S., Harris, J., Shaw, P., Jetten, M. and Strous, M. (2003). *Candidatus "Scalindua brodae"*, sp. nov., *Candidatus "Scalindua wagneri"*, sp. nov., two new species of anaerobic ammonium oxidizing bacteria. *System Appl. Microbiol.*, **26**, 529-538.

Schmidt, I. and Bock, E. (1997). Anaerobic ammonia oxidation with nitrogen dioxide by *Nitrosomonas eutropha. Arch. Microbiol.* **167**, 106-111.

Schmidt, I., Hermelink, C., Van de Pas-Schoonen, K., Strous, M., op den Camp, H.J., Kuenen, J.G. and Jetten, M.S.M. (2002). Anaerobic ammonia oxidation in the presence of nitrogen oxides (NOx) by two different lithotrophs. *Appl. Environ. Microbiol.* **68**, 5351~5357.

Schmidt, I., Sliekers, O., Schmid, M., Bock, E., Fuerst, J., Kuenen, J.G., Jetten, M.S.M. and Strous, M. (2003). New concepts of microbial treatment processes for the nitrogen removal in wastewater. *FEMS Microbiol.,* **27**, 481~492.

Schlotelburg, C., von Wintzinggerode, F., Hauck, R., Hegemann, W. and Gobel, U.B. (2000). Bacteria of an anaerobic 1,2-dichloroprapane- dechlorinating mixed culture are phylogenically related to those of other anaerobic dechlorinating consortia. *Int. J. Sys. Evol. Microbiol.*, **50**(4), 1505-1511.

Seyfried, C.F., Hippen, A., Helmer, C., Kunst, S. and Rosenwinkel, K.-H. (2001). One-stage deammonification: nitrogen elimination at low costs. *Wat. Sci. Tech.,* **1,** 71~80.

Sliekers, A.O., Derwort, N., Campos Gomez, J.L., Strous, M., Kuenen, J.G. and Jetten, M.S.M. (2002) Completely autotrophic nitrogen removal over nitrite in one single reactor. *Water Res* **36**:2475-2482

Sliekers, A.O., Third, K.A., Abma, W., Kuenen, J.G. and Jetten, M.S.M. (2003). CANON and Anammox in a ga-lift reactor. *FEMS Microbiology Letters*. **218**, 339~344.

Straub, K.L., Benz, M., Schink, B. and Widdel, F. (1996). Anaerobic, nitrate-dependent microbial oxidation of ferrous iron. *Appl. Environ. Microbiol.*, **62**, 1458-1460.

Strous, M. (2000). *Microbiology of anaerobic ammonium oxidation*. Ph.D Thesis, Kluyver laboratory, TU Delft University, Delft, The Netherlands.

Strous, M., Fuerst, J.A., Kramer, E.H.M., Logemann, S., Muyzer, G., Van de Pas-Schoonen, K.T., Webb, R., Kuenen, J.G. and Jetten, M.S.M. (1999a). Missing lithotroph identified as new planctomycete. *Nature,* **400**, 446~449.

Strous, M., Van Gerven, E., Kuenen, J.G. and Jetten, M.S.M. (1997a). Effects of aerobic and microaerobic conditions on anaerobi ammonium-oxidizing (anammox) sludge. *Appl. Environ. Microbiol.* **63**, 2446~2448.

Strous, M., van Gerven, Zheng, P., Kuenen, G.J. and Jetten, M.S.M. (1997b). Ammonium removal from concentrated waste streams with the anaerobic ammonium oxidation (anammox) process in different reactor configurations. *Water Res*. **31**, 1955~1962.

Strous, M., Heijnen, J.J., Kuenen, J.G. and Jetten, M.S.M., (1998). The sequencing batch reactor as a powerful tool for the study of slowly growing anaerobic ammonium-oxidizing microorganisms. *Appl. Microbiol. Biotechnol.* **50**, 589~596.

Strous, M., Kuenen, J.G. and Jetten, M.S.M. (1999b). Key physiology of anaerobic ammonia oxidation. *Appl. Environ. Microbiol.,* **65**, 3248-3250

Su, J.J., Liu, B.Y. and Liu, C.Y. (2001). Comparison od aerobic denitrification under high oxygen atmosphere by *Thiosphaera pantotropha* ATCC35512 and *Pseudomonas stutzeri* SU2 newly isolated from the activated sludge of piggery wastewater treatment system. *J. of App. Microbiol.*, **90**, 457-462.

Teske, A., Alm, E. Regan, J.M., Toze, S., Rittmann, B.E. and Stahl, D.A. (1994). Evolutionary relationship among ammonia- and nitrite-oxidizing bacteria. *J.Bacteriology*, **176**, 6623-6630.

Third, K.A., Slieker, A.O., Kuenen, J.G. and Jetten, M.S.M. (2001). The CANON system under ammonium limitation: interaction and competition between three groups of bacteria. *Sys. Appl. Microbiol.* **24**, 588~596.

Toh, S.K. and Ashbolt, N.J. (2002). Adaptation of anaerobic ammonium-oxidising consortium to synthetic coke-ovens wastewater. *Appl. Microbiol. Biotechnol.*, **59**, 344-352.

Udert, K.M., Fux, C., Münster, M., Larsen, T.A., Siegrist, H. and Gujer, W. (2003). Nitrification and autotrophic denitrification of source-separated urine. *Wat. Sci. Tech.,* **48**(1), 153~160.

U.S. EPA (1993). *Process Design Manual of Nitrogen Control.* EPA 625/r-93/010, Cincinnati, Ohio.

Van de Graaf, A.A., De Bruijn, P., Robertson, L.A., Jetten, M.S.M. and Kuenen, J.G. (1996). Autotrophic growth of anaerobic ammonium-oxidizing micro-organisms in a fluidzed bed reactor. *Microbiology UK.* **142**, 2187~2196.

Van de Graaf, A.A., Mulder, A., De Bruijn. P., Jetten, M.S.M., Robertson, L.A. and Kuenen, J.G.. (1995). Anaerobic oxidation of ammonium is a biologically mediated process. *Appl, Environ, Microbiol.* **61**, 1246~1251.

Van Dongen LGJM, Jetten MSM and van Loosdrecht MCM (2001) *The Combined Sharon/ ANAMMOX Process*. STOWA Report IWA Publishing London UK.

Van Loosdrecht, M.C.M. (2004). Recent development on biological wastewater nitrogen removal technologies. Presented in *International Conf. on Wastewater Treatment for Nutrient Removal and reuse 2004 (ICWNR'04),* AIT, Bangkok, Thailand, 26-29, Jan.

Von Wintzinggerode, F., Selent, B., Hegemann, W. and Gobel, U.B. (1999). Phylogenic analysis of an anaerobic trichlorobenzene-transforming microbial consortium. *Appl. Environ. Microbiol.*, **65**(1), 283-286.

Wagner, M., Rath, G., Koops, H.P., Floos, J. and Amann, R. (1996). *In situ* analysis of nitrifying bacteria in sewage treatment plants. *Wat. Sci. Tech.*, **34**, 237-244.

Watt, G.W. and Chrisp, J.D. (1952) A spectrophotometric method for the determination of hydrazine. *Analytical Chemistry* **24**:2006-2008.

Zart, D. and Bock, E. (1998). High rate of aerobic nitrification and denitrification by *Nitrosomonas eutropha* grown in a fermenter with complete biomass retention in the presence of gaseous NO_2 or NO. *Arch. Microbiol.*, **169**, 282-286.

Zumft, W.G. (1992). The denitrifying prokaryotes. *In* A Balows, H.G. Truper, M. Dworkin, W. Harder, and K.H. Schleifer (ed.), *The prokaryotes*. A handbook on the biology of bacteria: ecophysiology, isolation, identification, applications, 2nd ed., vol.1. Springer-Verlag, New York, N.Y. 554-582.

In: Research Progress in Materials Science
Editors: W. Olsson and F. Lindberg, pp. 103-129
ISBN: 978-1-60456-629-1

Chapter 4

FRP RETROFIT OF RECTANGULAR RC COLUMNS: A STATE OF THE ART

Mohamed ElGawady and Mesay Endeshaw
Civil and Environmental Engineering, Washington State University, USA

Abstract

Several earthquakes have demonstrated that many existing bridge columns are seismically vulnerable. A number of retrofitting and upgrading techniques including steel and concrete jacketing have been developed in order to reduce the seismic vulnerability of the highway system. Recently, researchers have been considering utilizing fiber reinforced polymers to replace traditionally used steel/concrete jackets. Extensive research has been conducted worldwide by a number of researchers to verify enhancements made by FRP retrofitting system, as well as to develop design guidelines for future implementations. This paper presents common failure modes of existing reinforced concrete columns as well as a brief discussion of the traditional retrofitting solutions. Developments in retrofitting of rectangular bridge columns using FRP are presented as well. Different international codes and guidelines for retrofitting of rectangular columns using FRP are reviewed as well. Investigations to date have demonstrated that the shear and flexural capacity of rectangular columns can be adequately enhanced by externally bonded FRP; whereas FRP retrofitting was less effective in improving the behavior of columns having deficient lap splices. In terms of design, several codes and guidelines present design equation for shear and confinement design. However, for rectangular columns having deficient lap splice, retrofitting using FRP is not recommended by several codes.

Introduction

The 1971 San Fernando earthquake and other recent earthquakes have demonstrated that reinforced concrete bridges built prior to modern seismic codes and design guidelines performed poorly under seismic loading. Insufficient shear transverse reinforcement and inadequate lap splice length of the main longitudinal column reinforcement dictate by the pre-1971 codes, coupled with poor detailing, including but not limited to poor anchorage of transverse reinforcement, insufficiently bent hoops, inadequate use of ties, lap splices located at high moment region, and inappropriate cut-off of longitudinal reinforcement created deficiencies in old columns. Figure 1 shows some of these deficiencies. Other deficiencies in the substructures of old bridges include inadequate development length of footing dowels (footing embedment), absence of reinforcement in the top of a footing, and inadequate footing support capacity (Kapur 2005).

Many of the interstate bridges in the United States were constructed in the 1950s and 1960s, and incorporated these deficiencies that make them vulnerable to damage or even collapse in a moderate to strong seismic events. If a major earthquake occurs in a metropolitan area, the damage to the highway system and other infrastructures will be severe. For example, a hypothetical earthquake with a magnitude of 7.0 on the Richter scale is estimated to cause a damage of $500 billion for the Los Angeles area (Folger 2007).

Since it is not financially feasible to replace most of the deficient bridges, retrofitting them is a viable option. Several retrofitting techniques such as reinforced concrete jacketing or steel jacketing have been developed for rehabilitation of seismically inadequate RC bridge columns. In the last decade or so, fiber reinforced polymer (FRP) has attracted the attention of researchers as an alternative material for retrofitting reinforced concrete (RC) structural elements. The objective of this paper is to review recent development in retrofitting of reinforced concrete columns having rectangular cross section using FRP.

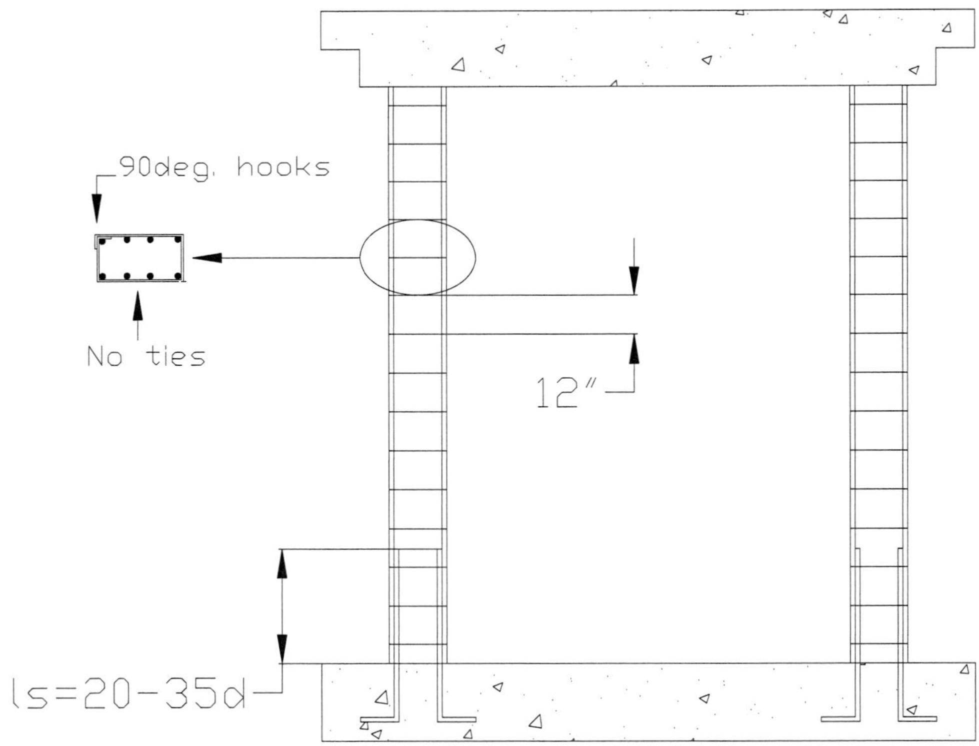

Figure 1. Typical deficiencies in existing bridges.

Failure Modes of Defiefient Columns

Bridge columns built prior to 1971 codes were designed for limited seismic forces, if any at all. Strong earthquakes often impose forces that are beyond the resisting capacity of these RC columns and cause the following modes of failure (Seible et al. 1997, Ghobarah and Galal 2004).

Shear Failure

The shear resistance of RC columns emanates from the transverse reinforcement and the core concrete. Columns having limited amount of transverse reinforcement lack sufficient shear resisting capaci y. Earthquakes often generate loads that are beyond a column's shear resisting capacity and cause brittle shear failure with a very limited ductility (Figure 2a, Figure 3a). In such cases, abrupt non-ductile deterioration with diagonal shear cracks around the section of maximum shear force occurs. With increasing load, the concrete cover spalls off and the transverse reinforcement ruptures/opens. Buckling of longitudinal bars follows or precedes lateral reinforcement failure depending on the longitudinal reinforcement size and hoop detailing.

For existing RC slender columns, the concrete core and the stirrups may develop shear strength higher than the flexural strength of the critical section, i.e., plastic hinge. However, the high bending moment at the plastic hinge regions will degrade the contribution of the concrete core component to the shear strength of the column resulting in a semi-ductile shear failure.

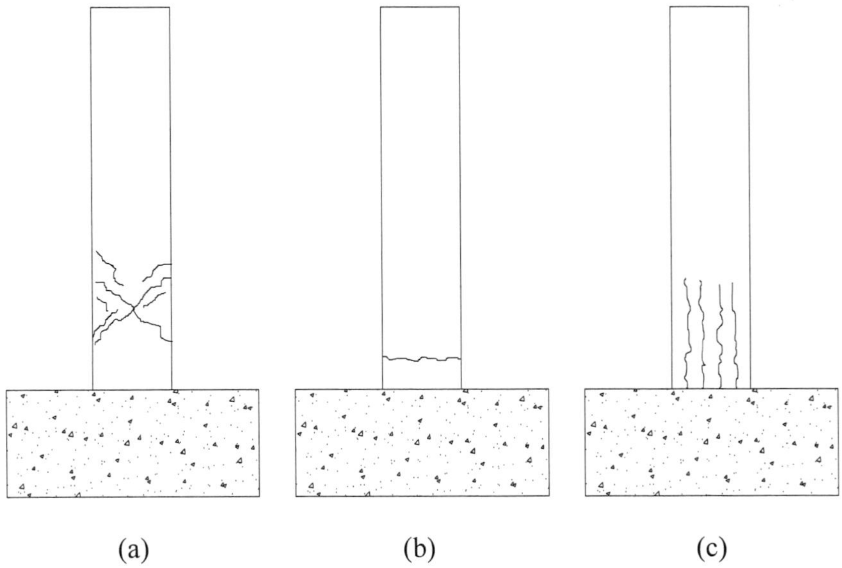

Figure 2. Typical failure modes of existing columns: (a) shear failure, (b) confining failure, (c) lap-splice failure.

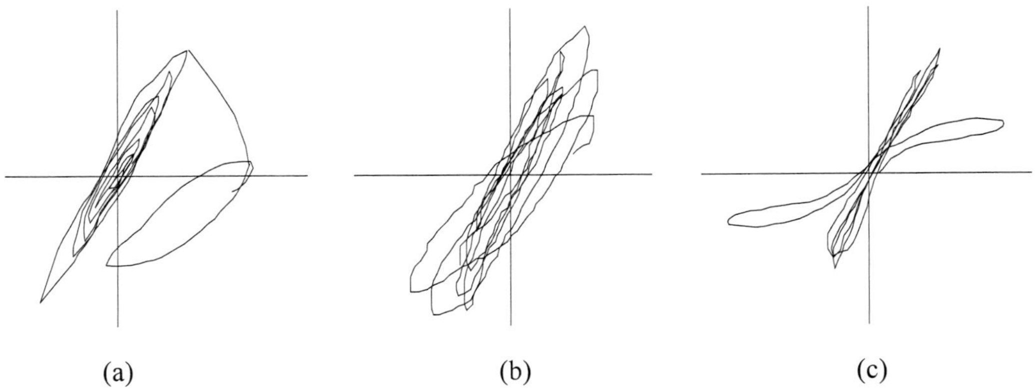

Figure 3. Sketch showing the hysteretic behavior of columns having the following modes of failure: (a) shear failure, (b) premature flexural failure, (c) lap splice failure

Premature Flexural Failure of the Plastic Hinge

Premature flexural failure (i.e., confinement failure) is associated with the lack of sufficient confining transverse reinforcement (Figure 2(b)). The transverse column reinforcement was typically No. 3 (D10) or 4 (D13) of grade 40 spaced at 12 in. independent of column size, strength, or deformation demands. Such column is incapable of providing adequate confinement to develop full flexural capacity and/or prevent buckling of the longitudinal reinforcement. Once the column reaches its flexural capacity and cracks were developed in the plastic hinge; the concrete cover will crush and spall, followed by longitudinal bars buckling. Hence, the column will not develop the required plastic deformations to dissipate the input seismic energy and the column strength will dramatically degrade (Figure 3b).

Lap Splices Failure

Lap splice failures occur when the longitudinal bars develop more force than the lap splice can resist. The pre-1971 construction practices used to lap-splice the longitudinal reinforcement rebar at the base of columns for ease of construction. The starter bars were extended 20 or 35 times the bar diameter from the footing face, which does not provide sufficient lap length to transfer the tensile force of the longitudinal reinforcement to the starter bars. Consequently, the bond fails in a brittle manner with splitting cracks (Figure 2c). In addition, inadequate confinement in the lap splice region fails to minimize the dilation of the concrete. Thus, a significant slippage of reinforcing bars occurs when the concrete around the reinforcing steel shears-off under reversed cyclic loads. Once the lap splice failed, the column will rock on its foundation (Saadatmanesh et al. 1997, Harries et al. 2006). Figure 3c shows a sketch for the hysteretic behavior of such failure.

Conventional Retrofit of Rectangular Columns

The performance of seismically vulnerable RC bridges can be upgraded using various techniques such as concrete jacketing, steel jacketing and active confinement by prestress wires. Flexural strength, ductility, and shear strength of columns can be improved by encasing the original column using a reinforced concrete Jacket. A sufficient number of dowels must be provided between the jacket and the existing column. The dowels will not significantly increase the debonding force; however, will increase the ultimate shear force that can be transferred between the old column and new jacket. Factors such as pre-wetting the interface between the old and new concrete jacket is quite controversial between different codes. Adding bonding agent between the old and new surfaces improves the performance if the surface is correctly roughened. Adding the concrete jacket will increase the moment arm between the original longitudinal reinforcement and the concrete compression block in the added concrete. This moment arm increase will increase the flexural strength of the column. Adding longitudinal reinforcement in the jacket will also increase the flexural strength of the column. However, the added reinforcement needs to be anchored into the existing pile cap/footing and other appropriate structural elements. The Seismic Retrofitting Manual for Highway Structures (FHWA 2006) proposes that the added reinforcement is anchored into newly constructed portions of the pile cap/footings. For improving the ductility of columns having rectangular x-section, it will be necessary to provide sufficient cross ties by drilling and grouting transverse reinforcement through the existing column (FHWA 2006, Bousias et al. 2006, Lehman et al. 2001, Priestley et al. 1996).

The direct cost of RC jackets is considered cheaper than other retrofit solutions. However, the indirect cost of the jacket construction should be considered. Also, in the case of bridge columns submerged in water, RC jackets are preferable over other retrofitting solutions. Although concrete jacketing enhances the performance of deficient columns, it considerably increases the mass and lateral stiffness of the rehabilitated columns. This increment will change the dynamic characteristics of the bridge, possibly, leading to higher seismic demand on the bridge (Prota et al.2005).

Steel jackets have been widely employed to retrofit seismically vulnerable bridge columns in the State of California, and elsewhere (Chang et al. 2002, Chai et al. 1994). For circular columns, two steel plate half-shells, which have been rolled to a radius equal to the column radius plus 13 to 25 mm for clearance, are positioned over the portion of the column to be retrofitted and the vertical seams are then welded. The gap between the jacket and the column is grouted with a pure cement grout, after flushing with water (Priestley et al. 1996, FHWA 2006). However, welding depends on workmanship and weather conditions. Other engagement joints have been successfully used to retrofit railway viaducts in Japan.

To avoid any significant increase in the column flexural strength, a gap of approximately 50 mm is typically provided between the end of the jacket and any supporting member (i.e. footing, cap beam, or girders) since at large drift angles the jacket can act as a compression member and bear against the supporting members (FHWA 2006). When the column cross section is rectangular, a steel jacket will be effective for enhancing shear and flexural performance. However, rectangular steel jacket is not effective for lap splice deficient columns (Chai et al 1990). In the later case, the rectangular cross section is often modified into oval/elliptical shapes before a steel jacket is applied using similar procedure as that of columns having circular cross section (Figure 4). Other researchers (Change et al. 2002) have shown that modifying rectangular cross sections to an octagonal cross section is effective as much as the oval cross section. However, the cost of construction of octagonal jacket (including grout) is 15% lower than that of the elliptical jacketing. The effectiveness of steel jackets has been demonstrated through laboratory experiments (e.g. Priestley et al. 1996, Chang et al. 2002) and field applications where some 50 steel-jacketed columns survived the 1994 Northridge earthquake without noticeable damage. Steel jackets, however, have several disadvantages including loss of space, increase of strength, corrosion, construction time and difficulties to develop bond between old structure and newly added parts.

Advanced Retrofitting of RC Columns Using FRP

Development in material science and manufacturing enabled researchers to come up with new materials such as fiber reinforced polymer (FRP). FRPs have several advantages including but not limited to: extremely low weight to strength ratio, high elastic modulus, immunity to corrosion, and ease of application which help to overcome the limitations of steel and concrete jackets. In addition, well designed FRP jacket improves column ductility without considerable increase in the column's stiffness; thereby, providing a mechanism to dissipate seismically induced energy while keep the bridge dynamic properties constant. The commonly-employed FRP materials are carbon fiber reinforced polymers (CFRP), glass fiber reinforced polymer (GFRP) and aramid fiber reinforced polymer (AFRP). FRP materials have essentially linear elastic behavior up to failure. In general, the elastic modulus of CFRP is nearly the same as normal mild steel; whereas, AFRP and GFRP have lower moduli of elasticity. In terms of tensile strength, CFRP has the highest strength followed by AFRP and GFRP. However, due to its lower cost GFRP is preferable for many civil engineering applications (Xiao et al. 2003, ACI 440.XR 2006).

In addition to the mechanical characteristics of FRP, durability is a crucial issue and must be considered by the design engineer. One of the disadvantages of applying GFRP wrapping is the freeze-thaw problem. Subjecting GFRP to low-temperature thermal cycles appears to have a greater deteriorative effect than constant immersion at a below-freezing temperature. Moisture can collect between the jacket and the concrete column and freeze resulting in expansive pressures that can cause a rapid deterioration of the concrete or the wrap. In addition, the moisture absorption of FRP causes plasticization and hydrolysis through attack on the ester linkage; both processes result in a higher induced molecular mobility, causing the deterioration of composites. Polyester epoxy matrix can provide a measure of protection against moisture absorption. Finally, CFRP are not affected by the moisture problem; however, the epoxy matrix used in fabrication of CFRP is quite sensitive to moisture absorption (Karbhari 2002, FHWA 2006). The Italian guidelines for design of FRP (CNR-DT 200/2004) apply an environmental reduction factor to the strength of the different types of FRP. This reduction factor depends on the exposure and type of FRP. For the same type of exposure, GFRP has the highest reduction in the design strength followed by AFRP, and CFRP, respectively.

Figure 4. Using an oval steel jacket for retrofitting of rectangular column.

Factors Affecting Retroftting of RC Columns Using FRP

To enhance the overall performance of RC columns, retrofit schemes must address all the possible failure modes since retrofitting for one deficiency can shift one mode of failure to another. Therefore, wrapping FRP around reinforced concrete columns is an effective means to retrofit since it addresses all aspects of the original column deficiencies. Externally bonded composite material complements the transverse reinforcement directly by providing additional shear-resistant capacity to the column. In addition, retrofitting using FRP significantly increases core concrete confinement, which results from the development of tensile force in the FRP as the cracked concrete dilate. A well confined concrete core remains intact even at higher displacement ductility and maintains the shear and flexural resistance. Moreover, the passive confining pressure helps to restrain the longitudinal reinforcement from buckling and provides a clamping force in the lap spliced region (Seible et al. 1997, Galal et al. 2005, Harries et al. 2006).

Different parameters affect the performance of RC columns after retrofitting. These parameters includes: geometry and size of the column's cross section, type and reinforcement ratio of the retrofitting material, applied normal force, applying prestressing to the FRP, and the state of the column before retrofitting/upgrading. In the next few paragraphs, these factors are discussed. Other factors such as the effect of the RC surface, characteristics of finishing mortar, and adhesive affect the performance as well (Fukuyama et al. 1999). However, these later factors are beyond the scope of this paper.

Effects of Column's Cross Section Geometry and Size

Column's cross section geometry and size significantly influence the confinement effect of FRP. Hence, the discussion below concern columns deficient in lap splice or confinement. There is no evidence that the column cross section geometry affecting columns deficient in shear.

Wrapping columns, having a circular cross section, results in a uniform confining pressure (Figure 5). In the case of columns having rectangular cross section, wrapping results in a non-uniform confining pressure. As illustrated in Figure 6, the confining of a rectangular cross section relies on the arching action. Such arching action is very effective in confining the corners of the section while it is less significant at the middle of the cross section. Thus, the effectiveness of the confinement reduces in rectangular columns. However, confining rectangular sections using FRP, results in improved performance under axial loading (Lam and Teng 2003). For high strength concrete, Zhao et al. (2001) have shown that FRP can increase the ductility but not the lateral flexural strength.

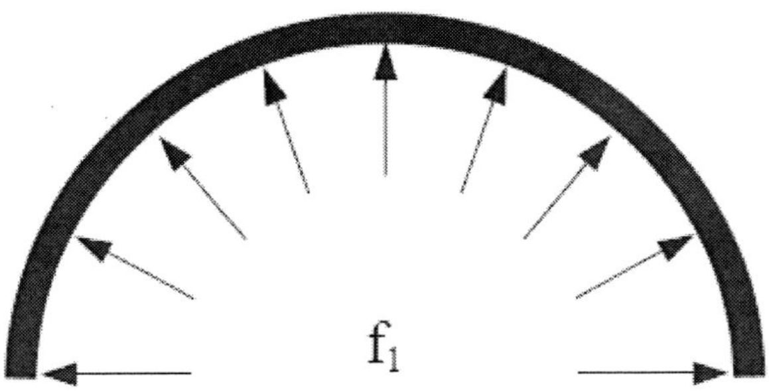

Figure 5. Uniform confining pressure of FRP in a circular column.

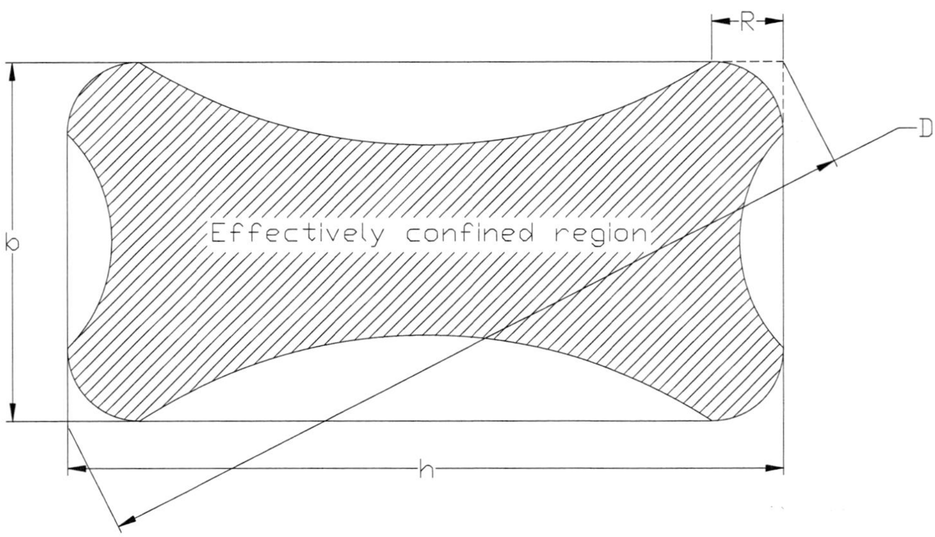

Figure 6. Effect of confinement and arching action in rectangular columns.

Increasing the cross section aspect ratio as well as the cross section dimensions will reduce the confinement efficiency (Lam and Teng 2003, Maalej et al. 2003). In order to consider retrofitting of rectangular cross sections using FRP, several codes and guidelines [CNR-DT 200/2004, S806 2002, Sieble et al 1997] recommended that the aspect ratio of the column's cross section should be equal to or smaller than 1.5. Other design recommendations [ISIS 2001] reduce this ratio to 1.1. The Italian Code for design of FRP (CNR-DT 200/2004) limits the size of the cross section to 600 x 900 mm. Seible et al. (1997) limits the columns size to 500 x 750 mm.

Another problem associated with confining rectangular cross sections is the stress concentration at the corners. In order to reduce such stress concentrations, the sharp corners of rectangular columns are typically rounded. FRP wrap will satisfactorily improve the columns confinement if the corners are rounded. Uniaxial compression tests have shown that the radius of the rounding has insignificant effect on initial stiffness while it has significant effect on the characteristics of the post-peak stress-strain relationship of these columns. Specimens with a rounding radius of approximately 2% D (see Figure 5) suffered from strain-softening phenomena. The softening was vanished and the stress strain relation became perfectly elasto-plastic by increasing the rounding radius. Finally, by increasing the rounding radius to 18%D, the post peak stiffness became positive with strain hardening (Yan et al. 2005, Sieble et al. 1997, Ghobarah and Galal 2004, Galal et al. 2005, Saadatmanesh et al. 1997, Al-Salloum 2007, Rochette and Labossiere 2000). Few other researchers (e.g. Paula and Silva 2002) reported that increasing the rounding radius will increase the ultimate compressive strength of the strengthened columns. More controversies exist on the effect of rounding radius on the ultimate strain. While some researchers (Al-Salloum 2007, Rochette and Labossiere 2000) reported that increasing the radius will increase the ultimate strain; others (e.g. Paula and Silva 2002) claimed that increasing the radius will decrease the ultimate strain. Both S806 (2002) and CNR-DT 200/2004 (2004) recommended that the minimum rounding radius be 20 mm. This value increased in ISIS design manual 4 (2001) to 35 mm or b/6 whichever is smaller.

In several cases, the rounding radius of the columns corner is restricted by the existence of reinforcing bars which limit the efficiency of the FRP jacketing. In such cases, a viable alternative to enhance confinement is to modify the cross section into an oval section. An oval section, similar to rounding, avoids stress concentrations; moreover, the effectiveness of confinement for oval section is higher than that of a rectangular section (Seible et al. 1997; Haroun et al. 2003; Saadatmanesh et al. 1997, Yan et al. 2005). Haroun et al. (2003) used a premold mortar blocks to rounded a square cross section. The rounding significantly improves the performance. However, modifying the cross sectional shape cuts down the advantages gained from a FRP retrofit, and the cost of retrofitting increases considerably.

Recently, several researchers came up with different hybrid solutions to overcome the confinement problem of rectangular cross sections. Most of these solutions are based on improving the arching action through using high stiffness materials such as steel plates or sandwich panels beneath the FRP layers. Chang et al. (2001) has shown that the hybrid steel-FRP solution is very successful in developing the required confining pressure for deficient lap splice columns.

Effect of the Axial Rigidity and Modulus of Elasticity of the FRP

In general, increasing the FRP axial rigidity, defined as the FRP reinforcement ratio times its modulus of elasticity, will improve the strength, deformation capacity, and ductility of the retrofitted columns (MacRae et al. 1997, Zhang et al. 1999, Ye at al. 2001, Iihoshi et al. 1999). However, researchers (Triantafillou and Antonopoulos 2000, ElGawady et al. 2006, Chaallal et al. 2002) have shown that there is an upper limit on the amount of FRP axial rigidity beyond this limit any increment in the FRP has no effect on the response of the retrofitted columns.

Besides the axial rigidity of the FRP, the type of FRP i.e. GFRP, CFRP, or AFRP may influence the behavior of the retrofitted columns. Generally, CFRP has relatively higher modulus of elasticity compared to AFRP and GFRP; nevertheless, comparable axial rigidity, can be obtained when a sufficient amount of GFRP is provided. Galal et al. (2005) have shown that RC column retrofitted using CFRP dissipated energy more than similar column retrofitted using GFRP. In addition, the column retrofitted using CFRP had a slightly higher strength and deformation capacity. However, the column retrofitted using CFRP had an axial rigidity 16% higher than the one that was retrofitted using GFRP. Moreover, the column that was retrofitted using GFRP had been tested as a reference specimen before retrofitting. This difference in the state of damage before retrofitting between the two columns significantly influenced the seismic response of the specimens as discussed later in this paper. Other researchers (Bousias et al. 2004) have shown that replacement of CFRP by equivalent (i.e. the same axial rigidity ratio) GFRP has a very limited effect on the response of the specimens. Specimens that were retrofitted using GFRP had a lower strength but higher deformation capacity. Furuta et al. (2003) have shown that for the same FRP reinforcement ratio, the axial stresses in AFRP are lower than the stresses in the CFRP.

Effect of Applied Normal Force

It is anticipated that applying higher normal forces will improve the performance of columns having deficient lap splice. However, by increasing the normal forces the seismic demand on the column confinement and bar buckling are increased since the normal force will shift the mode of failure from bar slippage to bar buckling which will require more FRP axial rigidity. In general, increasing the normal force will reduce the column ductility and energy dissipation. In terms of strength, the effect of the applied normal force depends on the level of the applied normal force, the column aspect ratio, and the axial rigidity of the applied FRP. Iihoshi et al. (1999) increased the axial load from 0.15 to 0.30 of the column ultimate compressive strength with a limited increase in the strength. While Ghosh and Sheikh (2007) when increased the axial load from 0.3 to 0.5 of the column ultimate compressive strength there was degradation in the ultimate strength of the column. In all cases, increasing the axial load leads to significant degradation in the post peak branch of the lateral load displacement relationship (Ghosh and Sheikh 2007, Iihoshi et al. 1999, Nagai et al. 1999, Wu et al. 2001).

For columns deficient in shear and retrofitted using FRP, increasing the applied normal force adversely affected the seismic response of the columns and increased the seismic demand on FRP Jacketing. Member and section ductility, energy dissipation, and columns strength will reduce with increasing the applied axial load. The effect of the higher axial load can be superseded by increasing the amount of FRP (Memon and Sheikh 2005, Iacobucci et al. 2003). However, other researchers (Furuta et al. 2003) examined the data in the literature of 65 columns tested in Japan. The conclusion was that there is no correlation between the stresses in the FRP and the amount of axial load. Finally, it should be noted that all the existing design formulae have not taken into consideration the effect of the applied normal force on the performance of the retrofitted columns.

Effect of Prestressing FRP

Most of the research carried out to date including applying FRP for improving the confinement or shear strength applying FRP without prestressing. Quite recently, few researchers applied prestressing to FRP before using it in confinement of existing columns. The prestressing applied in general by cutting the FRP to an appropriate length. Use the epoxy to glue one end of the FRP to the other end to make a closed loop. Approximately 100 mm length of the FRP was used as a splice length. Straighten the loop to form a two-ply FRP belt with eye hook at each end through which a crossbar can be installed. The straightened belt is wounded around the column and the ends are clamped together using a steel coupler and the installed crossbar. The steel coupler is manually screw driven using wrench to give the required pretension. The available results of applying prestressing for shear retrofitting are quite controversial. Satoh et al. (2001) used this technique to retrofit shear-deficient RC columns having rectangular cross section. This technique improved the strength, ductility, and energy dissipation of the tested columns. However, Nesheli and Meguro (2005) have shown that the performance of RC columns with pretension FRP is identical to RC columns retrofitted using normal FRP. More research is needed to confirm the effect of prestressing the FRP before retrofitting.

Retrofitting versus Upgrading

The effectiveness of using FRP in retrofitting of damaged columns is based on the extent of damage before retrofitting and the level of the applied normal force. Retrofitting of columns carrying small normal forces after being damaged by lateral loading is quite successful. The retrofitting is able to restore and improve the inherent seismic characteristics of the existing column. However, the seismic performance of specimens that have been upgraded using FRP directly after construction (i.e. without any prior damage) is significantly superior to those retrofitted after damage. The effect of damage before retrofitting using FRP is more pronounced under high normal forces. In general, if the damage was extensive such as beyond the yielding of the reinforcement, the retrofitting will be less successful in terms of strength and deformation. Finally, increasing the normal force and/or damage of the column before retrofitting will significantly increase the demand on FRP jacket (Bousias et al. 2004, Memon and Sheikh 2005, Iacobucci et al. 2003, Ghosh and Sheikh 2007).

Performance of Columns Retrofitted Using FRP

In the previous sections the parameters that affected the performance of the retrofitted columns were discussed. In the next few sections, a brief review of the performance of columns having different types of structural deficiencies and that have been retrofitted using FRP is discussed.

Retrofitting of Confinement Deficient Columns

Several researchers discussed the advantages, disadvantages, and limitation on confinement of columns having rectangular cross sections. Most of these tests carried out under uniaxial compression tests. A summary of these tests and models may be found in Lam and Teng (2003). Details discussion of confinement is beyond the scope of this paper; however, the main factors affecting the performance of the retrofitted column are the column size and cross section aspect ratio as discussed earlier in this paper. Increasing the section aspect ratio will decrease the ultimate strain and stress of the confined column (Lam and Teng 2003, Cole and Belarbi 2001).

Retrofitting of Lap-splice Deficient Columns

The goals of retrofitting lap splice deficient columns are to increase the ultimate bond stress between the splice bars and/or to keep this ultimate stress constant while the joint sustains high displacement. Limiting the lateral dilation of the concrete at the lap splice zone to approximately 0.001, by confining using FRP, can achieve these goals (Seible et al. 1997). For circular columns having lap splice length of at least 20 times the diameter of the longitudinal bars, the FRP retrofitting can increase the bond stress and prolonging bar slippage long enough to yield starter bars. Retrofitted columns are able to achieve ductility of eight (Seible et al. 1997, Xiao and Ma 1997). However, in some cases and after this high ductility is reached, the columns can still fail due to bar slippage (Haroun et al. 2003).

In the case of columns having rectangular cross sections, the limited confinement provided by the FRP is unable to improve the ultimate bond stress and to limit the hoop strains below 0.001; as a result, slippage of reinforcing bars occurred before high ductility is attained. However, the improved confinement may control concrete cracks and maintain the ultimate bond stress until at least double the ductility of the unretrofitted columns is reached (Seible et al. 1997; Harries et al. 2006, Ghosh and Sheikh 2007, Yalcin el al. 2006, Prota et al. 2005, Sause et al. 2004, Elnabelsy and Saatcioglu 2004) Saadatmanesh et al. 1997, Haroun et al. 2003; Dagher and Harajli 2007). Based on the limited improvement in the ductility of the retrofitted, some researchers proposed not to use FRP for confinement of lap splice deficient columns (Elnabelsy and Saatcioglu 2004).

One issue with the retrofitted columns is that the amount of energy dissipation is limited. Even if the displacement ductility of the retrofitted columns is improved, pinching of hysteretic loops at relatively lower displacement ductility lead to limited energy dissipation. To improve both ductility and energy dissipation, it is recommended to modify rectangular cross sections into an oval shape before the FRP jacket is installed as discussed earlier. An alternative solution to substandard lap splices may be to relocate the plastic hinge away from the column bottom, a high moment region, by strengthening columns using steel spikes or other feasible materials (Prota et al. 2005).

The improvement in the lap splice performance due to FRP confinement is associated with deformed steel bars as the bar ribs crush against the surround concrete keys. Using the appropriate FRP thickness will help this mechanism. With continuous cyclic loading, however, the concrete between the ribs shears-off resulting in slippage of the longitudinal reinforcement. For undeformed bars, FRP confining will not improve the lap splice performance due to the lack of a mechanism to delay slippage in the lap splice region since the bars rely solely on adhesion and friction with the surrounding concrete (Harries et al. 2006, Yalcin el al. 2006).

Retrofitting of Shear Deficient Columns

Shear failure in columns is very brittle in nature and the columns may fail before even a ductility of one. Well designed and applied FRP improve the shear strength and may shift the mode of failure to a very ductile flexural failure. For both circular and rectangular columns FRP is quite successful and it is able to reach a ductility of up to nine (Haroun et al. 2003). The retrofitting improved the energy dissipation as much as ten times the reference deficient columns. As discussed earlier in this paper, increasing the axial rigidity of FRP significantly improves the performance of the retrofitted specimens. Moreover, if the FRP is well designed and applied the upgraded columns may achieve a performance superior to columns designed according to ACI-318 (Memon and Sheikh 2005, Iacobucci et al. 2003).

Failure Modes of Retrofitted Columns

Failure of existing columns is initiated by crushing of the concrete at low compressive strain of 0.003 (Parvin et al. 2002). This implies that failure occurs instantaneously after ultimate strength is achieved without undergoing much inelastic deformation. The absence of such inelastic deformations leads to a poor energy dissipation and seismic performance. For retrofitted columns, a number of different modes of failure were observed. In lap spliced rectangular columns, slippage of bars governed column failure before the wrapping FRP rupture. In the case of circular columns, where confinement using FRP is quite significant, lap spliced deficient columns fail due to low-cycle fatigue rupture of starter bars (Haroun et al. 2003, Seible et al. 1997). For FRP having higher axial rigidity, extreme concrete crushing in the plastic hinge initiates failure with no signs of tensile failure of the FRP. On the other hand, sudden rupture of the FRP at the rounded corners of rectangular columns is observed for jackets having relatively small axial rigidity FRP. The rupture strain of the FRP approximately less than 50% of the actual ultimate strain obtained from the tensile coupon tests (Haroun et al. 2005, Ghobarah et al. 2004, Xiao et al. 2003, and Triantafillou et al. 2000).

FRP Retrofit Design

As discussed in previous sections, several factors influence the performance of the retrofitted columns. However, most of the current design equations have not taken into considerations all these factors. Only a few of them such as the FRP axial rigidity are introduced into the design equations. Recently, few researchers attempted to use finite element models to determine parameters affecting the performance of the retrofitted columns. Galal et al. (2005) have shown that higher concrete compressive strength, yield strength of longitudinal reinforcement, longitudinal reinforcement content, and high axial load require thicker FRP jackets to accomplish targeted displacement ductility (μ_Δ). On the other hand, an increase in the transversal reinforcement yield strength, shear span-to-depth ratio, and transverse reinforcement percentage require fewer FRP wraps to develop a given displacement ductility. Therefore, such parameters have to be considered in the development of any future design model for FRP retrofit.

Shear Retrofit Design

The effects of continues fiber reinforcement are expressed by substituting fiber stress for steel reinforcement. Evaluation of shear capacity based on yield strength is possible for steel reinforcement as steel reinforcement yield with an increasing deformation of a member. However, in the case of FRP, the stress strain relation is perfectly linear elastic up to failure, evaluation of the strain and stress in the FRP based on the failure mechanism is necessary.

Design models for shear retrofit design are limited. For columns having rectangular cross sections, the shear strength of the retrofitted columns can be calculated as follows:

$$V = V_{RC} + V_{FRP} \tag{1}$$

Where V_{RC} is the shear strength of the existing reinforced concrete column, and V_{FRP} is the contribution of FRP to the shear strength. The contribution of the existing RC column is calculated based on an appropriate RC design code. The contribution of FRP can be calculated as follows:

$$V_{FRP} = 2 f_{FRP} d t_j \tag{2}$$

Where d = column dimension in the loading direction, t_j = is the total thickness of the FRP, f_{FRP} = is the strength of the FRP which is smaller than the ultimate strength of the FRP. The value of the f_{FRP} depends on the selected model as discussed in the next few paragraphs.

In order to maintain the contribution of the concrete core to the V_{RC} component, Seible et al. (1997) proposed an upper limit of 0.4% on the strain in the FRP. Beyond this limit, the strength of the existing concrete decrease due to degradation in the aggregate interlock mechanism. In other words the stress in the FRP can be calculated as follows:

$$f_{FRP} = \varepsilon_{FRP} E_{FRP} \tag{3}$$

Where E_{FRP} = the FRP modulus of elasticity, ε_{FRP} = an upper limit on the axial strain on the FRP = 0.004. Substituting for V_{FRP} and V_{RC}, the FRP thickness can be calculated as follows:

$$t_j = \frac{\frac{V}{\phi_v} - V_{RC}}{2 f_{FRP} d} \tag{4}$$

Where V is the column shear demand based on full flexural over strength in the potential plastic hinges, Φ_v is shear capacity reduction factor.

Equations 1and 2 have been recommended by several codes and standards. Calculating V_{RC} is different from code to code; however, the difference in calculating V_{RC} is beyond the scope of this chapter. The focus of this paper is that how these codes calculate V_{FRP}. The Canadian Standard for design of FRP (S806 2002) recommended that:

$$f_{FRP} = 0.4 f_{FRP,U} \tag{5}$$

Where $f_{FRP,u}$ is the ultimate tensile strength of FRP. However, if equation 5 is greater than equation 3 with ε_{FRP} = 0.4%, the contribution of the concrete component V_C to V_{RC} should be neglected. Also, V_{FRP} in equation 2 should be multiplied by the appropriate resistance factor for FRP. Finally, the S806 (2002) uses the following upper limit on the strength of the retrofitted column.

$$V \leq V_C + 0.6 \lambda \varphi_c \sqrt{f'_c} bd \tag{6}$$

Where $\lambda = 1$ for normal weight concrete, and $\phi_c = 0.6$.

Another Canadian Guidelines for retrofitting of RC structures using FRP is the ISIS manual 4 (ISIS 2001). ISIS uses procedure similar to equations 1, 2, and 3 with an upper limit on the strain in the FRP (ε_{FRP}) of 0.2% and a resistance factor of 0.75.

The ACI 440 (2002) uses equations similar to equation 1 to 4 with two main differences. First, in equation 4, the value of f_{FRP} should not exceed $0.75\varepsilon_{FRP,u}E_{FRP}$ where $\varepsilon_{FRP,u}$ = the ultimate strain of the FRP. Second, beside the strength reduction factor, there is an additional reduction factor of 0.95 is applied to the contribution of V_{FRP}.

The Japanese Guidelines for design of upgrading of concrete structures using FRP (2001) uses equation 1 with V_{FRP} calculated as follows:

$$V_{FRP} = K.[t_f . f_{FRP,U} .z] / \gamma_b \tag{7}$$

$$K = 1.68 - 0.67R \leq 0.8 \tag{8}$$

$$R = \left(\frac{t_j E_{FRP}}{b}\right)^{\frac{1}{4}} \left(\frac{f_{FRP,u}}{E_{FRP}}\right)^{\frac{2}{3}} \left(\frac{1}{f'_c}\right)^{\frac{1}{3}} \leq 2.0 \tag{9}$$

$$z = \frac{d}{1.15} \tag{10}$$

However, K and R should be greater than or equal to 0.4 and 0.5, respectively.

The Italian Code for design of FRP (CNR-DT 200/2004) recommended that only complete wrapping, with the FRP fiber oriented perpendicularly to the column longitudinal axis, be allowed for retrofitting of rectangular RC columns. In the Italian Code, Equation 2 is replaced by Equation 11.

$$V_{FRP} = \frac{1}{\gamma_{Rd}}.\min\{0.9d, h\}.f_{fed}.2.t_j \tag{11}$$

$$f_{fed} = f_{fdd}[1 - \frac{1}{6}.\frac{l_e}{\min\{0.9d, h\}}] + \frac{1}{2}(\Phi_R.f_{fd} - f_{fdd})[1 - \frac{l_e}{\min\{0.9d, h\}}] \tag{12}$$

$$f_{fdd} = \frac{1}{\gamma_{f,d}\sqrt{\gamma_c}}\sqrt{\frac{2E_{FRP}\Gamma_{FK}}{t_f}} \tag{13}$$

$$f_{fd} = \min\{\eta_a.\frac{f_{FRPk}}{\gamma_f}, f_{fdd}\} \tag{14}$$

$$\Gamma_{FK} = 0.03k_b\sqrt{f_{ck}f_{ctm}} \tag{15}$$

$$k_b = \sqrt{\frac{1}{1+\frac{b_f}{400}}} \geq 1 \tag{16}$$

$$\Phi_R = 0.2 + 1.6\frac{r_c}{b_w} \tag{17}$$

$$l_e = \sqrt{\frac{E_{FRP}.t_j}{2f_{ctm}}} \tag{18}$$

Where f_{FRPk} = the characteristic strength of FRP, γ_{Rd} = partial safety factor for resistance models = 1.2 (for shear design), $\gamma_{f,d}$ = partial safety factor for FRP debonding = 1.2, γ_c=

partial safety factor for concrete, γ_f = partial safety factor for FRP rupture=1.1, n = an environmental factor depends on the FRP type and environmental conditions and ranging from 0.5 to 0.95, b_f = width of FRP = min {0.9d,h} for complete wrapping, $\Box_{FK}$ = the specific fracture energy of FRP concrete interface, f_{ck} = characteristic concrete strength, f_{ctm} = the average concert tensile strength, l_e = the bond length, k_b = geometric factor, Note that the units of the stresses in Equation 15 are N/mm^2.

As presented the existing design models and codes with one exception (i.e. the Japanese Guidelines) are consistently used a constant value usually 0.4% for the strain in the FRP. Other researchers proposed that the strain in the FRP is a percentage of its ultimate strain. Mutsuyoshi et al (1999) and Masuo (1999), based on calibration of experimental data, proposed that the strain in the FRP is 80% and 67% of the FRP ultimate strain, respectively. However, several experiments on columns failed in shear have shown that the FRP may fail before it reaches a strain of 0.4% due to the variations in the modes of failure, applied FRP axial rigidity, types of FRP, and strength of the substrate material (Triantafillou and Antonopoulos 2000, Masukawa et al. 1997). To account for such differences, empirical expressions to calculate effective strain of the FRP at shear failure is proposed based on tests conducted on simply supported RC shallow beams (Triantafillou and Antonopoulos 2000). Later on, Furuta et al. (2003), based on analyzing the test data of 65 specimens, recommended that the stresses in FRP can be calculated as follows:

$$f_{FRP} = \sqrt{5.9 \times 10^{-3} \times \frac{E_{FRP}}{\rho_{FRP}}} \tag{19}$$

The average ratio of the experimental strength to the calculated strength, using equations 1 and 19, is 1.16 and with a standard deviation of 11%. However, this model does not take into considerations the effect of different types of FRP, and the level of the applied normal force.

Some experimental evidences (Masukawa et al. 1997, Ye et al. 2002) have indicated that for retrofitting using small amount of FRP, adding the ultimate strength of the existing column to the contribution of the FRP as indicated by Equation 1 is justified as the hoop yield before failure of FRP. However, at higher FRP reinforcement ratio, adding the ultimate strength of existing column to the contribution of FRP is not justified as the hoop will not yield until the FRP failure. Therefore, applying Equation 1 is not always correct. Using basics of mechanics, ElGawady et al. (2006) have proven the same idea is true for masonry walls retrofitted using FRP. To take into consideration this effect, Ye et al. (2002) proposed the following expression to calculate the shear strengths based on its own experimental data:

$$V_{\mathrm{Ret}} = V_{RC} + V_{FRP} = V_{RC} + \left(\frac{2t_j}{b}\right)\frac{1.639(0.403 - 1.053\eta + 0.176a/h)}{1.207 + \sqrt{\frac{2t_j}{b}\frac{f_{FRP}}{f_t}}} f_{FRP} bh \tag{20}$$

Where a = shear span = column height for cantilever column, η = applied normal force normalized by f'_c and column cross section dimensions, and f_t = concrete tensile strength.

Masukawa et al. (1997) proposed to multiply the V_{RC} in equation 1 by a factor of 0.8 to take into consideration the lag in strength between the FRP and the existing column.

Another factor that affects the strength of the retrofitted column is the degradation in the component V_{RC} at the plastic hinge zone with increasing the column ductility. To account for such degradation, Ghobarah A and Galal K E. (2004) proposed the following expression to calculate the shear strength of the existing columns V_{RC}:

$$V_{RC} = v_c + v_P + v_S \tag{21}$$

$$v_c = 0.3\sqrt{f'_{cc}}A_e \tag{22}$$

$$v_P = K_p \frac{Pt}{2H} \tag{23}$$

$$v_S = \frac{A_v f_{vy} d}{s} \tag{24}$$

Where $A_e = 0.8bh$, and f'_{cc} = the confined concrete strength, v_C = concrete contribution to the shear strength, v_P= applied normal force contribution to the shear strength, v_S=stirrups contribution to the shear strength, $K_p = 1$ for columns in double curvature and 0.5 for columns in single curvature, P = applied normal force, H = column height, f_{vy} = yield strength of the stirrups, S= spacing of the stirrups, A_v = total area of the stirrups. Equation 21 is valid till ductility of 2, after that equation 21 should be read as follows:

$$V_{RC} = \frac{1}{3}(v_c + v_P) + v_S \qquad for\ \mu = 4 \tag{25}$$

$$V_{RC} = v_S \qquad for\ \mu = 6 \tag{26}$$

Comparing Equations 21, 25, and 26 there is degradation in the shear strength of the existing column. At ductility of 6, the concrete and applied normal force contributions to the shear strength are vanished.

Calculation of the ductility of the retrofitted columns has been discussed rarely in the past. Mutsuyoshi et al (2001) proposed the following expression to calculate the ductility of the retrofitted columns:

$$\mu_{FRP} = 3.35 + 1.06 * (V_{FRP}.a / M_{FRP}) \tag{27}$$

Where μ_{FRP} = ductility factor for the retrofitted column; M_{FRP} = flexural strength of the retrofitted member. As discussed earlier, the performance of the retrofitted columns depends on whether it is a retrofitting of damaged column or upgrading of undamaged column. To take this factor into considerations, Ye et al. (2003) proposed the following expression to calculate the column ductility

$$\mu_{FRP} = \frac{\sqrt{1+30\alpha\lambda_e}}{0.045+1.4n} \tag{28}$$

$$\lambda_e = \lambda_{SV} + e_{FRP}\lambda_{FRP} \tag{29}$$

$$\lambda_{SV} = \frac{2A_{SV}}{Sb}\frac{f_{ySV}}{f_c'} \tag{30}$$

$$\lambda_{FRP} = \frac{2t_j}{b}\frac{f_{FRP}}{f_c'} \tag{31}$$

Where α = factor representing the hoop type = 1 for rectangular hoop; e = efficiency factor = 0.473 for strengthening and 0.387 for repair.

Plastic Hinge Confinement Retrofit Design

Designing a FRP retrofit system for confinement of the plastic hinge of a rectangular column requires an accurate quantification of confined concrete strength and ultimate strain. Correlation between experimental data and existing confining models has shown that current models can closely predict the ultimate strength whereas it fails in predicting the ultimate strain (Bhagat et al. 2006). Quantifying confined concrete properties of rectangular columns pose a greater challenge since the confining pressure is non-uniform throughout the cross section. Typically, the confined rectangular section is simplified into two distinct sections—a uniformly confined part and the unconfined section (Lam and Teng 2003, Mander et al. 1988). Based on the ratio of the confined to unconfined parts, most researchers tried to define an efficiency factor that approximates the properties of a given rectangular cross section from an equivalent circular cross section (Mander et al. 1988). For rectangular columns having a cross sectional aspect ratio less than 1.5 and confined using FRP, Sieble et al. (1997) recommended that designers should define an equivalent circular column and design this circular one. Once this equivalent column is designed for certain target displacement ductility, the rectangular column should be retrofitted using double the amount of this FRP. The equivalent diameter can be calculated as follows:

$$D_e = \sqrt{\left(\frac{h}{2k_1}\right)^2 + \left(\frac{b}{2}\right)^2\left(\frac{1}{k_1}+k_1\right)} \tag{32}$$

Where

$$k_1 = \left(\frac{h}{b}\right)^{\frac{2}{3}} \tag{33}$$

For a given displacement ductility μ_Δ, Seible et al. (1997) design the equivalent circular column as follows:

$$t_j = 0.09\frac{D_e(\varepsilon_{cu} - 0.004)f'_{cc}}{\Phi_f f_{FRP,u}\varepsilon_{FRP,u}} \tag{34}$$

Where

$$\varepsilon_{cu} = \phi_y \mu_\phi c_u \tag{35}$$

$$\mu_\phi = \frac{\phi_u}{\phi_y} = \frac{\mu_\Delta - 1}{3\frac{L_p}{L}\left(1 - 0.5\frac{L_p}{L}\right)} + 1 \tag{36}$$

$$L_p = 0.08L + 0.022 f_{sy} d_b \tag{37}$$

and Φ_f = flexural strength reduction factor (typically 0.9), L_p = length from the plastic hinge to the contra-flexure point, d_b = diameter of longitudinal bars, f_{sy} = yield strength of longitudinal reinforcement, μ_ϕ = curvature ductility, ϕ_y = section yield curvature, ϕ_u = section ultimate curvature, c_u = neutral axis depth. Both the neutral axis depth and the ultimate section curvature can be estimated form a sectional moment curvature analysis. The confinement jacket should extend over a column height of L_j, at the plastic hinge zone, where L_j is the larger of D/2 or L/8. This region of retrofitting is called the primary region. To avoid the possibility that the plastic hinge emigrate higher than the retrofitted region, another region of length L_j should be retrofitted using one half the jacket thickness used in the primary region. This later region is called the secondary region.

The S806 (2002) uses the following equation which relates the required jacket thickness to the drift demand.

$$t_j = 2D\frac{f'_c}{f_{FRP}}\frac{\delta}{\sqrt{k_c}}\frac{P_f}{P_{ro}} \tag{38}$$

Where D = length of the column cross section in the loading direction, i.e., either b or h; f_{FRP} = $0.004E_{FRP}$; δ = column drift; k_c = confinement efficiency factor = 1 for circular columns and 0.25 for rectangular and square columns; P_f/P_{ro} = the ratio of the applied axial load to the ultimate axial capacity of the column and should be equal to or greater than 0.2. The primary region confining length is the larger of the plastic hinge length, D/2, or L/8. In the secondary region, a FRP jacket having one-half the thickness calculated by equation 38 should be used for a length of D/2 or L/8 whichever is larger. Finally, for rectangular cross section, the S806 (2002) recommended that h/b be smaller than or equal to 1.5.

ACI440 (2002) calculates the lateral pressure due to confining using FRP as follows:

$$f_l = \frac{k_h \rho_{FRP} \varepsilon_{FRP} E_{FRP}}{2} \tag{39}$$

$$k_h = 1 - \frac{(b-2r)^2 + (h-2r)^2}{3A_g(1-\rho)} \tag{40}$$

$$\rho_{FRP} = \frac{2t_j(b+h)}{bh} \tag{41}$$

The limits on the strain in the FRP are similar to those described in the shear retrofitting. If b/h is greater than 1.5 or h is greater than 900 mm (36 in) the effect of FRP on confinement should be neglected.

ISIS (2001) allows using FRP for confinement of columns having a cross section aspect ratio smaller than or equal to 1.1. In this case the lateral pressure due to the FRP can be calculated as follows:

$$f_{FRP} = \frac{2t_j \Phi_{FRP} E_{FRP} \varepsilon_{FRP} (b+d)}{bd} \tag{42}$$

Where $\varepsilon_{FRP} = 0.002$, Φ_{FRP} = Environmental reduction factor based on the type and exposure of FRP.

Several finite element codes and models have been developed to model the response of concrete columns confined using FRP (Parvin et al. 2002, Mirmiran et al. 2000). A more simplified numerical model approach using the principle of continuum damage mechanics has been developed by Perera (2005). The ultimate strength and strain are reasonably modeled in this approach; whereas, it failed to account for pinching effects and stiffness degradation of the hysteretic loops. Recently, Binici and Mosalam (2007) were able to model the ultimate strength, strain, and stiffness of circular columns confined using FRP. The model was developed using a discrete fiber frame elements with a variable confining pressure.

Lap Splice Retrofit Design

Very limited design equations and models are available for lap splice retrofitting. Sieble et al. (1997) recommended using shape modification for columns having rectangular cross section. The FRP thickness can be calculated as follows:

$$t_j = 500\frac{D_e(f_l - f_h)}{E_j} \tag{43}$$

Where f_h = the horizontal stress level provided by the existing hoop reinforcement in a circular column at a strain of 0.1%; and f_l = the lateral clamping pressure over the lap splice, which can be determined as follows:

$$f_l = \frac{A_s * f_{sy}}{\left[\frac{p}{2n} + 2(d_b + cc)\right]L_s} \tag{44}$$

Where p = the perimeter line in the column cross section along the lap-spliced bar locations, n = the number of spliced bars along p, cc = the concrete cover, A_s = the area of one main column reinforcement with diameter d_b. The retrofitting will extend over a primary region having a length equal to the lap splice length.

The Italian Guidelines (CNR-DT 200/2004) uses equations similar to equation 43 and 38 to calculate the FRP jacket thickness. However, in the Italian code, D_e is replaced by D where D is the maximum of b and d. Finally, the confinement effect of FRP shall be reduced by a factor k_h which can be calculated as follows:

$$k_h = 1 - \frac{b'^2 + d'^2}{3A_g} \tag{39}$$

Where A_g = the gross sectional area.

The S806 (2002) does not allow using of FRP for retrofitting of columns having rectangular cross-section which has a deficient lap splice. For columns having circular cross section, S806 (2002) uses equation 38 to calculate the FRP jacket thickness. However, in the lap splice confinement, $f_{FRP} = 0.002E_{FPR}$. The primary region height is equal to the lap splice length where the secondary region length is equal to L/8 or D/2, whichever is larger.

Conclusion

Recent progress in FRP retrofitting of rectangular bridge column is presented in this chapter. The research reviewed has shown that FRP retrofit system is a potential substitute for conventional steel or concrete jacketing. Experimental studies have demonstrated that the performance of FRP wrapped rectangular columns deficient in shear are superior to columns designed according to recent reinforced concrete codes such as ACI-318.

Confining rectangular columns are able to improve the seismic performance as long as the column cross section aspect ratio is less than 1.5. The effectiveness of the FRP jacket increase with decreasing the cross section aspect ratio. To improve the seismic performance of the retrofitted columns, the corners should be rounded with a radius greater than approximately 25 mm (1 in.).

For columns having deficient lap splice, the improvement in the seismic performance is not too much. The FRP jacket is not able to keep the lateral dilation of the concrete smaller than 0.1% to delay slippage. Therefore, further research is needed to come up with innovative ways to delay slippage of bars. However, the FRP jacket prevents spalling of the concrete cover and applies limited pressure which improves to certain extend the ductility of the retrofitted columns.

Regarding design of the jacket, several design models exist for shear design and the predictions of these models are close enough to some experimental data. For confined columns, several models are available. However, the models predict quite closely the strength but fail in predicting the ultimate strain of the confined concrete. In addition, there has been limited research regarding rectangular columns having large cross section or large cross section aspect ratio. For lap splice deficient columns, only a single model is developed. While the model proves to predict correctly the strength and displacement of circular columns, it needs some experimental verification in the case of rectangular columns.

References

Al-Salloum Y. “A. Influence of edge sharpness on the strength of square concrete columns confined with FRP composite laminates”, *Composites Part B: Engineering,* Vol. 38(5-6), pp. 640-650.

American Concrete Institute (ACI). Report on fiber-reinforced polymer (FRP) reinforcement for concrete structures. *ACI 440.XR,* 2006, Farmington Hills, Mich.

American Concrete Institute (ACI). "Guide for the design and construction of externally bonded FRP systems for strengthening concrete structures." *ACI 440.2R-02,* 2002, Farmington Hills, Mich.

Bhagat SR, Tesfaye A, Bhargava P, Kaushik SK. Strength and ductility of CFRP confined concrete. *International Conference on Earthquake Engineering.* 2006

Binici B., and Mosalam K. (2007). “Analysis of reinforced concrete columns retrofitted with fiber reinforced polymer lamina” *Composites:* Part B, Vol. 38, pp 265–276.

Bousias S, Fardis M., Spthis A., and Biskinis D. (2006). “Concrete of FRP jacketing of concrete columns for seismic retrofitting” *Advances in earthquake Engineering for Urban Risk Reduction* S, Wasti T. and Ozcebe G. (Ed.), Springer.

Chai YH, Priestley MJN, Seible F. Analytical model for steel jacketed RC circular bridge columns. *Journal of Structural Engineering. ASCE* 1994;120(8):2358-76.

Chaallal, O., Shahawy, M., Hassan, M. (2002). “Performance of reinforced concrete T-girders strengthened in shear with GFRP fabrics,” *ACI structural Journal,* **99**(3), pp. 335-343.

Chang K., Chung L., Lee B., Li Y., Tsai K., Hwang J., and Hwang S. Seismic retrofit study of RC bridge columns. *International Training Programs for Seismic Design of Building Structures,* Taiwan. 2002.

CNR-Advisory Committee on Technical Recommendations for Construction, *Guide for the design and construction of externally bonded FRP systems for strengthening existing structures* (CNR-DT 200/2004), Italy, 2004

Cole C. and Belarbi A. (2001). *“FRP jacketed reinforced concrete columns” Department of Civil Engineering,* University of Missouri Rolla, Report number CIES 99-13.

Dagher F. and Harajli M. (2007). "Seismic strengthening of bond-critical regions in rectangular RC columns using external FRP jackets" *Research report,* Department of Civil and Environmental Engineering, American University of Beirut, Lebanon.

ElGawady M., Lestuzzi P., and Badoux M. (2006). "Analytical model for the in-plane shear behavior of URM walls retrofitted with FRP" *Composites Science and Technology,* Vol. 66(3-4), pp. 459-474.

Elnabelsy G. and Saatcioglu M. (2004). "Seismic retrofit of circular and square bridge columns with CFRP jackets" *4th Advanced Composite Materials in Bridges and Structures,* Calgary, Canada

Federal Highway Administration FHWA (2006). *"Seismic retrofitting manual for highway structures: Part 1-Bridges"* Publication Number FHWA- HRT-06-032

Folger P. Earthquakes: risk, monitoring, notification, and research. *CRS Report for Congress* 2007.

Fukuyama H., Suzuki H., and Nakamura H. (1999). "Seismic retrofit of reinforced concrete columns by fiber sheet wrapping without removal of finishing mortar and side wall concrete" 4th Intl Symposium - Fiber Reinforced Polymer Reinforcement for Reinforced Concrete Structures, *ACI SP188,* American Concrete Institute, Detroit

Furuta T., Kanakubo T., and Fukuyama (2003). "Evaluation of shear capacity of RC columns strengthened by continuous fiber" *FRPRCS-6,* Singapore, 8-10 July.

Galal K, Arafa A, Ghobarah A. Retrofit of RC square short columns. *Engineering structures* 2005;27:801-813

Galal K. (2006). Effect of design variables on displacement ductility of FRP-rehabilitated RC squat columns. *8th US National Conference on Earthquake Engineering,* San Francisco, CA

Ghobarah A, & Galal K E. Seismic rehabilitation of short rectangular RC columns. *Journal of Earthquake Engineering.* Vol. 8, No. 1 (2004) 45-68

Haroun M & Elsanadedy H, Behavior of cyclically loaded squat reinforced concrete bridge columns upgraded with advanced composite-material jackets. *Journal of Bridge Engineering* (ASCE) 2005, vol. 10, No.6,

Haroun MA, Mosallam AS, Feng MQ, Elsanadedy HM. Experimental investigation of seismic repair and retrofit of bridge columns by composite jackets. *Journal of Reinforced Plastics and Composites* 2003;22(14):1243

Harries K, Ricles J, Pessiki S & Sause R. (2006). "Seismic retrofit of lap splices in nonductile square columns using carbon fiber-reinforced jackets". *ACI Structural Journal,* V. 103(6)

Iihoshi C., Fukuyama H., Matsumoto Y., and Abe S. (1999). "Strengthening effect of reinforced concrete elements with polyacetal fiber sheets" *4th Intl Symposium - Fiber Reinforced Polymer Reinforcement for Reinforced Concrete Structures, ACI SP188*, American Concrete Institute, Detroit

Iacobucci R., Sheikh S., and Bayrak O. (2003). "Retrofit of Square Concrete Columns with Carbon. Fiber-Reinforced Polymer for Seismic Resistance" *ACI Strcutral Journal,* Vol. 100(6), pp. 785-794

ISIS Design Manual #4 (2001) *"Strengthening Reinforced Concrete Structures with Externally-Bonded Fibre Reinforced Polymers"* Canada

"Japanese Recommendations for Upgrading of Concrete Structures with Use of Continuous Fiber Sheets" (2001) Concrete Engineering Series, Japan

Karbhari, V. M. Response of fiber-reinforced polymer confined concrete exposed to freeze and freeze-thaw regimes, 2002, *J. Compos. Constr.*, Vol. 6(1), 35–40.

Ghosh K. and Sheikh S. (2007). "Seismic Upgrade with Carbon Fiber-Reinforced Polymer of. Columns Containing Lap-Spliced Reinforcing Bars" *ACI Structral Journal,* Vol. 104(2), pp. 227-236.

Kapur J. Washington State's bridge seismic retrofit program. *21st US-Japan Brdige Engineering Workshop,* October 3-5, 2005.

Lam L and Teng J G. Design-oriented stress-strain model for FRP-confined concrete in rectangular columns. *Journal of Reinforced Plastics and Composites,* V. 22, No. 13 (2003)

Lehman D., Elhin S., Nacamuli A., Moehle J. Repair of earthquake damaged bridge columns. *ACI Structural Journal,* 2001, Vol. 98(2).

Lu Y, Gu Xiaoming, Guan J. Probablilistic drift limits and performance evaluation of reinforced concrete columns. *Journal of Structural Engineering* (ASCE) 2005; 131(6): 966-978

Maalej M, Tanwongsval S, Paramasivam. Modelling of rectangular RC columns strengthened with FRP. *Cement & Concrete Composite* 2003; 25: 263-276

MacRae G.A., Nosho K., Stanton J., Myojo T. Carbon Fiber Retrofit of Rectangular RC Gravity Columns in Seismic Regions. *Third International Symposium on Non-Metalic (FRP) Reinforcement for Concrete Structures* (FRPRCS-3) 1997, Sapporo, Japan.

Mander JB, Priestley MJN, Park R. Theoretical stress-strain model for confined concrete. *ASCE Journal of Structural Engineering* 1988;114(8): 1804-26.

Masukawa J., Akiyama H., and Saito H. (1997). "Retrofit of existing concrete piers by using carbon fiber sheet and aramid fiber sheet" Non-Metalic (FRP) Reinforcement for Concrete Structures, *3rd international symposium.*

Masuo K. (1999). "Seismic retrofitting of reinforced concrete columns with wing walls using carbon fiber reinforced plastic walls". *4th Intl Symposium - Fiber Reinforced Polymer Reinforcement for Reinforced Concrete Structures, ACI SP188,* American Concrete Institute, Detroit

Mirmiran A, Zagers K, Wenquing Y. Nonlinear finite element modeling of concrete confined by fiber composites. *Finite Elements in Analysis and Design* 2000;35: 79-96

Memon M. and Sheikh S. (2005). "Seismic resistance of square concrete columns retrofitted with glass fiber-reinforced polymer" *ACI Structural Journal,* pp 774-783.

Mutsuyoshi H., Ishibashi T., Okano M., Katsuki F. (1999). "New design method for seismic retroit of bridge columns with continuous fiber sheet-performance-based-design" *4th Intl Symposium - Fiber Reinforced Polymer Reinforcement for Reinforced Concrete Structures, ACI SP188*, American Concrete Institute, Detroit

Nagai H., Kanakubo T., Jinno Y., Matsuzaki Y., and Morita S. (1999). "Study on strcutral performance of reinforced concrete columns with waist-high walls strengthened by carbon fiber reinforced plastic sheets" 4th Intl Symposium - Fiber Reinforced Polymer Reinforcement for Reinforced Concrete Structures, *ACI SP188*, American Concrete Institute, Detroit

Nesheli K., and Meguro K.(2005). "External prestressing concrete columns with fibrous composite belts" *7th Intl Symposium - Fiber Reinforced Polymer Reinforcement for Concrete Structures, ACI SP230*, American Concrete Institute, Detroit

Rochette P. and Labossiere P. (2000) "Axial testing of rectangular column models confined with composites", *J Compos Construct, ASCE,* **4**(3), pp. 129–136.

Satoh H., Yamakawa T., Nesheli K., Inaba H. (2001). "Seismic and/or emergency retrofit of extremely short RC columns utilizing prestressed aramid fiber belts and steel angles" *FRP Composites in Civil Engineering,* Hong Kong, china.

S806-02 *"Design and construction of building components with fiber-reinforced polymers"* Canadian Standards Association, Ontario, (2002).

Saadatmanesh H, Ehsani MR, Jin L. Repair of earthquake-damaged RC columns with FRP warps. *ACI Structural Journal* 1997; 1(2):52-62

Saadatmanesh H, Ehsani MR, Jin L. Seismic Retrofitting of rectangular bridge columns with composite straps. *Eng Res Inst J SPECTRA* 1997 13(2):1977

Sause R., Harries K., Walkup S., Pessiki S., and Ricles, J. (2004). "Flexural behavior of concrete columns with carbon fiber composite jackets" *ACI Structural Journal,* Vol. 101 (5), pp. 708-716

Seible F, Priestley MJN, Hegermier GA & Innamorato D. Seismic retrofit of RC columns with of RC columns with continuous carbon fiber jackets. *Journal of Composite for Construction* ASCE 1997; 1(2) 52-62

Triantafillou TC & Antonopoulos CP. Design of Concrete flexural member strengthened in shear with FRP. *Journal of Composite for Construction* (ASCE) 2000: 4(4): 198-205.

Parvin A, Wang Wei. Concrete columns confined by fiber composite wraps under combined axial and cyclic lateral loads. *Composite Structure* **58** (2002) 539-549

Perera R. (2006) "A numerical model to study the seismic retrofit of RC columns with advanced composite jacketing" *Composites:Part B,* Vol. 37, pp 337-345

Priestley MJN, Seible F and Calvi GM. *Seismic design and retrofit of bridges.* New York: Wiley; 1996

Prota A, Manfredi G, Balsamo A, Nanni A & Cosenza E. Innovative technique for seismic upgrade of RC square columns. *7th Int. Symposium FRP Reinforcement for Concrete Structures* (C.K. Shield, J.P. Busel, S.L. Walkup, D.D. Gremel Eds.), SP-230, American Concrete Institute, Vol. 1, Farmington Hills (Miissouri, USA), pp. 1289-1303, 2005.

Paula, R. and Silva M.. Sharp edge effects on FRP confinement of RC square columns. *3rd International Conference on Composites in Infrastructure,* San Francisco, Ca, 2002.

Wu G., Zhitao L., and Jianbiao J. (2001). "Seismic behavior of RC columns strengthened with CFS" *FRP Composites in Civil Engineering,* Hong Kong, china.

Xiao Y, Ma R. Seismic retrofit of RC circular columns using prefabricated composite jacketing. *Journal of Structural Engineering ASCE* 1997; 123(10): 1357-64

Xiao Y, Wu H. Compressive behavior of concrete confined by various types of FRP composite jackets. *Journal of Reinforced Plastics and Composites* 2003;22(13): 1187

Yalcin C, Kaya O & Sinangil M. Seismic retrofitting of R/C columns having plain rebars using *CFRP sheets for improved strength and ductility.* Construction and Building Materials 2006

Yan Z, Pantelides CP, Reaveley LD. Shape modification with expansive cement concrete for confinement with FRP composites. *7th International Symposium on Fiber-Reinforced (FRP) Polymer Reinforcement for Concrete Structures,* Missouri, 2005

Ye L., Zhang K., Zhao S., and Feng P. (2001). "Experimental study on seismic strengthening of RC columns with wrapped CFRP sheets" *FRP Composites in Civil Engineering,* Hong Kong, china.

Ye L., Zhang K., Zhao S., and Feng P. (2003). "Experimental study on seismic strengthening of RC columns with wrapped CFRP sheets" *Construction and Building Materials,* vol. 17.

Ye L., Yue Q., Zhao S., and Li Q. (2002). "Shear strength of reinforced concrete columns strengthened with carbon-fiber-reinforced plastic sheet" ASCE, *Journal of Structural Engineering,* Vol. 128 (12).

Yoshimura K., Kikuchi K., Kuroki M., Ozawa K., and Masuda Y. (2001)."Experimental study on seismic behavior of R/C short columns strengthened by carbon fiber sheets" *6th ASCCS International Conference on Steel-Concrete Composite Structures,* Los Angeles, California.

Zhang A., Yamakawa T., Zhong P., and Oka T. (1999). "Experimental study on seismic performance of reinforced concrete columns retrofitted with composite-materials jackets" *4th Intl Symposium - Fiber Reinforced Polymer Reinforcement for Reinforced Concrete Structures, ACI SP188,* American Concrete Institute, Detroit

Zhao T., Liu M., and Xie J. (2001). "Research on the new method of improving the ductility of high-strength concrete columns with continuous carbon fiber sheet" *FRP Composites in Civil Engineering,* Hong Kong, china.

In: Research Progress in Materials Science
Editors: W. Olsson and F. Lindberg, pp. 131-148
ISBN: 978-1-60456-629-1

Chapter 5

CARBON NANOTUBE GROWTH BY SURFACE DECOMPOSITION OF SiC

Takahiro Maruyama[1,2,*], Shigeya Naritsuka[1,2] and Michiko Kusunoki[3,**]

[1]Department of Materials Science and Engineering, Meijo University, Tempaku, Nagoya 468-8502, Japan

[2]Meijo University 21st CENTURY COE program "Nano Factory", Nagoya 468-8502, Japan

[3]Japan Fine Ceramics Center, 2-4-1 Mutsuno, Atsuta, Nagoya 456-8587, Japan

Abstract

The formation process of carbon nanotubes (CNTs) on a thermally treated 6H-SiC(000-1) carbon face substrate was investigated using X-ray photoemission spectroscopy (XPS), atomic-resolution ultrahigh-vacuum scanning tunneling microscopy (UHV-STM) and transmission electron microscopy (TEM). STM observations showed that carbon networks formed on the surface of carbon nanocap structures, which appear in the initial stage of CNT growth. This indicates that the carbon nanocaps are crystalline when they are formed on the SiC surface. By suppressing surface oxidation during heating, details of the formation process of carbon nanocaps were successfully observed using UHV-STM. Based on the experimental results, we proposed a model that describes the formation of CNT, and also a model that explains why the chirality of the CNTs is the zigzag-type. We also investigated the effect of ambient oxygen gas on the SiC surface and the phase diagram of nanocap formation was obtained. Our results show that both active and passive oxidation transitions occurred depending on the oxygen pressure. They also show that carbon nanocaps formed at about 1200°C, irrespective of the ambient pressure, when the oxygen pressure was under 10^{-2} Pa.

---

* E-mail address: takamaru@ccmfs.meijo-u.ac.jp. Corresponding author: Department of Materials Science and Engineering, Meijo University, 1-501 Shiogamaguchi, Tempaku, Nagoya 468-8502, Japan

** Present address: EcoTopia Science Institute, Nagoya University

1. Historical Overview

Since the discovery of carbon nanotubes (CNTs) by Iijima [1], CNTs have been used in numerous fields. To date, three main growth techniques have been employed for CNT growth: the carbon-arc discharge technique, laser ablation technique and chemical vapor deposition (CVD) [2]. Among these, the CVD technique is the most suitable for organizing CNTs precisely over a large surface and is widely investigated at present. However, accurate control of the chirality of CNTs has never been achieved, even though improvement in uniformity of the diameter, linearity, alignment and structure of CNTs is critical to developing various CNT-based electronic devices.

As distinct from these conventional growth techniques, Kusunoki et al. have reported a novel CNT growth technique: surface decomposition of SiC This technique was discovered by Kusunoki et al. in 1997 [3]. After heating a 3C-SiC single crystal at about 1700°C in the transmission electron microscope (TEM), they observed a very fine, stringy contrast, oriented perpendicular to the <111> plane of SiC, as shown in Fig. 1(a). From magnified electron micrographs and diffraction patterns, they identified these strings to be a number of overlapped tubes of 2-5 nm width (Figs. 1(b) and (c)). Their results suggested that CNTs tended to grow along the [111] direction on the SiC(111) surface. Soon thereafter, Takikawa et al. also reported that CNTs grew on SiC powder surface after heating at 1600-1700°C in vacuum [4].

A few years after, Kusunoki et al. reported that CNT growth proceeded even after heating 6H-SiC(000-1) substrates in a vacuum electric furnace [5]. Through TEM observations, they found that aligned CNT films 0.3 μm in thickness were formed perpendicular to the surface after heating at 1700°C for 30 min, as shown in Fig. 2 (a). High-resolution TEM (HRTEM) images showed that two to five layered concentric circles 2-5 nm in diameter were adjacent to each other, indicating that multi-walled CNTs were densely grown on the SiC surface [6]. In addition, electron diffraction pattern analysis showed that most of the CNTs were of the zigzag type [7]. This is characteristic of this growth technique because, at present, it is still difficult to grow CNTs with unique chirality. The automatic alignment of the chirality of CNTs will be useful for numerous purposes.

It is well known that SiC shows polytypes of various stacking sequences with the same chemical composition [8]. Among the many polytypes, the technologically important ones are 3C-, 4H-, and 6H-SiC. For technological applications, Si-terminated faces of SiC are more important and have been widely investigated. However, to date, aligned CNT growth perpendicular to the SiC surface by surface decomposition has been reported for 6H-SiC (000-1) [5-7] 3C-SiC(-1-1-1) [3, 9] and 4H-SiC(000-1) [10], which are all carbon-terminated faces (C-faces) of SiC. Recently, CNT growth was also observed on the Si-faces of SiC [11, 12]. However, upon heating the Si-faces of SiC, CNTs generally grow parallel to the surface, as shown in Fig. 2(b), which varies markedly with the polarity of the SiC surface, indicating that the surface structure of SiC strongly influences the CNT formation by surface decomposition.

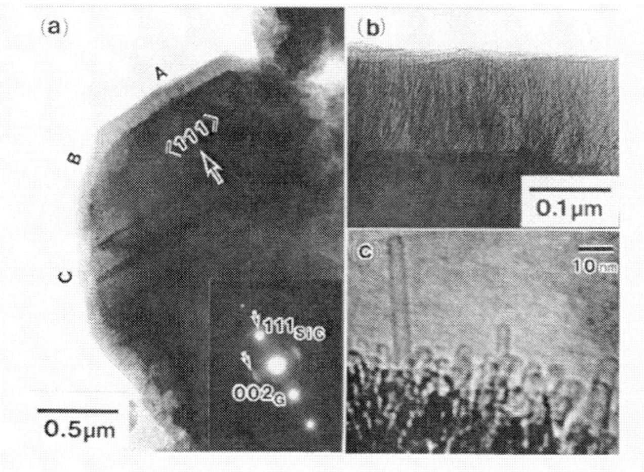

Figure 1. (a) TEM micrograph of a β-SiC single crystal particle heated to about 1700°C for 3 min. (b) Micrograph of the well-oriented strings in area A. (c) Blowup of a part of the surface in (b). Caps 2-5 nm in width are observed.

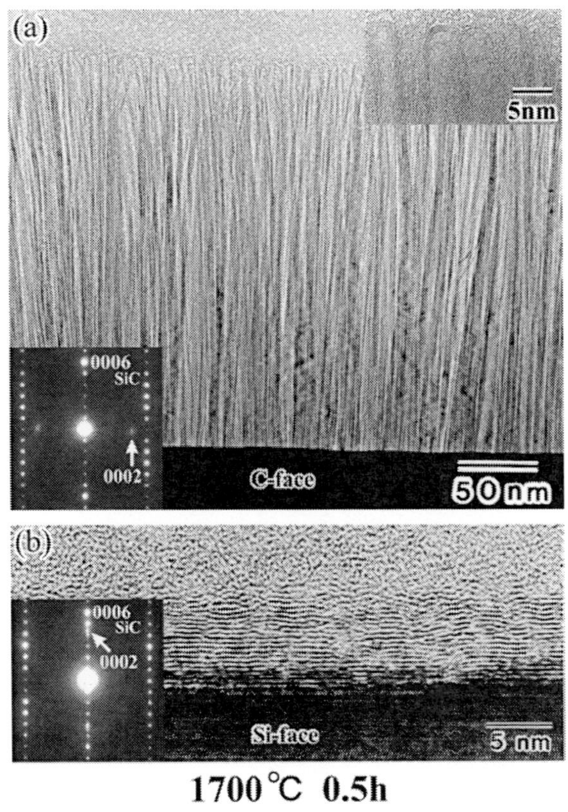

Figure 2. Cross-sectional TEM images and the diffraction patterns of the surface of SiC single-crystal wafers heated to 1700°C for half an hour on (a) the C and (b) Si-faces.

2. Characterization of CNT

CNTs grown by surface decomposition of SiC have mainly been characterized by TEM observations. Kusunoki et al. used the TEM to determine the structure of grown CNTs. They observed that the aligned CNTs grew perpendicular to the SiC(000-1) surface after heating 6H-SiC(000-1) crystals above 1500°C and roughly estimated the density of the CNTs to be 3×10^{12} cm^{-2} [6]. A selected-area electron diffraction pattern obtained from a CNT film and SiC(000-1) substrate (circled area in Fig. 3(a)), showed both six fold 10-10* spots from the horizontal graphene net plane and 0002* and 0004* spots from vertical planes of the aligned CNTs [7]. From the orientation relationship between these 0002* and 10-10* spots, they concluded that most of the grown CNTs were zigzag-type, as shown in Fig. 4.

So far, TEM observations have revealed several important characteristics of the grown CNTs. In the case of CNT growth from SiC(000-1) single crystals, 1) most of the CNTs are multi-walled nanotubes, 2) the density of the CNTs is high ($\sim10^{12}/cm^2$) compared to that obtained by CVD growth, and 3) most of the CNTs are zigzag-type. These characteristics are unique to CNTs grown by this technique. In addition, a recent TEM study showed the presence of single-wall CNTs (SWNTs) with a diameter of 1 nm after introduction of hydrogen gas during heating (see below) [13].

In contrast to TEM observations, Raman measurements have not been reported until recently, even though Raman spectroscopy is a very important technique for identifying the presence of CNTs, especially for determining their diameter [14]. This is presumably because the diameters of most CNTs are rather large (3-5 nm), which makes Raman spectroscopy difficult since conventional Raman measurements can only detect a Raman shift in the region above ~100 cm^{-1}. Thus, it is difficult to observe CNTs greater than 2.5 nm in diameter using Raman spectroscopy.

Recently, our group succeeded in Raman observation of CNTs formed on a 6H-SiC (000-1) using a 1064 nm Nd:YAG laser as excitation source [15]. Spectra in the range from 130 to 400 cm^{-1} for the sample after heating at various temperatures are shown in Fig. 5. Before and after heating at 1250 and 1350°C, only a peak that corresponds to the E_2 transverse acoustic phonon mode of 6H-SiC was observed at 150 cm^{-1} [16]. After heating at 1700°C, the peak at 150 cm^{-1} disappeared and a small peak at 338 cm^{-1} appeared, which was a radial breathing mode (RBM) peak due to double-walled CNTs and/or the innermost wall of MWNTs. On the basis of a simple approximation between the mode frequency ω_{RBM} (cm^{-1}) and the diameter d (nm) for SWNTs [17, 18], the RBM peaks at 189 and 338 cm^{-1} correspond to a diameter of 1.3 and 0.7 nm, respectively. It should be noted that these diameters are quite a bit smaller than the average tube diameter of the CNTs grown by surface decomposition of SiC. TEM observations showed that the CNTs detected by Raman measurements were small-diameter CNTs formed during heating, and were partially distributed among the larger-diameter CNTs and the percentage of small-diameter CNTs was not so high. On the basis of the Kataura plot [19], these peaks observed in the Raman spectra could be assigned to the E_{11} and E_{22} electronic transitions of semiconducting tubes, respectively. Our Raman measurement results indicate that the quality of CNTs formed by surface decomposition of SiC is as good as that of CNTs grown by other growth techniques.

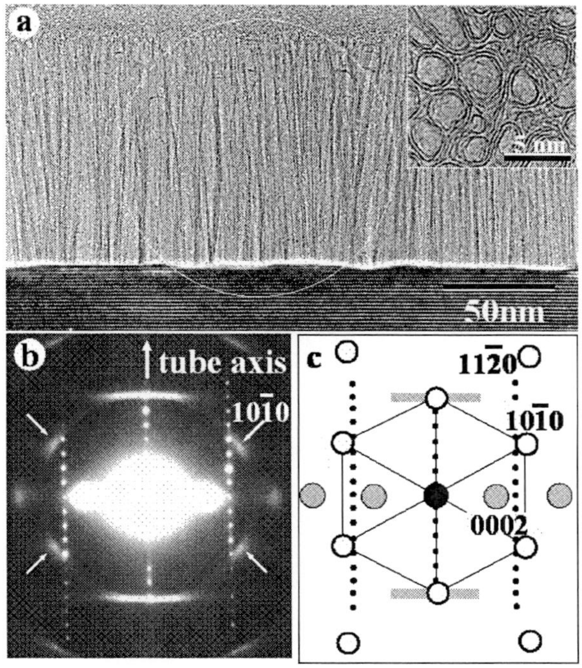

Figure 3. Aligned CNT film on the SiC(000-1) C-face formed by surface decomposition of SiC. (a) TEM micrographs of the cross-section and in the plan-view direction (upper right). (b) A selected-area electron diffraction pattern obtained from the circled area in (a). (c) A schematic of the diffraction pattern in (b). The intensity distribution is classified into three types of reflections: solid circles – a SiC crystal (electron beam // {1-102}SiC); small dots – horizontal plane of CNTs; gray circles – vertical planes of CNTs.

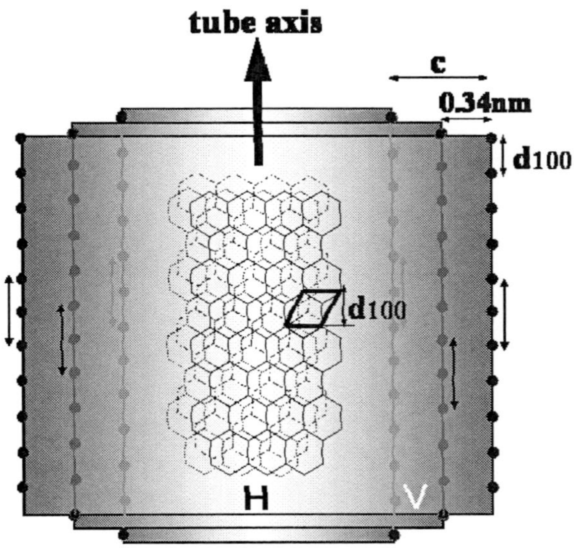

Figure 4. Schematic of the helical and vertical planes for zigzag-type MWNT with the walls shifted with respect to each other along and perpendicular to the CNT axis.

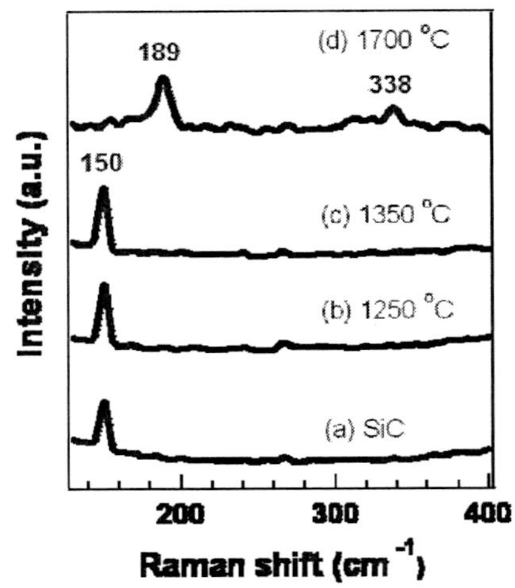

Figure 5. Raman spectra in the 130-400 cm^{-1} range for 6H(000-1) samples (a) before heating, and after heating (b) at 1250°C for 30 min, (c) at 1350°C for 30 min, and (d) at 1700°C for 2 h.

3. Mechanism of CNT Growth

3.1. Early Studies

TEM studies have revealed the appearance of cap-like structures prior to the formation of CNTs. In early studies, Kusunoki et al. showed that arced sheets of carbon with an outer diameter of 5 nm and height of 1-2 nm appeared on the SiC surface after heating at 1250°C for 30 min, as shown in Fig. 6(a). After heating at 1300°C, two- to three-layered carbon nanocaps 3-5 nm in diameter and 3-5 nm in height were generated in a dense formation on the surface (Fig. 6(b)). These carbon nanocaps corresponded to the top of CNTs, which grew from below the carbon nanocaps [5]. Thus, the carbon nanocaps are thought to be strongly related to the structure of the grown CNTs, namely, the diameter, number of walls and chirality. To elucidate the formation mechanism of the carbon nanocaps, several studies have been carried out.

The formation process in the initial stage of CNT growth has been investigated by several groups. Watanabe et al. carried out TEM observation of 3C-SiC(-1-1-1) samples that were heated in the TEM [9]. After heating at 1170°C, a strong diffraction contrast was observed at the nanometer scale due to local strains of the 3C-SiC. At 1360°C, amorphous carbon layers with high porosity were observed, which was presumably caused by the additional evaporation of Si. After successive heating for several minutes, the amorphous carbon changed into graphene. Finally, a ring contrast was observed in TEM images, suggesting the appearance of CNTs whose axis was perpendicular to the surface. They also carried out scanning tunneling microscope (STM) observations and observed the appearance of convex structures with diameters of 3-7 nm after heating at 1360°C for 30 min. They believe that these corresponded to the tops of newly growing CNTs. On the basis of these results, they proposed a model for the initial growth process of CNTs. According to this model, the amorphous carbon that forms on the SiC surface by the evaporation of Si crystallizes into graphene flakes aligned parallel to the SiC surface. Then, carbon nanocaps are formed by lifting of part of the graphene. They concluded that, first, pentagons and heptagons were generated inside the graphene and that the heptagons moved down along the graphene surfaces as the graphene was lifted, forming the nanocaps. As the temperature increased, other graphene layers were formed below the nanocaps, leading to double-walled or multi-walled CNTs through lifting of the graphene layers. However, if their model is correct, the diameter of the CNTs seems to increase as the CNTs elongate, which makes it difficult to explain the high density of the grown CNTs.

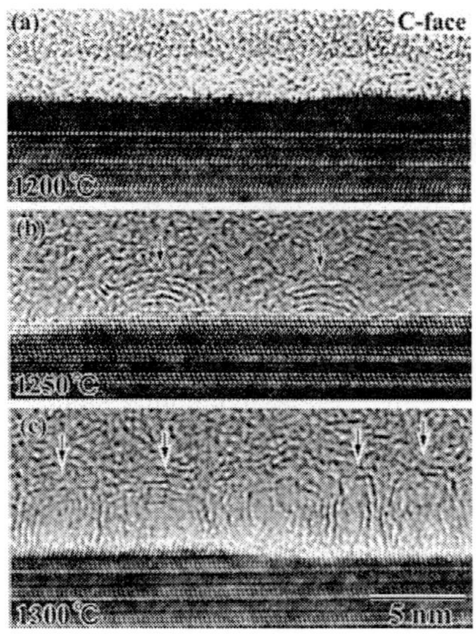

Figure 6. Cross-sectional HREM images showing surfaces of the C-face heated at (a) 1250°C, (b) 1300°C, and (c) the Si-face heated at 1350°C for half an hour. The amorphous contrasts on the upper side of each micrograph were due to the glue used in TEM sample preparation.

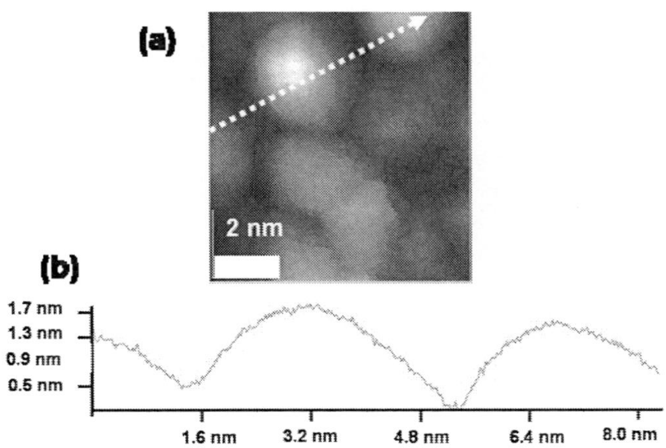

Figure 7. (a) STM image of the cap structures produced on 6H-SiC(000-1) substrates. The image was taken at room temperature. (b) Height profile of cap structures in the direction marked by arrow in (a).

STM of the formation process of CNTs by surface decomposition of SiC was also reported by Hisada et al [10]. They annealed both a 4H-SiC(0001) Si-face and (000-1) C-face in an ultra-high vacuum (UHV) system and carried out STM observations. Their results also showed that many protrusions corresponding to the tops of CNTs were generated on the C-face upon annealing at 1800°C.

Recently, our group carried out high-resolution STM observations of 6H-SiC(000-1) samples and succeeded in obtaining atomic-resolution STM images of samples after annealing at 1250°C in a vacuum electric furnace [20]. Figure 7(a) shows the blowup of an STM image of a cap structure produced on a 6H-SiC(000-1) substrate. We observed convex structures, 3-5 nm in diameter and 1-2 nm in height, exhibiting cobweb-like patterns, as shown in Fig. 7(b). Figures 8(a) and (b) show one portion of the surface of a carbon nanocap. These two figures are the same except for the markers shown by open circles in Fig. 8(b). Figure 8(a) clearly reveals the cobweb-like pattern, which corresponds to carbon networks, on the surface of carbon nanocaps. In Fig. 8(b), two large circles and nine small circles can be seen. There are six small open circles around the right large open circle, which confirms that this pattern is a carbon network composed of hexagons. In contrast, the left large circle is surrounded by only five small circles, which indicates that the portion of network represented by the left large open circles is a pentagon on the nanocap surface. It is well known that a convex carbon network contains several pentagons, as observed on the surface of fullerenes. Given the convex shape of the nanocap, it is reasonable to observe a pentagon on the surface. These STM observations confirmed the presumption that the carbon nanocap had been already crystallized when it was formed on the SiC surface. From the STM results, only the nanocap surface could be observed, however, we believe that crystallization simultaneously occurs in the inner layers of the nanocaps if the nanocaps are composed of several layers. This speculation is reasonable because each layer should have the same composition.

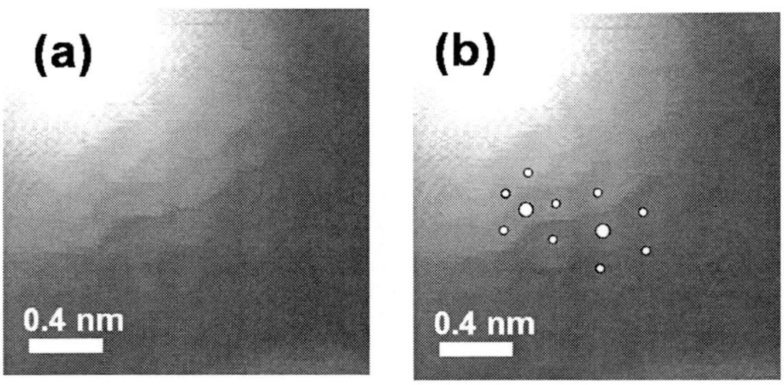

Figure 8. Magnified feature of STM images of the cap structure produced on 6H-SiC(000-1) substrates. (a) and (b) are exactly the same image, except for the open circle markers in (b). Two sizes of open circles are observed in (b): two large circles and nine small ones. The large circle on the right is surrounded by six small circles, while the large circle on the left is surrounded by five small circles.

Assuming the nanocap structure is maintained during CNT growth, this observation is important because it means that the carbon nanocap structure is strongly related to that of the structure of the CNTs grown: the diameter and number of walls of the nanocap and the CNT growing from below it must be the same; in addition, the peripheral bonds of the carbon nanocap must be connected smoothly to the bonds of the zigzag-type CNT. Through STM observation, we also compared the carbon nanocap size between 1250 and 1300°C, and showed that there was no distinct difference between them, which indicates that the nanocap diameter is maintained during the temperature increase [20]. This is consistent with the TEM observation in Fig. 6, which also points to the preservation of the nanocap size. Thus, it is necessary to construct another model that can explain the preservation of CNT diameter during growth.

3.2. Mechanism of Carbon Nanocap Formation

As mentioned above, one of the most important characteristics of CNTs grown by surface decomposition of SiC is that they mostly exhibit zigzag-type chirality. Since the structure of the carbon nanocap presumably correlates strongly to that of the grown CNT, it is important to clarify the formation mechanism of the carbon nanocaps, which may enable control of the CNT structure. STM observation during CNT growth provides an effective method of observing the formation of carbon nanocaps directly. However, our previous study showed that surface oxidation occurred below 1200°C upon heating in a vacuum electric furnace, which made it difficult to observe the formation process of the carbon nanocap directly [21]. Figure 9 shows the deconvoluted Si 2p X-ray photoelectron spectra (XPS spectra) for 6H-SiC(000-1) before and after heating at 1100°C in a vacuum electric furnace. Before heating, the SiO_2 component was negligible, however, after heating, a satellite peak related to SiO_2 appeared. This indicates that the SiO_2 layer formed below the temperature of nanocap formation. In this experiment, the pressure in the vacuum electric furnace was about 10^{-2} Pa. In general, oxidation of the SiC surface is strongly dependent on the ambient pressure during heating. To prevent oxidation, SiC must be heated under lower pressures.

To prevent the surface oxidation induced by reaction with residual gases, our group carried out STM observation of a SiC(000-1) substrate annealed under UHV conditions [22]. To suppress oxidation, the pressure was kept below 10^{-6} Pa during heating. Figure 10 shows the deconvoluted Si 2p and C 1s XPS spectra for 6H-SiC(000-1) after heating at 1100°C under UHV. It should be noted that no SiO_2 layer was observed even after heating, indicating that surface oxidation could be suppressed by annealing under UHV.

Topological STM images of 6H-SiC(000-1) obtained in the subsequent stage of the annealing process under UHV are shown in Fig. 11. After heating at 1100°C, densely packed grains, around 1-2 nm in size, were observed over the entire surface. These were nanosized carbon particles that were formed mainly by carbon sp^2 bonding. After heating at 1150°C, clusters of several particles were formed on the surface, though the size of individual particles seemed to remain unchanged. The cluster size ranged from around 3 to 5 nm. After heating at 1160°C, the particles in some clusters began to coalesce, as shown in Fig. 11(c). The high porosity that formed on the surface of these clusters may be defects caused by the redistribution of carbon atoms in the cluster.

STM images of the 6H-SiC(000-1) after heating at 1200°C under UHV are shown in Fig. 12. It can be seen that the SiC surface was covered by a large number of grains ranging in size from around 3 to 5 nm. The height profile showed that all grains showed a convex shape, with a height of 1-2 nm. The blowup in Fig. 12(b) shows that a cobweb-like or striped pattern covered the convex surfaces and that the periods of these patterns were 0.2-0.3 nm, which was the same as that of a hexagonal carbon network in a graphene sheet, i.e., 0.24 nm. The shape and period indicate that these convex structures correspond to carbon nanocaps, which represent the initial state of CNT growth on SiC.

On the basis of the present experimental results, we propose a model for CNT growth on SiC. Schematic images are shown in Fig. 13. (1) At 1100°C, graphitization induced by Si evaporation occurs on the SiC surface, and carbon atoms isolated from the SiC surface form chemical bonds with each other, yielding carbon particles 1-2 nm in diameter (Fig. 13(a)). These carbon particles are mainly composed of sp^2 hybrid bonds. It is thought that carbon particles were generated on the C-face of SiC to reduce the surface energy in the C/SiC(000-1) system. (2) Upon heating at 1150°C, carbon particles start migrating on the SiC surface as a result of the thermal energy, forming clusters of several tens of carbon nanoparticles to reduce the surface energy further (Fig. 13(b)). Then, the redistribution of carbon atoms begins within each nanoparticle cluster, and the nanoparticles begin to coalesce (Fig. 6(c)). (4) At 1200°C, coalescence proceeds within the clusters, and crystallization occurs, forming CNT nanocaps (Fig. 13(d)). In the crystallization, the size of clusters is presumably almost the same as that of the formed nanocaps. Thus, we believe that the size of clusters determines the nanocap size. (5) Above 1200°C, CNT growth into SiC proceeds by the sublimation of Si atoms from the SiC(000-1) surface (Fig. 13(e)). It is thought that as the CNTs grow into SiC, crystallization occurs by utilizing the crystal structure of SiC (see the next section, in addition to the structure of nanocaps).

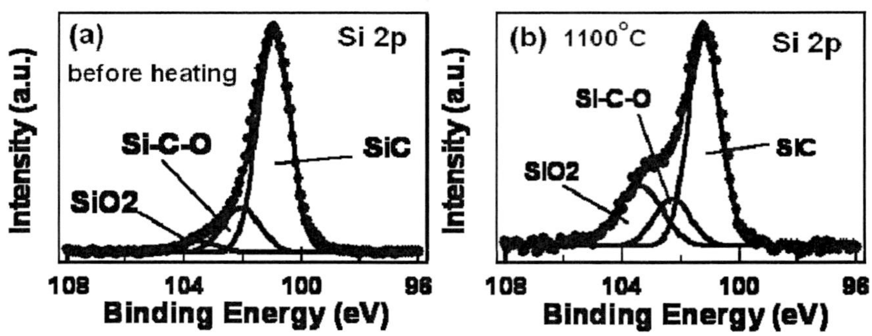

Figure 9. Si 2p XPS spectrum for 6H-SiC(000-1) (a) before heating and (b) after heating at 1100°C in a vacuum electric furnace.

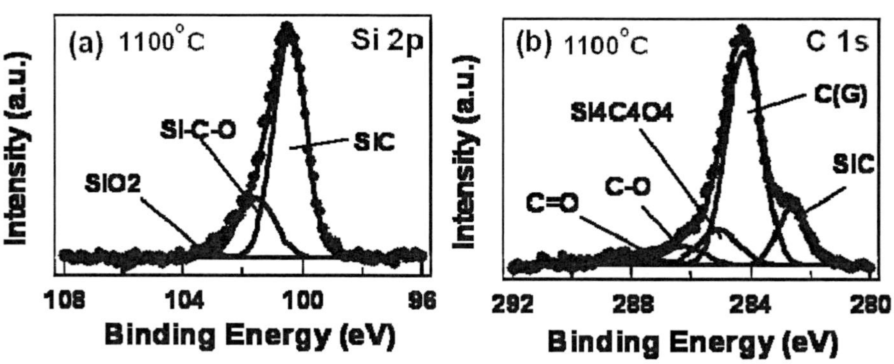

Figure 10. (a) Si 2p and (b) C 1s XPS spectra after heating 6H-SiC(000-1) at 1100°C under UHV.

Figure 11. STM images of 6H-SiC(000-1) substrate obtained in subsequent stages of the annealing process under UHV: (a) after annealing at 1100°C, (b) 1150°C and (c) 1160°C.

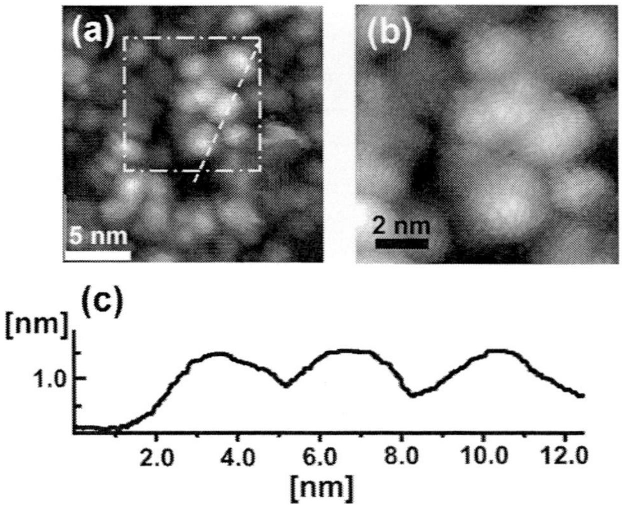

Figure 12. (a) An STM image of a 6H-SiC(000-1) substrate after annealing at 1200°C under UHV. (b) A blowup of the area indicated by the dotted-and-dashed line in (a). (c) Height profile of the cap structures for the direction marked by the arrow in (a).

Our results indicated that amorphous carbon atoms generated by both decomposition of the SiC surface and sublimation of Si are crystallized, forming carbon nanocaps on the surface. Since most of CNTs that grow from below the nanocaps are the zigzag type, the bonds of the bases of the nanocaps are thought to be perpendicular to the SiC surface. The dangling bonds of the SiC(000-1) C-face may be involved in the formation of such bonds, because their direction is completely perpendicular to the surface. Recently, Irle et al. reported high-temperature quantum chemical molecular dynamics simulations [23]. Their result showed that graphene sheets tend to warp on the surface of the SiC(000-1) C-face, forming small diameter nanocaps. In contrast, graphene sheets attached on the Si-face did not readily warp, keeping largely flat on the SiC surface. They pointed out that carbon dangling bonds perpendicular to the SiC C-face were essential to forming zigzag-type CNTs because a zigzag-type CNT favors perpendicular C-C bonds for connecting the C-C bonds between the nanocaps and the SiC surface. These results explain, to some extent, the experimental observation that carbon nanocaps are more easily formed on the SiC C-face than on the Si-face. In brief, dangling bonds of the SiC surface are thought to be an important factor in forming only zigzag-type CNTs.

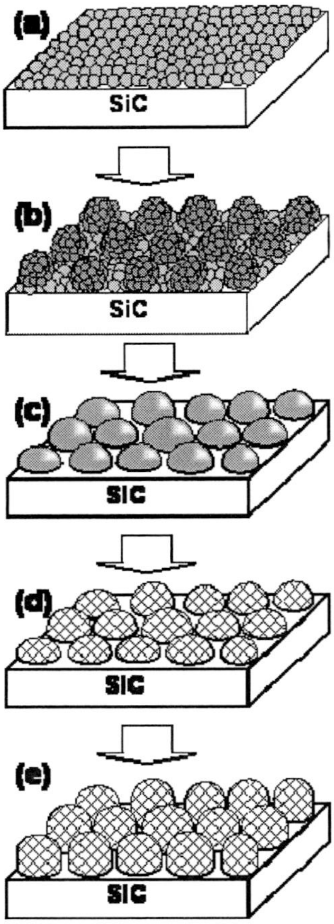

Figure 13. Schematic model of the carbon nanocap formation process during surface decomposition of SiC. (a) formation of carbon nanoparticles by surface decomposition of SiC, (b) aggregation of carbon nanoparticles, (c) coalescence, (d) crystallization and (e) CNT growth into SiC.

3.3. Mechanism of Chirality Alignment

STM observations have revealed that carbon nanocaps formed in the initial stage are already composed of carbon hexagon networks, suggesting the existence of dangling bonds perpendicular to the SiC surface at the base of the nanocaps. Cross-sectional TEM observations showed that vertical planes of CNT walls connect directly to the SiC crystal with no amorphous layer at the interface, and the lattice distance of the graphite (0002) lattice within a CNT is 0.34 nm, which roughly matches that of the (10-12) and (10-1-2) planes of 6H-SiC [7]. These results suggest that, the CNT walls are formed immediately after decomposition and sublimation of Si by inheriting the atomic structure of the original SiC through dangling bonds.

Kusunoki et al. also pointed out that the {10-12} planes of 6H-SiC are composed of networks of slightly wavy hexagons consisting of three Si atoms and three C atoms. As a monolayer network structure of these Si-C hexagons exits on the (10-12) plane, they proposed a model in which a carbon honeycomb structure is formed with the outward diffusion of Si atoms and the collapse of carbon atoms from the inner side of the CNT wall to the Si positions (Fig. 14) [7]. The Si-C distance projected on the {10-12} plane is 0.166 nm, which is slightly longer than the C-C bond length of 0.142 nm. Thus, a Si-C network has to shrink at a rate of 14% to form a graphene sheet, which may be caused by the {10-12} planes becoming perpendicular to the surface, since the (10-12) plane is tilted by 19.5° with respect to the (10-10) plane along the CNT axis. The {10-10} SiC plane has six-fold symmetry along the c axis. Thus, we believe that the graphene sheets easily warped cylindrically along the periphery of the carbon nanocaps, forming CNTs from below the nanocaps. If the carbon nanocap has several inner caps, the corresponding number of walls are considered to be cylindrically formed, leading to a multi-walled nanotube. Given that the tubes were empty, it is thought that the vacancies formed at the Si sites were filled with carbon atoms from the inner side of the walls. If our assumption is correct, hexagons contained in the graphene sheets formed by shrinkage of the {10-12} planes will have a pair of C-C bonds parallel to the c-axis. We believe that all these processes occur during the formation of CNTs, and that this is why only zigzag-type CNTs grow by this technique. In our model, CNTs walls are formed along the periphery of carbon nanocaps, so both the diameter and density of the nanocaps are maintained during CNT growth, which is consistent with experimental observations. In addition, the density of CNTs can be high in this model because carbon nanoparticles cover almost the entire surface after sublimation of Si (Fig. 11(a)).

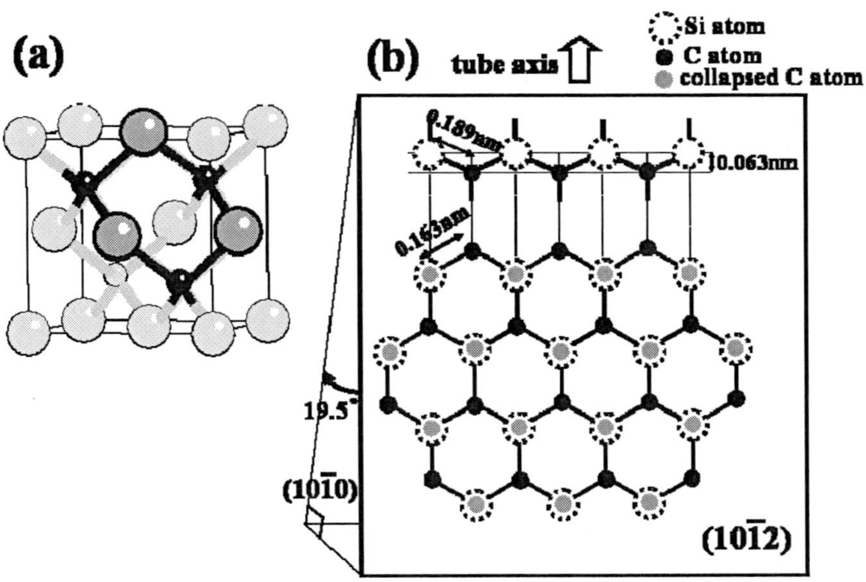

Figure 14. Schematic representations of (a) one of the Si-C hexagons in the 3C-SiC unit cell and (b) the formation mechanism for a zigzag-type CNT along the c-axis. A slightly wavy honeycomb network of Si-C hexagons on the (10-12) plane changes into a part of the graphene of a CNT wall with a zigzag edge through selective evaporation of Si atoms and occupation of the Si positions with free carbon atoms.

4. Effect of Ambient Gas

Previously, most CNT growth by surface decomposition of SiC was carried out under medium pressure, e.g., in a vacuum electric furnace [5-7], and the effect of the ambient gas on CNT growth has not been investigated in detail. So far, attempts have been made to determine the effect of oxygen and hydrogen gas on CNT growth. The effects of oxygen on the formation of carbon nanocaps and CNTs were first reported by Kusunoki et al. [6]. Comparing the length of CNTs grown under different pressures, they suggested that the pressure during heating is a major factor determining the CNT film thickness. In addition, their electron energy loss spectroscopy (EELS) results showed the segregation of oxygen atoms at the interface between the CNT film, suggesting that residual oxygen in the chamber plays an important role for CNT growth.

Recently, we carried out X-ray photoelectron spectroscopy (XPS) measurements on several samples annealed under various oxygen pressures [21]. Figure 15 shows the C 1s XPS spectra for samples annealed at 1270°C under UHV and different oxygen partial pressures (10^{-4} and 10^{-3} Pa). The main peaks around 284.3 eV are due to carbon sp^2 bonding, and the shoulder peaks around 282.7 eV correspond to SiC substrates. The annealing conditions indicate that the carbon sp^2 bonding arose from carbon nanocaps or CNTs. All spectra were normalized to the carbon sp^2 peaks. It should be noted that the ratio of the peak intensity of the SiC component to that of the carbon sp^2 component decreased as the oxygen partial pressure increased. This indicates that oxygen enhances the decomposition of the SiC surface, promoting CNT growth.

To determine the effect of oxygen on SiC, surface oxidation has also been widely investigated. It is well known that there are two modes in oxidation of the SiC surface: passive oxidation, in which a SiO_2 layer is formed on the SiC surface, and active oxidation, in which the silicon-containing products are gaseous, such as Si or SiO, leading to etching of the SiC surface [24]. The Si 2p XPS spectra shown in Figs. 9(a) and (b) correspond to active and passive oxidation, respectively. Thus, it is interesting to determine how oxygen affects the SiC surface, through passive oxidation, active oxidation or formation of nanocaps and CNTs.

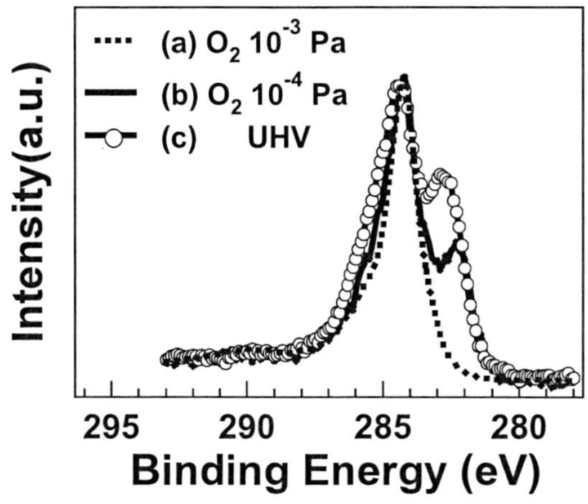

Figure 15. C 1s XPS spectra for samples after heating at 1270°C for 30 min under UHV and various oxygen partial pressures (10^{-4} and 10^{-3} Pa).

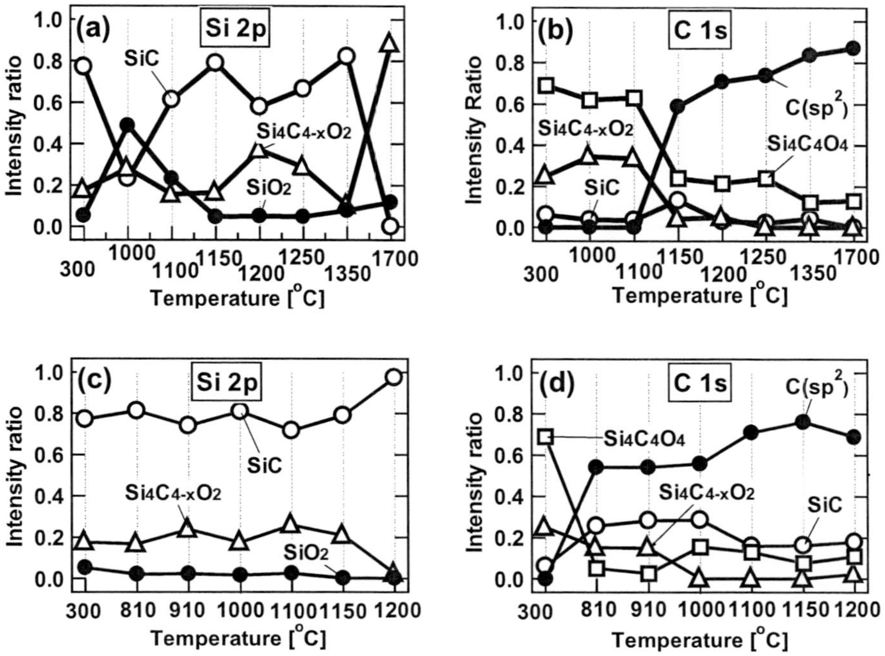

Fig. 16. Variation of surface composition of 6H-SiC(000-1) after each treatment. (a) Silicon- and (b) carbon-derived compositions before and after heating at various temperatures in a vacuum electric furnace. (c) Si- and (d) C-derived compositions before and after heating at various temperatures under UHV. The Si- and C-derived compositions were estimated by the deconvolution of Si 2p and C 1s XPS spectra after each treatment.

Figure 16 shows the variation of surface composition after annealing at various temperatures both in a vacuum electric furnace and under UHV. The surface compositions were determined by the deconvolution of Si 2p and C 1s XPS spectra. As mentioned before, the amount of SiO_2 before heating was negligible. After heating in a vacuum electric furnace, a SiO_2 layer formed on the SiC surface between 1000 and 1100°C, as shown in Fig. 16(a). No SiO_2 layer was formed between 810 and 1200°C when the SiC was annealed under UHV. These results confirm that the oxygen partial pressure strongly affects the surface oxidation mode of SiC. In the case of the vacuum electric furnace, a carbon sp^2 component corresponding to elemental carbon arising from the decomposition of SiC appeared at 1150°C, the temperature where the SiO_2 layer desorbed completely. Elemental carbon was observed even below 1000°C in the case of annealing under UHV, which suggests that the surface SiO_2 layer suppresses the decomposition of Si-C bonds. This indicates that from the standpoint of nanocap formation, the presence of oxygen has an adverse effect on the SiC surface, depending on the oxygen pressure and temperature. Thus, it is important to investigate the relationship between the formation of the SiO_2 layer and the formation of carbon nanocaps in order to determine the effect of oxygen on the surface and CNT growth. Figure 17 shows a phase diagram of the products (nanocap, elemental carbon, SiO_2) generated by the interaction between O_2 and the SiC surface. The diagram shows the active-passive oxidation boundary determined from previous experimental results [25], although this was estimated from data in the pressure range between 10^{-1} and 10^2 Pa. The critical oxygen pressure at which the active-passive transition occurred was estimated to be $P_C(O_2, T)$

= $P_0\exp(-\Delta E/k_BT)$, where $P_0 = 2.7\times10^{15}$ Pa and ΔE=4.8 eV. In spite of the difference in oxygen pressure, SiO_2 formation was observed in the passive oxidation region, indicating that passive oxidation proceeds via the same chemical reactions. In the case of heating under UHV, formation of SiO_2 was not observed even in the lower temperature region owing to the negligible presence of oxygen during heating. It should be noted that formation of carbon nanocaps occurred at about 1200°C, irrespective of the ambient pressure. This indicates that the heating at 1200°C is necessary to coalesce and crystallize carbon atom clusters to form nanocaps. A previous study reported that crystallization begins above 1200°C for certain kinds of coals composed of amorphous carbons, which supports our experimental observation [26].

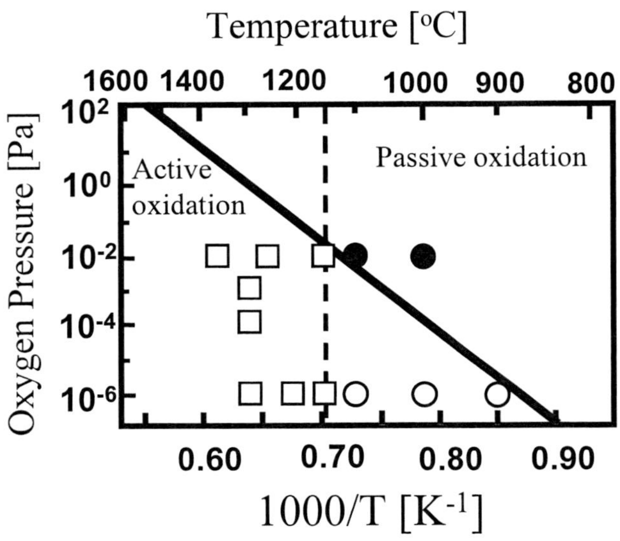

Figure 17. Phase diagram for the interaction of O_2 with the 6H-SiC(000-1) face showing the active-passive boundary. Our experimental results are indicated as well: the solid circles, open circles and open squares represent the surfaces covered with SiO_2, elemental carbon (unformed nanocaps) and carbon nanocaps, respectively.

Recently, the effect of hydrogen gas during CNT growth was investigated by Zhang et al [13]. They heated 6H-SiC(000-1) C-face at 1600°C for 5 h in a furnace into which H_2 gas was introduced at a pressure of 1×10^{-4} Torr. Their TEM observation suggested that the addition of H_2 increased the yield of SWNTs and decreased the SWNT diameter. These results suggest that addition of H_2 may enable the growth of very thin SWNTs by the surface decomposition technique.

5. Surface Polarity Dependence

In this paper, we have focused on CNT growth from the SiC C-face by surface decomposition, which shows several interesting characteristics. Heating under similar conditions yields graphite layers on the Si-face of SiC, as shown in Fig. 2 (b) [5]. As mentioned, one of the important reasons for this difference is thought to be the dangling bonds on the SiC surface. Recently, the formation of CNTs was also observed on the SiC Si-face [12]. These CNTs were aligned parallel to the surface of the SiC Si-face, unlike in the case of the C-face. In addition, the CNTs on the Si-face were formed on the graphite layers. On the basis of their in situ STM observations, Hayashi et al. proposed that local swelling of the graphene sheet, possibly resulting from the compressive stress, gives rise to a CNT nucleus, followed by CNT formation by continuous and undirected prolongation and rolling-up. They maintained that these processes may be the result of relaxation of the compressive stress in the graphite layer on the SiC surface, but the exact mechanism remains unclear. It is very interesting that the mechanism of CNT formation on the SiC Si-face differs completely from that on the C-face.

6. Summary

The present state of research on CNT growth by surface decomposition of SiC was reviewed and the CNT formation mechanism was discussed on the basis of data from our group. This growth technique has several characteristics: automatic alignment of chirality, relatively uniform CNT diameter, and high-density CNT growth. In addition, in this technique, CNTs grow directly from the semiconductor substrate. We believe that this is a unique characteristic that will benefit the fabrication of CNT devices on semiconductor substrates, leading to fusion between the semiconductor and CNT in the future. However, there are still several issues that need to be addressed before CNT devices can be fabricated. These include improvement of wall quality, improvement and control of uniformity of both the diameter and number of walls, removal of graphitic impurities in the CNT layer.

Acknowedgements

The authors thank Dr. Ito, Mr. Kawamura, Ms. Fujita and Mr. Tanioku for assisting with our study on CNT growth by surface decomposition of SiC. We are also grateful to Prof. Bandow in Meijo University for Raman measurements and Prof. Yokoyama and Dr. Nakagawa in the Institute of Molecular Science for XPS measurements. This work was partly supported by the Ministry of Education, Culture, Sports and Technology, Japan, through the 21st century COE program, the Grant-in-Aid for Scientific Research (C) (#17560015) and the nanotechnology support project. We also appreciate the funding from Meijo University.

References

[1] Iijima, S. *Nature*. 1991, vol. 354, 56-58.

[2] Moravsky, A. P.; Wexler, E. M. Loutfy, R. O. & Meyyappan, M. In *Carbon Nanotubes Science and Applications*; Meyyappan, M.; Ed.; CRC Press: Boca Raton, FL, 2005; pp.65-116.

[3] Kusunoki, M.; Rokkaku, M. Suzuki, T. *Appl. Phys. Lett.* 1997, vol.71, 2620-2622.

[4] Takikawa, H.; Miyano, R.; Yatsuki, M.; Sakakibara, T. *Jpn. J. Appl. Phys*. 1998, vol. 37, L187-L189.

[5] Kusunoki, M.; Suzuki, T.; Hirayama T.; Shibata N. *Appl. Phys. Lett.* 2000, vol.77, 531-533.
[6] Kusunoki, M.; Suzuki, T.; Kaneko K.; Ito M. *Phils. Mag. Lett.* 1999, vol.79, 153-161.
[7] Kusunoki, M.; Suzuki, T.; Honjo, C.; Hirayama, T.; Shibata, N. *Chem. Phys. Lett.* 2002, vol. 366, 458-462.
[8] Verma, A.R.; Krishna, P. *Polymorphism and Polytypism in Crystals*; Wiley; New York, NY 1966.
[9] Watanabe, H.; Hisada, Y.; Mukainakano, S.; Tanaka, N. *J. Microscopy* 2001, vol. 203, 40-46.
[10] Hisada, Y.; Mitsuoka, Y.; Mukainakano, S. *e-J. Surf. Sci. Nanotech.* 2004, vol. 2, 8-16.
[11] Derycke. V.; Martel, R.; Radosavljevic, M.; Ross, F.M.; Avouris, Ph. *Nano Lett.* 2002, vol. 2, 1043.
[12] Hayashi, K.; Mizuno, S.; Tanaka, S.; Toyoda, H.; Tochihara, H.; Suemune, I. *Jpn. J. Appl. Phys.* 2005, vol. 44, L803-L805.
[13] Zhang, G.; Hashimoto, Y.; Nakamura, A.; Tanaka, A.; Temmyo, J.; Matsuki, Y. *Jpn. J. Appl. Phys.* 2007, vol. 46, L53-L56.
[14] Saito, R.; Dresselhaus, G.; Dresselhaus, M. S. *Physical Properties of Carbon Nanotubes*; Imperial College Press: London, 1998; pp.183-206.
[15] Maruyama, T.; Shiraiwa, T.; Fujita, N.; Kawamura, Y.; Naritsuka, S.; Kusunoki, M. *Jpn. J. Appl. Phys.* 2006, vol. 45, 7231-7233.
[16] Nakashima, S.; Harima, H.; Tomita, T.; Suemoto, T. *Phys. Rev. B. 2000*, vol. 62, 16605-16611.
[17] Saito, R.; Dresslhaus, G.; Dresselhaus, M. S., *Phys. Rev. B. 2000,* vol. 61, 2981-2990.
[18] Jorio, A.; Saito, R.; Hafner, J. H.; Liever, C.M.; Hunter, M.; McClure, T.; Dresselhaus, G.; Dresselhaus, M.S. *Phys. Rev. Lett.* 2001, vol. 86, 1118-1121.
[19] Kataura, H.; Kumazawa, Y.; Maniwa, Y.; Umezu, I.; Suzuki, S.; Ohtsuka, Y.; Achiba, Y. *Synth. Met.* 1999, vol. 103, 2555-2558.
[20] Bang, H.; Ito, Y.; Kawamura, Y,; Hosoda, E.; Yoshida, C.; Maruyama, T.; Narituska, S.; Kusunoki, M. *Jpn. J. Appl. Phys.* 2006, vol. 45, 372-374.
[21] Maruyama, T.; Bang, H.; Fujita, N.; Kawamura, Y.; Naritsuka, S.; Kusunoki, M., *Diamond Relat. Mater.* 2007, vol. 16, 1078-7081.
[22] Maruyama, T.; Bang, H.; Kawamura, Y.; Fujita, N.; Tanioku, K.; Shiraiwa, T.; Hozumi, Y. Shigeya, N.; Kusunoki, M. *Chem. Phys. Lett.* 2006, vol. 423, 317-320.
[23] Irle, S.; Wang, Z.; Zheng, G.; Morokuma, K.; Kusunoki, M. J. Chem. Phys. 2006, vol. 125, 044702/1-044702/5.
[24] Schneider, B.; Guette, A.; Naslain, R.; Cataldi, M.; Costecalde, A. *J. Mater. Sci.* 1998, vol. 33, 535-547.
[25] Song, Y.; Smith, F. W. *J. Am. Ceram. Soc.* 2005, vol. 88, 1864-1869.
[26] Tsubouchi, N.; Xu, C.; Ohtsuka, Y. *Energy & Fuels* 2003, vol. 17, 1119-1125.

In: Research Progress in Materials Science ISBN 978-1-60456-629-1
Editors: W. Olsson and F. Lindberg pp. 149-163

Chapter 6

COMPARATIVE STUDY OF STATISTICS OF COOPER'S ELECTRON PAIRS IN LOW-TEMPERATURE SUPERCONDUCTORS AND COUPLED HOLES PAIRS IN HIGH T_c CERAMICS

I.G. Kaplan* and O. Navarro
Instituto de Investigaciones en Materiales, Universidad Nacional Aut ónoma de México, Apartado Postal 70-360, 04510, México, D. F

Abstract

It is well known that the system of Cooper's pair is described by boson symmetric wave functions, but Cooper's pair operators are bosons only when the moments **k** are different and they are fermions for equal **k**. The analysis of trilinear commutation relations for the Cooper pair (pairon) operators reveals that they correspond to the modified parafermi statistics of rank $p = 1$. Two different expressions for the Cooper pair number operator are presented. We demonstrate that the calculations with a Hamiltonian expressed via pairon operators is more convenient using the commutation properties of these operators without presenting them as a product of fermion operators. This allows to study problems in which the interactions between Cooper's pairs are also included. The problem with two interacting Cooper's pairs is resolved and its generalization in the case of large systems is discussed.

It is shown that in site representation, the hole-pair operators obey the same commutation relations (paulion) as the Cooper pair operators in impulse representation, although the latter describe delocalized quasiparticles. In quasi-impulse representation, the hole-pair operators are also delocalized and their exact commutation relations correspond to a modified parafermi statistics of rank M (M is the number of sites in a "superlattice" formed by the centers of mass of each hole pair). From this follows that one state can be occupied by up to M pairs. Even in the absence of dynamic interaction, the system of hole pairs is characterized by some immanent interaction, named after Dyson as kinematic interaction. This interaction appears because of the deviation of the quasiparticle statistics from the Bose (Fermi) statistics and its magnitude

*E-mail address: kaplan at iim.unam.mx(Corresponding author)

depends on the concentration of hole pairs. In spite of the non-bosonic behavior, there is no statistical prohibition on the Bose-Einstein condensation of coupled hole pairs.

Keywords: Cooper's pair commutation relations, Pairing interactions, Strongly correlated electron systems.

1. Introduction

The theory of the low temperature superconductivity was created by Bardeen, Cooper, and Schrieffer (BCS) [1] only after Cooper [2] had shown that two electrons interacting above the Fermi sea of non-interacting electrons can couple in a stable pair, if the interaction resulting from virtual exchange of phonons is attractive near the Fermi surface. As was demonstrated in a more sophisticated study [3], in a full agreement with the Cooper assumption, the largest binding energy of the Cooper pair corresponds to electrons with the opposite momenta and spins. In the second quantization formalism, the operators of creation, b_k^+, and annihilation, b_k, of Cooper's pair in a state $(\mathbf{k}\alpha, -\mathbf{k}\beta)$, are defined as simple products of the electron creation and annihilation operators, $c_{k\sigma}^+$ and $c_{k\sigma}$, satisfying the fermion commutation relations,

$$b_k^+ = c_{k\alpha}^+ c_{-k\beta}^+; \qquad b_k = c_{-k\beta} c_{k\alpha}. \tag{1}$$

Let us call these operators, following Schrieffer [3], as "pairon" operators.

The Cooper pair has the total spin $S = 0$. Hence, in accordance with the Pauli principle, the wave functions describing the Cooper pair system have the boson permutation symmetry, that is, they are symmetric under permutations of pairs. But the pairon operators (1) do not obey the boson commutation relations. It is easy to show by direct calculation. Namely,

$$\left[b_k, b_{k'}^+\right]_- = \left[b_k^+, b_{k'}^+\right]_- = \left[b_k, b_{k'}\right]_- = 0 \qquad \text{for } k \neq k', \tag{2}$$

$$\left[b_k, b_k^+\right]_- = 1 - \hat{n}_{k\alpha} - \hat{n}_{-k\beta}, \tag{3}$$

$$\left(b_k^+\right)^2 = \left(b_k\right)^2 = 0, \tag{4}$$

where $\hat{n}_{k\alpha} = c_{k\sigma}^+ c_{k\sigma}$ is the electron number operator. As follows from Eqs. (2)–(4), for $k \neq k'$ the Cooper pairs are bosons, while for $k = k'$ they do not obey the boson commutation relations, although they obey the Pauli principle and have the fermion occupation numbers for one-particle states.

Thus, the pairon operators may not be considered neither as the Bose operators, nor as the Fermi operators. This is the reason that the problem with the model Hamiltonian of the BCS theory

$$H = \sum_k \varepsilon_k b_k^+ b_k + \sum_{k',k} V_{kk'} b_{k'}^+ b_k \tag{5}$$

cannot be directly solved by transforming Hamiltonian (5) to the diagonalized form

$$H = \sum_n \varepsilon_k' B_k^+ B_k \tag{6}$$

by means of some unitary transformation

$$B_n = \sum_k u_{nk} b_k, \qquad B_n^+ = \sum_k u_{nk}^* b_k^+. \tag{7}$$

The unitary transformation is canonical only for the Bose or Fermi operators. In general case, it is not canonical; it does not preserve the commutation properties of the operators transformed. Therefore, practically all calculations in the BCS approach were performed using the fermion properties of electron operators forming the Cooper pair.

In 1953 Green [4] and, independently, Volkov [5] have shown that the boson and fermion commutation relations do not exhaust all physically possible commutation relations for field operators in the second quantization formalism. The field operators satisfying the requirements of causality, relativistic invariance and positivity of energy can obey more general commutation relations than the boson and fermion ones. This new statistics was called parastatistics, it is shortly described in next section.

In section 3 we analyze the commutation relations for the pairon operators and show that the trilinear commutation relations correspond to the modified parafermi statistics of rank $p = 1$. In section 4, we demonstrate that the calculations with a Hamiltonian expressed via the pairon operators is more convenient to perform using the commutation properties of these operators without presenting them as a product of fermion operators. This allows to study problems in which the interactions between Cooper's pairs are also included [6]. The solution of the simplest problem with two interacting Cooper's pairs is presented.

Although, all elementary particles known at present are bosons or fermions, the parastatistics can be realized for quasiparticles. As was shown by one of the authors [7], the quasiparticles in a periodical lattice (the Frenkel excitons and magnons) obey a modified parafermi statistics of rank M, where M is the number of equivalent lattice sites within the delocalization region of collective excitations. Later on, the results [7] for the Frenkel excitons and magnons where extended to polaritons [8, 9], defectons in quantum crystals [10] and to the Wannier-Mott excitons [11]. The statistics and some properties of a system of noninteracting holes was studied in ref. [12]. Below, in section 5 and 6, we discuss in detail the statistics and properties of the coupled hole pairs [13].

2. Short Account of Parastatistics

Green [4] introduced in 1953 a generalized method of field quantization with trilinear commutation relations, which include the boson and fermion commutation relations as particular cases,

$$\begin{aligned} \left[[a_k^+, a_{k'}]_\pm, a_{k''} \right]_- &= -2\delta_{kk''} a_{k'}, \\ \left[[a_k^+, a_{k'}]_\pm, a_{k''}^+ \right]_- &= 2\delta_{k'k''} a_k^+, \\ \left[[a_k^+, a_{k'}^+]_\pm, a_{k''}^+ \right]_- = \left[[a_k, a_{k'}]_\pm, a_{k''} \right]_- &= 0. \end{aligned} \tag{8}$$

The relations (8) with upper sign at the inner brackets are called the paraboson commutation relations and with lower sign are called the parafermion commutation relations.

These names are connected with the fact that the ordinary boson and fermion operators also obey the relations (8). But the above relations are fulfilled by a more general type of operators given by the so-called Green Ansatz [4]:

$$a_k = \sum_{\rho=1}^{p} d_{\rho k}, \tag{9}$$

having an infinite set of solutions, labeled by the integers $p(p = 1, 2, ..., \infty)$.

For parabosons, the operators $d_{\rho k}$ obey the following commutation relations:
(a) $\rho = \rho'$, as for bosons

$$\left[d_{\rho k}, d^{+}_{\rho k'}\right]_{-} = \delta_{kk'} \ ; \ \left[d_{\rho k}, d_{\rho k'}\right]_{-} = \left[d^{+}_{\rho k}, d^{+}_{\rho k'}\right]_{-} = 0 \tag{10}$$

(b) $\rho \neq \rho'$, as for fermions

$$\left[d_{\rho k}, d^{+}_{\rho' k'}\right]_{+} = \left[d_{\rho k}, d_{\rho' k'}\right]_{+} = \left[d^{+}_{\rho k}, d^{+}_{\rho' k'}\right]_{+} = 0. \tag{11}$$

For parafermions:
(a) $\rho = \rho'$, as for fermions

$$\left[d_{\rho k}, d^{+}_{\rho k'}\right]_{+} = \delta_{kk'} \ ; \ \left[d_{\rho k}, d_{\rho k'}\right]_{+} = \left[d^{+}_{\rho k}, d^{+}_{\rho k'}\right]_{+} = 0 \tag{12}$$

(b) $\rho \neq \rho'$, as for bosons

$$\left[d_{\rho k}, d^{+}_{\rho' k'}\right]_{-} = \left[d_{\rho k}, d_{\rho' k'}\right]_{-} = \left[d^{+}_{\rho k}, d^{+}_{\rho' k'}\right]_{-} = 0. \tag{13}$$

As follows from Eqs. (10) and (12), the solution of the Green Ansatz (9) for $p = 1$, reduces to the usual boson and fermion operators in paraboson and parafermion cases, respectively. It is worth-while to note that such mixed boson-fermion behavior of the operators $d_{\rho k}$ is well known in solid state physics and called as *paulion*, see the book by Davydov [14]. The value of p in the Green Ansatz is called the rank of parastatistics. The application of operators (9) to the vacuum state is easily found using Eqs. (10)-(13). So,

$$a_k|0\rangle = 0, \qquad a_k a^{+}_{k}|0\rangle = p\delta_{kl}|0|\rangle \qquad \text{for all } k, l. \tag{14}$$

For the parafermi statistics, the maximum occupation numbers of one state is equal to the rank of parastatistics p

$$\left(a^{+}_{k}\right)^{N}|0\rangle \neq 0 \qquad \text{for } N \leq p, \tag{15}$$

$$\left(a^{+}_{k}\right)^{p+1}|0\rangle = 0. \tag{16}$$

At $p = 1$, the parafermi statistics is reduced to the Fermi-Dirac statistics. For parabose statistics there is no restriction in the occupation numbers, as in the Bose-Einstein statistics.

The particle number operator in parastatistics is defined as

$$\hat{N}_k = \frac{1}{2}\left[a^{+}_{k}, a_k\right]_{\pm} \mp \frac{1}{2}p, \tag{17}$$

where the upper sign for parabosons and the lower sign for parafermions. It satisfies the general property of the particle number operators

$$\left[\hat{N}_k, a_l^+\right]_- = \delta_{kl} a_k^+ \tag{18}$$

and in the Bose (Fermi) case turns to

$$\hat{N}_k = a_k^+ a_k. \tag{19}$$

As was shown in Refs. [15, 16], the parafermion and paraboson algebras are the Lie algebras of the orthogonal and symplectic groups, respectively. For example, the algebra for ν parafermion operators a_r and their adjoints $a_r^+ (r = 1, 2, ..., \nu)$, which constitutes a parafermi ring, is the Lie algebra of the orthogonal group $SO_{2\nu+1}$ in $2\nu + 1$ dimensions; more details about the parastatistics is presented in the book by Ohnuki and Kamefuchi [17].

In spite of numerous studies of all known elementary particles, see Refs. [18, 19], the elementary particles obeying the parastatistics were not revealed. On the other hand, as discussed in Refs. [20, 21], the ordinary fermions, which differ by some internal quantum numbers but are similar dynamically, can be described by the parafermi statistics. In this case, fermions with different internal quantum numbers are considered as non-identical distinguishable particles. The parafermi statistics of rank p describes systems with p different types of fermions. As a result, *quarks* with 3 colors obey the parafermi statistics of rank $p = 3$; *nucleons* in nuclei (isotope spin $1/2$) obey the parafermi statistics of rank $p = 2$.

Let us stress, the fermions characterized by the same value of internal quantum number obey the Fermi-Dirac statistics. It is the system of p fermion groups with different internal quantum numbers that obey the parafermi statistics of rank p.

3. Statistics of Cooper's Pairs

As was discussed in the Introduction, the pairon operators, Eq. (1), obey boson commutation relations only in the case of different momenta. For equal momenta, the right-hand part of commutation relation (3) contains the products of fermion operators that reflects the fermion structure of pairon operators.

$$\left[b_k, b_k^+\right]_- = 1 - c_{k\alpha}^+ c_{k\alpha} - c_{-k\beta}^+ c_{-k\beta}. \tag{20}$$

To operate with the pairon operators, the commutation relations for these operators do not have to include other kinds of operators. One of the ways to achieve this goal is to calculate trilinear commutation relations, as it is formulated in the parastatistics [4, 22].

The direct calculation leads to the following trilinear commutation relations

$$\left[\left[b_k^+, b_{k'}\right]_-, b_{k''}^+\right]_- = 2\delta_{kk'}\delta_{kk''} b_k^+, \tag{21}$$

$$\left[\left[b_k^+, b_{k'}\right]_-, b_{k''}\right]_- = -2\delta_{kk'}\delta_{kk''} b_k. \tag{22}$$

These relations coincide with the trilinear commutation relations (8) for the parafermi statistics for $k = k' = k''$. For different k, k', and k'' the relations are different. In the parafermi statistics in relations (8) corresponding to Eq. (21) instead of the two presented Kronecker symbols there is one, namely, $\delta_{k'k''}$; and in relation corresponding to Eq. (22), $\delta_{kk'}$ is absent. Thus, the pairon operators satisfy some modified parafermi statistics of the rank $p = 1$. The latter follows from Eq. (4), since the parastatistics of rank p satisfy Eqs. (15) and (16).

As follows from the definition of the particle number operator $\hat{N}_k$

$$\hat{N}_k|N_k\rangle = N_k|N_k\rangle. \tag{23}$$

For the boson and fermion number operators the well-known expression $\hat{N} = a_k^+ a_k$ is valid. But it is quite not evident that the same expression is valid for the pairon number operator. In the parafermi statistics, the particle number operator is defined as

$$\hat{N}_k = \frac{1}{2}\left(\left[a_k^+, a_k\right]_- + p\right), \tag{24}$$

for pairons $p = 1$ and

$$\hat{N}_k = \frac{1}{2}\left(\left[b_k^+, b_k\right]_- + 1\right). \tag{25}$$

As follows from the trilinear commutation relations (21) and (22), the operator (25) satisfies the commutation relations for the particle number operator that were established earlier for bosons and fermions, see Ref. [23],

$$\left[\hat{N}_k, b_k^+\right]_- = b_k^+, \tag{26}$$

$$\left[\hat{N}_k, b_k\right]_- = -b_k. \tag{27}$$

It is easy to check that for fermions, Eq. (25) is equivalent to the standard expression $\hat{n}_k = c_k^+ c_k$. Let us study it in the pairon case.

Using Eq. (3), the expression for the pairon number operator (25) can be written as

$$\hat{N}_k = \frac{1}{2}\left(\hat{n}_{k\alpha} + \hat{n}_{-k\beta}\right). \tag{28}$$

This is quite natural that the number of Cooper's pairs is two times less than the number of electrons forming pairs. It can be proved that from Eq. (28) follows that the expression $\hat{N}_k = b_k^+ b_k$ may be also used for the pairon number operators. Let us do it.

The product $b_k^+ b_k$ is equal to the product of the fermion number operators $\hat{n}_{k\alpha}\hat{n}_{-k\beta}$, but in the general case

$$\hat{n}_{k\alpha}\hat{n}_{-k\beta} \neq \frac{1}{2}\left(\hat{n}_{k\alpha} + \hat{n}_{-k\beta}\right). \tag{29}$$

Pairons operators possess the fermions occupation numbers, $n_{k\alpha}$ and $n_{-k\beta}$ equal to 0 or 1, in this case, and only in this case, the left-hand part of Eq. (29) is equal to its right-hand part. Thus,

$$\frac{1}{2}\left(\hat{n}_{k\alpha} + \hat{n}_{-k\beta}\right) = b_k^+ b_k \tag{30}$$

and from Eqs. (28) and (30) follows that although Cooper's pairs are neither bosons nor fermions, for the operators of their number, the traditional form $\hat{N}_k = b_k^+ b_k$ can be used.

If one substitutes the equality

$$\hat{n}_{k\alpha} + \hat{n}_{-k\beta} = 2b_k^+ b_k, \tag{31}$$

into the commutation relation (3), it transforms into

$$\left[b_k, b_k^+\right]_- = \left(1 - 2b_k^+ b_k\right) \tag{32}$$

or

$$\left[b_k, b_k^+\right]_+ = 1. \tag{33}$$

Thus, for equal **k**, the pairon operators obey the fermion commutation relations, while for different **k**, they obey the boson commutation relations.

Despite the fact that each Cooper's pair has the total spin $S = 0$, the pairons are not bosons, because for equal momenta **k** they behave as fermions. However, for different **k**, the pairons obey the Bose-Einstein statistics and can occupy one energy level, that is, they can undergo the phenomenon of the Bose-Einstein condensation. However, in this case all electrons composed into the condensed Cooper pairs must have different momenta **k**.

The Eqs. (2) and (32) can be combined into one commutation relation

$$\left[b_k, b_{k'}^+\right]_- = \delta_{kk'}\left(1 - 2b_k^+ b_k\right). \tag{34}$$

The application of pairon operators to the vacuum state follows from their definition, Eq. (1),

$$b_k|0\rangle = 0, \qquad b_k b_{k'}^+|0\rangle = \delta_{kk'}|0\rangle. \tag{35}$$

The relations (34) and (35) are sufficient for performing calculations using only the pairon operators.

4. Generalized Model Hamiltonian and the Problem of Two Interacting Cooper's Pairs

Let us add to the BCS Hamiltonian the term describing the interaction among Cooper's pairs. The generalized model Hamiltonian is

$$H = 2\sum_k \epsilon_k b_k^\dagger b_k - \sum_{k,k'}{}' V_{k,k'} b_{k'}^\dagger b_k + \frac{1}{2}\sum{}'\sum_{k_1',k_2'k_1'',k_2''}{}' V_{k_1'k_2',k_1''k_2''} b_{k_1''}^\dagger b_{k_2''}^\dagger b_{k_1'} b_{k_2'}, \tag{36}$$

where primes in sums denote that $k \neq k'$, $k_1' \neq k_2'$ and $k_1'' \neq k_2''$. The other restriction is concern with the potential energy of interpair interaction in which $k_1'' \neq k_1'$ and $k_2'' \neq k_2'$. While according to the Cooper model, $V_{k,k'} > 0$, the sign of $V_{k_1'k_2',k_1''k_2''}$ is not restricted, it can be both positive for a repulsive interpair interaction and negative for an attractive interaction.

In general, the variational wave function of the system with N pairs can be presented as

$$|\Psi(1,2,...,N)\rangle = \sum_{k_1,k_2,...,k_N}{}' \alpha(k_1,k_2,...,k_N) b_{k_1}^+ b_{k_2}^+ ... b_{k_N}^+|0\rangle, \tag{37}$$

where $k_1 \neq k_2 \neq ... \neq k_N$ because of the fermion condition (4), or it can be presented in the BCS form, which is not as precise as the variational function (37) but easier for calculations

$$|\Psi(1,2,...,N)\rangle = \prod_{k=1}^{N}(u_k + v_k b_k^{+})|0\rangle. \tag{38}$$

As an illustration of operations with the Cooper pair operators using their properties (34) and (35), we consider the model of two interacting pairs described by the BCS wave function

$$\begin{aligned} |\Psi(1,2)\rangle &= \prod_{k_j=1}^{2}\left(u_{k_j} + v_{k_j} b_{k_j}^{\dagger}\right)|0\rangle \\ &= \left(u_{k_1}u_{k_2} + u_{k_1}v_{k_2}b_{k_2}^{\dagger} + u_{k_2}v_{k_1}b_{k_1}^{\dagger} + v_{k_1}v_{k_2}b_{k_1}^{\dagger}b_{k_2}^{\dagger}\right)|0\rangle. \end{aligned} \tag{39}$$

Following the BCS theory, we assume in the general Hamiltonian (36) that the interaction energies do not depend upon the value of moments

$$H = 2\sum_k \epsilon_k b_k^{\dagger} b_k - V_0 \sum_{k,k'}{}' b_{k'}^{\dagger} b_k + \frac{V_1}{2}\sum{}'\sum_{k_1',k_2'k_1'',k_2''}{}' b_{k_1''}^{\dagger} b_{k_2''}^{\dagger} b_{k_1'} b_{k_2'}. \tag{40}$$

Using the properties of Cooper's pair operators, Eqs. (34) and (35), we calculate the expectation value for the energy $W = \langle\Psi(1,2)|H|\Psi(1,2)\rangle$ with the Hamiltonian (40) and wave function (39). The results is

$$W = 2v_{k_1}^2\epsilon_{k_1} + 2v_{k_2}^2\epsilon_{k_2} - 2V_0 u_{k_1}v_{k_1}u_{k_2}v_{k_2} + V_1 v_{k_1}^2 v_{k_2}^2. \tag{41}$$

By a minimization procedure respect to v_{k_1} and v_{k_2} using the Lagrange multiplicators, the following definitions

$$\Delta_{k_1} \equiv V_0 u_{k_2} v_{k_2}; \qquad \Delta_{k_1}' \equiv V_1 v_{k_2}^2, \tag{42}$$

$$\Delta_{k_2} \equiv V_0 u_{k_1} v_{k_1}; \qquad \Delta_{k_2}' \equiv V_1 v_{k_1}^2, \tag{43}$$

and introducing

$$E_{k_1}' = \sqrt{\left(\epsilon_{k_1} + \frac{\Delta_{k_1}'}{2}\right)^2 + \Delta_{k_1}^2}, \qquad E_{k_2}' = \sqrt{\left(\epsilon_{k_2} + \frac{\Delta_{k_2}'}{2}\right)^2 + \Delta_{k_2}^2}; \tag{44}$$

we obtain that

$$v_{k_1}^2 = \frac{1}{2}\left(1 - \frac{\epsilon_{k_1} + \frac{\Delta_{k_1}'}{2}}{E_{k_1}'}\right), \qquad u_{k_1}^2 = \frac{1}{2}\left(1 + \frac{\epsilon_{k_1} + \frac{\Delta_{k_1}'}{2}}{E_{k_1}'}\right), \tag{45}$$

$$v_{k_2}^2 = \frac{1}{2}\left(1 - \frac{\epsilon_{k_2} + \frac{\Delta_{k_2}'}{2}}{E_{k_2}'}\right), \qquad u_{k_2}^2 = \frac{1}{2}\left(1 + \frac{\epsilon_{k_2} + \frac{\Delta_{k_2}'}{2}}{E_{k_2}'}\right), \tag{46}$$

and

$$u_{k_1}v_{k_1} = \frac{1}{2}\left\{1 + \frac{\left(\epsilon_{k_1} + \frac{\Delta'_{k_1}}{2}\right)}{E'^2_{k_1}}\right\}^{1/2} = \frac{\Delta_{k_1}}{2E'_{k_1}}. \tag{47}$$

The calculation of the quasiparticle excitation energy results in

$$E'_k = \sqrt{\left(\epsilon_k + \frac{\Delta'_k}{2}\right)^2 + \Delta_k^2}. \tag{48}$$

Thus, the quantities introduced by Eq. (44) and entered into Eqs. (45)-(47) have the physical sense as the quasiparticle excitation energies.

Using, as in the BCS theory, the approximations

$$\begin{aligned} \Delta_{k_1} &= \Delta_{k_2} = \Delta \\ \Delta'_{k_1} &= \Delta'_{k_2} = \Delta' \\ \epsilon_{k_1} &\simeq \epsilon_{k_2} = \epsilon \end{aligned} \tag{49}$$

we obtain the explicit expressions for the parameters Δ' and Δ

$$\Delta' = V_1 \frac{V_0 - 2\epsilon}{2V_0 + V_1}, \tag{50}$$

$$\Delta = \sqrt{\left(\frac{V_0}{2}\right)^2 - \left\{\epsilon + \frac{V_1}{2}\left(\frac{V_0 - 2\epsilon}{2V_0 + V_1}\right)\right\}^2}. \tag{51}$$

In the expression for E'_k and u_k^2 and v_k^2, as well, the electron energy ϵ_k enters with the additive term $\Delta'_k/2$. According to Eq. (50), this term depends upon the interaction energies V_1 and V_0, and it disappears when $V_1 = 0$. Thus, one can say that the interpair interaction leads to the renormalization of the electron energy. This renormalization is proportional to the interaction potential V_1 between pairs. On the other hand, the interpair interaction leads to an augmentation of the gap, which is according to Eq. (48) equal to

$$\sqrt{\Delta_k^2 + (\Delta'_k/2)^2}. \tag{52}$$

The interpair interaction increases the energy of the quasiparticle excitation.

For comparison with the BCS theory, we have to neglect the interaction between pairs, $V_1 = 0$. With this condition, expressions for u_k^2, v_k^2 and E_k are reduced to the BCS expressions. But in the case of Δ, (Eq. 51), it is not so, it is reduced to

$$\Delta = \sqrt{\left(\frac{V_0}{2}\right)^2 - \epsilon^2}. \tag{53}$$

This formula differs from the exponential expression for the energy gap in the BCS theory. The difference is connected with the fact that in a system with a finite number of particles, the distances between the energy levels are also finite, there is a discrete set of energy levels.

The approximation $\epsilon_{k_1} = \epsilon_{k_2} = \epsilon$, Eq. (49), is valid only for systems with $N \gg 1$, for which the energy spectrum is continuous and after integration one obtains the exponential dependence as in the BCS theory.

Finally, it is important to mention that the two-pairon system was considered as an illustration of applying the commutation relation for pairons, Eq. (34). The employment of pairon operators b_k^+, b_k in place of presenting them as product of fermion operators reduces twice the number of operators in the Hamiltonian. For instance, for the two-pairon system, the interaction term contains four pairon operators instead of eight fermion operators.

Let us note that the problem of two interacting pairons has been solved exactly. The real physical problem for N interacting pairons with $N \to \infty$ is very complicated and can be solved only approximately. For its solution one should, similar to the BCS theory, consider the interacting pairons in the nearest neighborhood of the Fermi level and take into account an approximation similar to that given by Eq. (49), so that the summation over k can be replaced by an integration.

5. Statistics and Properties of Coupled Hole Pairs

At present, it is well established that the conductivity in high-T$_c$ ceramics has a hole origin with charge of carriers equal to $+2e$. Here, we present the results of our study of statistics and some physical properties of the hole-pair system [13]. But before we shortly consider the properties of isolated holes in superconducting ceramics.

Usually, holes are described as fermions. It came from atomic physics: the closed electronic shell after one electron is knocked out (the "hole" formation) has the same angular and spin momentum properties as the electronic shell with one electron. But in general case, the holes can have different values of spin S. For example, in the CuO_2 planes in high-T$_c$ ceramics where the hole conductivity is revealed, all spins are paired, the so-called Zhang-Rice singlet [24, 25] is realized. The holes in high-T$_c$ ceramics (at least in the CuO_2 planes) can be considered as spinless positive charged quasiparticles. On the CuO_2 plane, the hole is delocalized among Cu and four O coupled by covalent bonding [25].

In second quantization formalism in the site representation, the model Hamiltonian for one type of spinless holes is

$$H = \epsilon_0 \sum_n b_n^\dagger b_n + \sum_{nn'} M_{nn'} b_n^\dagger b_{n'} + \sum_{nn'} V_{nn'} b_n^\dagger b_{n'}^\dagger b_{n'} b_n, \tag{54}$$

where ϵ_0 is the energy for the hole creation in a lattice, $M_{nn'}$ is the so-called hopping integral and $V_{nn'}$ is the hole-hole interaction term. As we showed in [12], the hole creation, $b_n^\dagger$, and hole annihilation, b_n, operators are characterized by the paulion properties:

$$[b_n, b_{n'}^\dagger]_- = [b_n, b_{n'}]_- = [b_n^\dagger, b_{n'}^\dagger]_- = 0 \qquad \text{for } n \neq n', \tag{55}$$

$$[b_n, b_n^\dagger]_+ = 1; \quad [b_n, b_n]_+ = [b_n^\dagger, b_n^\dagger]_+ = 0, \tag{56}$$

the operators acting on different sites obey the boson commutation relations, while the operators acting on one site obey the fermion commutation relations.

Suppose that the hole-hole interaction term in (54) is attractive. In this case under some conditions, the coupled hole pairs can be formed.

The operators of creation and annihilation of the hole pair are defined as usual:

$$\begin{aligned} a_t^\dagger &= b_n^\dagger b_m^\dagger, \\ a_t &= b_m b_n, \end{aligned} \tag{57}$$

where t denotes the localization point of the center of mass of the coupled hole pair. In high-T$_c$ ceramics, the hole-pair localization region is not large: the correlation length in the CuO_2 planes is of the order of (10 – 12)Å.

It is easy to verify that the hole-pair operators $a_t^\dagger$ and a_t obey the same paulion commutation relations (55) and (56) as the hole operators. Let us note that the Cooper pair operators also obey the paulion commutation relations, see Eqs. (2) and (33). However, there is an essential difference: the Cooper pair operators are defined in the impulse space and so they are completely delocalized, on the other hand, the hole-pair operators $a_t^\dagger$ and a_t are defined in the site representation and are localized at some regions of the lattice. As we show below, in the quasi-impulse representation, the statistics of hole pairs radically changes.

Let us assume that all hole pairs have the same size and the region of the hole pair localization can be repeated in crystal so that the points t form a "superlattice" with M sites. The model Hamiltonian for hole pairs can be presented as

$$H = \sum_t \epsilon_p a_t^\dagger a_t + \sum_{tt'} M_{tt'} a_t^\dagger a_{t'}, \tag{58}$$

where $\epsilon_p = 2\epsilon_0 + V_0$ is the energy of the coupled hole pair, V_0 is the attractive potential between holes which we assume to be the same for all pairs, as in the BCS approach, $M_{tt'}$ is the hopping integral for a hole pair moving as a whole entity. The Hamiltonian (58) can be transformed by some unitary transformation:

$$A_{\mathbf{q}} = \frac{1}{\sqrt{M}} \sum_{t=1}^{M} u_{\mathbf{q}t} a_t, \quad A_{\mathbf{q}}^\dagger = \frac{1}{\sqrt{M}} \sum_{t=1}^{M} u_{\mathbf{q}t}^* a_t^\dagger \tag{59}$$

to the diagonalized form in the quasi-impulse space,

$$H = \sum_{\mathbf{q}} \epsilon_{\mathbf{q}} A_{\mathbf{q}}^\dagger A_{\mathbf{q}}. \tag{60}$$

For simple lattices with one site per cell, the unitary transformation (59) is completely determined by the translation symmetry of the lattice and the coefficients $u_{\mathbf{q}n} = \exp(-i\mathbf{q}r_n)$. The self-energy of the diagonalized Hamiltonian (60) is equal to

$$\epsilon_{\mathbf{q}} = \epsilon_p + \sum_{t'(\neq t)} M_{tt'} \exp[i\mathbf{q}\cdot(\mathbf{r}_t - \mathbf{r}_{t'})]. \tag{61}$$

Since the operators (57) obey neither the boson nor the fermion commutation relations, the unitary transformation in general case is not canonical; this means that it does not preserve the commutation properties of the operators transformed. In particular, the operators (59) do not describe the paulion quasiparticles. As we showed in [12] for holes in quasi-momentum space, such operators obey the modified parafermi statistics of rank M (where

M is the number of lattice sites at which the hole can be created) with the values of quasi-momentum defined from the quasi-momentum conservation law. This statistics has been introduced by Kaplan [7] in 1976 for the Frenkel excitons and magnons.

For lattices, diagonalized by an exponential unitary transformation, the operators (59) obey the following trilinear commutation relations:

$$[[A^\dagger_{\mathbf{q}}, A_{\mathbf{q}'}]_-, A_{\mathbf{q}''}]_- = -2M^{-1}A_{\overline{\mathbf{q}}}, \qquad \overline{\mathbf{q}} = \mathbf{q}' + \mathbf{q}'' - \mathbf{q} \tag{62}$$

$$[[A^\dagger_{\mathbf{q}}, A_{\mathbf{q}'}]_-, A^\dagger_{\mathbf{q}''}]_- = 2M^{-1}A^\dagger_{\overline{\mathbf{q}}}, \qquad \overline{\mathbf{q}} = \mathbf{q} - \mathbf{q}' + \mathbf{q}''. \tag{63}$$

Eqs. (62) and (63) have not the Kronecker symbols, as the parafermi trilinear commutation relations (8), the value of $\overline{\mathbf{q}}$ is determined by the quasi-momentum conservation law. Thus, the commutation relations (62) and (63) correspond to the modified parafermi statistics [7]. It can be proved that the rank of parastatistics is equal to the number of sites, M, in the superlattice. This means that one state can be occupied by up to M hole pairs:

$$(A^+_{\mathbf{q}})^N |0\rangle \neq 0, \quad N \leq M \tag{64}$$

$$(A^+_{\mathbf{q}})^{M+1} |0\rangle = 0. \tag{65}$$

The state with N noninteracting pairs, each with the same $\mathbf{q}$, is defined by the usual expression

$$|N_{\mathbf{q}}\rangle = C_N (A^\dagger_{\mathbf{q}})^N |0\rangle, \tag{66}$$

where the normalization factor C_N can be found by the induction method using the operator equation obtained from the commutation relation (63)

$$\begin{aligned} A_{\mathbf{q}}A^\dagger_{\mathbf{q}'}A^\dagger_{\mathbf{q}''} &= A^\dagger_{\mathbf{q}'}A_{\mathbf{q}}A^\dagger_{\mathbf{q}''} + A^\dagger_{\mathbf{q}''}A_{\mathbf{q}}A^\dagger_{\mathbf{q}'} - A^\dagger_{\mathbf{q}''}A^\dagger_{\mathbf{q}'}A_{\mathbf{q}} - \frac{2}{M}A^\dagger_{\overline{\mathbf{q}}}, \\ \overline{\mathbf{q}} &= \mathbf{q}' + \mathbf{q}'' - \mathbf{q}. \end{aligned} \tag{67}$$

The expression for C_N differs from that for a Bose system and is given by:

$$C_N = \left[N!\left(1 - \frac{1}{M}\right)\left(1 - \frac{2}{M}\right)\cdots\left(1 - \frac{N-1}{M}\right)\right]^{-\frac{1}{2}}. \tag{68}$$

Now, it is easy to find

$$A^\dagger_{\mathbf{q}}|N_{\mathbf{q}}\rangle = \sqrt{(N_{\mathbf{q}}+1)\left(1 - \frac{N_{\mathbf{q}}}{M}\right)}\,|N_{\mathbf{q}}+1\rangle \tag{69}$$

$$A_{\mathbf{q}}|N_{\mathbf{q}}\rangle = \sqrt{N_{\mathbf{q}}\left(1 - \frac{N_{\mathbf{q}}-1}{M}\right)}\,|N_{\mathbf{q}}-1\rangle. \tag{70}$$

As $M \longrightarrow \infty$, relations (69) and (70) turn into the well known relations for bosons. From Eqs. (69) and (70) it follows that

$$A^\dagger_{\mathbf{q}}A_{\mathbf{q}}|N_{\mathbf{q}}\rangle = N_{\mathbf{q}}\left(1 - \frac{N_{\mathbf{q}}-1}{M}\right)|N_{\mathbf{q}}\rangle. \tag{71}$$

Thus, the operator $A_{\mathbf{q}}^{\dagger}A_{\mathbf{q}}$ is not a particle number operator in a state $\mathbf{q}$, as in the case of boson, fermion and paulion operators. It can be proved that for the modified parafermi statistics, the operator of particle number in a state $\mathbf{q}$ does not exist, see the Refs. [7, 12]. This is the consequence of the absence of the Kronecker symbols in the commutation relations (62) and (63). What can be defined is the operator of the total number of hole pairs, $\hat{N}$. For the commutator, the following relation is valid [7, 12]:

$$[A_{\mathbf{q}}, A_{\mathbf{q}}^{\dagger}]_{-} = 1 - \frac{2\hat{N}}{M}. \tag{72}$$

Only for small concentrations, $\langle\hat{N}\rangle/M \ll 1$, the hole pairs satisfy the Bose statistics.

6. Some Properties of the Hole-Pair System

As we showed above, the operator $A_{\mathbf{q}}^{\dagger}A_{\mathbf{q}}$ is not the hole-pair number operator; so, the diagonalized Hamiltonian (60) does not describe the ideal gas of the hole pairs. The latter does not exist in principle. Even in the absence of dynamical interactions, some immanent interaction in the hole-pair system is always present. The origin of this interaction, which called after Dyson [26] the kinematic interaction, is in the deviation of the hole-pair statistics from the Bose (Fermi) statistics.

Let us estimate the magnitude of the kinematics interaction in the state (66) with N noninteracting hole pairs, each pair with energy $\epsilon_{\mathbf{q}}$ (61). Using the equation (67) for shifting the operator $A_{\mathbf{q}}$ to the right, after straightforward although cumbersome calculations, we obtain [7, 12]

$$\begin{aligned} E(N_{\mathbf{q}}) &= \left\langle N_{\mathbf{q}} \left| \sum_{\mathbf{q}'} \epsilon_{\mathbf{q}'} A_{\mathbf{q}'}^{\dagger} A_{\mathbf{q}'} \right| N_{\mathbf{q}} \right\rangle \\ &= N\epsilon_{\mathbf{q}} \left(1 - \frac{N-1}{M}\right) + \frac{N}{M-1}\frac{N-1}{M} \sum_{\mathbf{q}'(\neq \mathbf{q})} \epsilon_{\mathbf{q}'} \\ &= N\left[\epsilon_{\mathbf{q}} + \frac{N-1}{M}(\bar{\epsilon} - \epsilon_{\mathbf{q}})\right], \end{aligned} \tag{73}$$

where

$$\bar{\epsilon} = \frac{1}{M-1} \sum_{\mathbf{q}'(\neq \mathbf{q})} \epsilon_{\mathbf{q}'} \tag{74}$$

is the mean energy of the hole-pair band. The second term in the last line of Eq. (73) is the kinematic interaction. It is proportional to the concentration of hole pairs and its magnitude is larger the larger is the difference between the $\epsilon_{\mathbf{q}}$ and the mean energy of the hole pair band. According to Eq. (73), there is an immanent coupling among all states of the hole pair band. Therefore, we cannot define the independent quasi-particles in some particular state. As we mentioned above, the ideal gas of the hole pairs does not exist fundamentally. It can exist only in the low concentration limit in which the kinematic energy becomes small and we get the case of the Bose statistics, cf. Eq. (72).

In high T_c superconducting ceramics, the maximum T_c is achieved for a hole concentration in CuO_2 planes equal to $0.2 - 0.25$ per CuO_2 unit [27, 28]. The same order of magnitude has to be for the hole-pair concentration because the latter is counted not per CuO_2 units, but per the number of sites M in the superlattice. Thus, the deviations from the Bose statistics for the hole-pair system are not negligible and have to be taken into account.

As we showed above, the hole pairs obey the modified parafermi statistics of rank M, so, one state can be occupied by up to M hole pairs. The number of hole pairs cannot exceed the number of sites M in the superlattice. This means that, in spite of the non-boson behavior of the hole-pair system, there is no statistical prohibition for the Bose-Einstein condensation. On the other hand, the hole-pair system is always non-ideal (because of the kinematic interaction). For a rigorous study of the Bose-Einstein condensation phenomenon, one has to include also a dynamic interaction and consider an interplay between kinematic and dynamic interactions to study the stability of the Bose condensate, as was done for the molecular exciton system in Ref. [29].

Acknowledgement

This work was partially supported by grants from CONACYT (México) 41226-F, 46770-F and from UNAM by IN102203, IN107305.

References

[1] J. Bardeen, L.N. Cooper, and J.R. Schriefer, *Phys. Rev.* **106**, 162 (1957); Ibid. 108,1175 (1957).

[2] L.N. Cooper, *Phys. Rev.* **104**, 1189 (1956).

[3] J.R. Schrieffer, *Theory of Superconductivity*, Addison-Wesley, Redwood City, California, 1988.

[4] H.S. Green, *Phys. Rev.* **90**, 270 (1953).

[5] D.V. Volkov, *Sov. Phys. JETP* **9** 1107 (1959); Ibid. 11 375 (1960).

[6] I.G. Kaplan, O. Navarro and J.A. Sanchez, *Physica C* **419**, 13 (2005).

[7] I.G. Kaplan, *Theor. Math. Phys.* **27** , 466 (1976).

[8] A.N. Avdyugin, Yu. D. Zavorotnev and L.N. Ovander, *Sov. Phys. Solid State* **25** , 1437 (1983).

[9] B.A. Nguen, *J. Phys.: Condensed Matter* **1**, 9843 (1989).

[10] D.I. Pushkarov, *Phys. Status Solidi* **(b) 133** , 525 (1986).

[11] B.A. Nguen and N.C. Hoang, *J. Phys.: Condens. Matter* **2**, 4127 (1990).

[12] I.G. Kaplan and O. Navarro, *J. Phys.: Condens. Matter* **11**, 6187 (1999).

[13] I.G. Kaplan and O. Navarro, *Physica C* **341-348**, 217 (2000).

[14] A.S. Davydov, *Theory of Molecular Excitons* McGraw-Hill, New York, 1962.

[15] S. Kamefuchi and Y. Takahachi, *Nucl. Phys.* **36**, 177 (1962).

[16] C. Rayan and E.C.G. Sudarshan, *Nucl. Phys.* **47**, 207 (1963).

[17] Y. Ohnuki and S. Kamefuchi, *Quantum Field Theory and Parastatistics* Springer-Verlag, Berlin, 1982.

[18] O.W. Greenberg and A.M. Messiah, *Phys. Rev.* **138**, B1155 (1965).

[19] R.C. Hilborn and G.M. Tino, Eds., *Spin-Statistics Connection and Commutation Relations*, *AIP Conf. Proc.* No. **545**, AIP, Melville, New York, 2000.

[20] N.A. Chernikov, *Acta Physica Polonica* **21**, 52 (1962).

[21] A.B. Govorkov, *Sov. J. Part. Nucl.* **14**, 520 (1983).

[22] I.G. Kaplan, in *Fundamental World of Quantum Chemistry. A Tribute Volume to the Memory of Per-Olov Lowdin*, Eds. E.J. Brandas and E.S. Kryachko, Kluwer Academic Publ., Dordrecht, 2003, pp.183-220.

[23] S.S. Schweber, *An Introduction to Relativistic Quantum Field Theory*, Row Petersen, New York, 1961.

[24] F.S. Zhang and T.M. Rice, *Phys. Rev. B***37** , 3759 (1988).

[25] I.G. Kaplan, J. Soullard, J. Hernández-Cobos, and R. Pandey, *J. Phys.: Condens. Matter* **11** , 1049 (1999).

[26] F. Dyson, *Phys. Rev.* **102** , 1217 (1956).

[27] M.W. Shafer et al., *Phys. Rev. B***39** , 2914 (1989).

[28] H. Zhang and H. Sato, *Phys. Rev. Lett.* **70** , 1697 (1993).

[29] I.G. Kaplan and M.A. Ruvinskii, *Sov. Phys. JETP* **44** , 1127 (1976).

In: Research Progress in Materials Science
Editors: W. Olsson and F. Lindberg, pp. 165-171

ISBN: 978-1-60456-629-1

Chapter 7

APPLICATION OF RARE EARTH TRIFLATES AS EFFICIENT CATALYSTS IN FINE CHEMICALS INDUSTRY: THE EXAMPLE VITAMIN E

Werner Bonrath, Thomas Netscher and Jan Schütz
Research and Development, DSM Nutritional Products, P.O. Box 3255,
CH-4002 Basel, Switzerland

Abstract

The development and application of catalytic processes is one of the fundamental issues for fine chemicals industry where many transformations are still based on stoichiometric reactions from the area of organic chemistry, and often large amounts of waste are formed. Therefore there is a need for more selective, especially catalytic processes. Major activities in industrial process research and development are directed to environmental aspects, raw material costs, technology, and selectivity of a reaction. In this article, new industrially important catalytic processes for the application to the synthesis of the vitamin α-tocopherol and its acetate derivative are presented. Protocols with rare earth catalysts provide the advantages of high selectivity and low catalyst loading for a variety of reactions. Here we report on state of the art in this area, and on recent results obtained in our laboratories. For example, the known processes for the synthesis of α-tocopherol suffer from disadvantages like formation of by-products and waste-water problems. To overcome these drawbacks new types of catalysts have been developed for the synthesis of α-tocopherol and its acetate derivative, starting from trimethylhydroquinone or its 1-acetate, and phytol, isophytol or derivatives, e. g. halides or esters like alkanoates. Catalysts used are various rare earth metal triflates or triflimides. This new approach to vitamin E or derivatives thereof has several advantages, based on high yields and selectivities. Furthermore the reactions can be carried out in continuous or batch-wise modes. When using catalyst-to-substrate ratios from 1:100 to 1:10'000, yields of above 94% can be achieved.

Introduction

Lewis acids are important reagents for organic synthesis as strongly activating mediators for a variety of transformations. In general, typical examples like BF_3, $SnCl_4$, or $AlCl_3$ must be handled under strictly anhydrous conditions. Small amounts of water may interfere with the reactions because conventional Lewis acids react generally very quickly with water and are immediately deactivated leaving the substrates inactivated [1]. Although many of such reagents are utilized in industry most transformations need nearly stoichiometric or even higher amounts of the Lewis acid, causing often serious problems in production due to waste formation and corrosion of equipment. In sharp contrast, lanthanide(III) salts [Ln(III)X, Ln = Sc, Y, and the elements La to Lu, X = anion] have found wide use as versatile water tolerant Lewis acid catalysts in organic synthesis [2], applied in really catalytic amounts.

Lanthanide compounds act as strong Lewis acids because of their hard character [3]. The acidity of lanthanide(III) triflates is higher compared to salts containing other anions such as halides, due to the strong electron withdrawing effect of the triflyl group [4]. Baes and Mesmer predicted in 1976 the hydrolysis of lanthanide(III) triflates [$Ln(OTf)_3$, Tf = trifluoromethanesulfonyl] to be slow based on their physical constants [5]. Triflates show not only a high stability in aqueous media but in some cases the catalyses are even faster in the presence of water. In 1987 $Ln(OTf)_3$ was used for the first time in organic synthesis in the addition of amines to nitriles under anhydrous conditions [6].

Ln(III) salts, and in particular $Ln(OTf)_3$, are finding increasing use as catalysts in organic synthesis, for example in Aldol reactions, Diels-Alder and aza Diels-Alder reactions, 1,3-dipolar cycloadditions, Mannich-type reactions, allylation reactions, Strecker-type reactions [4], and Friedel-Crafts reactions [7-9]. Kobayashi et al. showed that one could substitute stoichiometric amounts of e.g. $AlCl_3$ with the catalytic system $Ln(OTf)_3$-$LiClO_4$ (down to 0.1 mol% $La(OTf)_3$; 4 equiv. $LiClO_4$) to give high yields in Friedel-Crafts acylation reactions. $Sc(OTf)_3$ proved to be the most active catalyst whereas $LiClO_4$ is necessary to enhance the catalytic activity [7].

The recovery of $Ln(OTf)_3$ is easily and fully achieved because the solubility of these salts in water is considerably higher than in organic solvents. Reactions are typically quenched with water and the organic compounds are extracted with an organic solvent.

Recent developments include solid phase synthesis applying Ln(III) salts [10]. Substrates are immobilized for instance in Eu(III)-complex catalyzed heterocycloadditions [11], in $La(OTf)_3$ catalyzed Baylis-Hillman reactions [12], in $Yb(OTf)_3$ catalyzed Mannich-Michael reactions [13] and aza Diels-Alder reactions [14], or in $Yb(OTf)_3$ catalyzed aldol condensations in peptide synthesis [15]. There are also examples for the immobilization of the Ln-catalysts. $Yb(OTf)_3$ was immobilized through an ionic polymer [16] and La(III)- and Yb(III)-BINOL complexes were immobilized on polystyrene [17].

Another recent trend is the application of so-called green media in $La(OTf)_3$ catalyzed reactions [18]. Since the catalysts are air and water stable, they can be easily recycled and reused. Various green media such as supercritical CO_2, ionic liquids, and water, as well as solvent-free conditions have been applied.

In this article, the application of rare earth compounds, in particular triflates, to some industrially relevant transformations from the field of vitamins and fine chemicals is described. The examples chosen concern the synthesis of α-tocopherol and α-tocopheryl acetate, representing the economically important group of fat-soluble antioxidants.

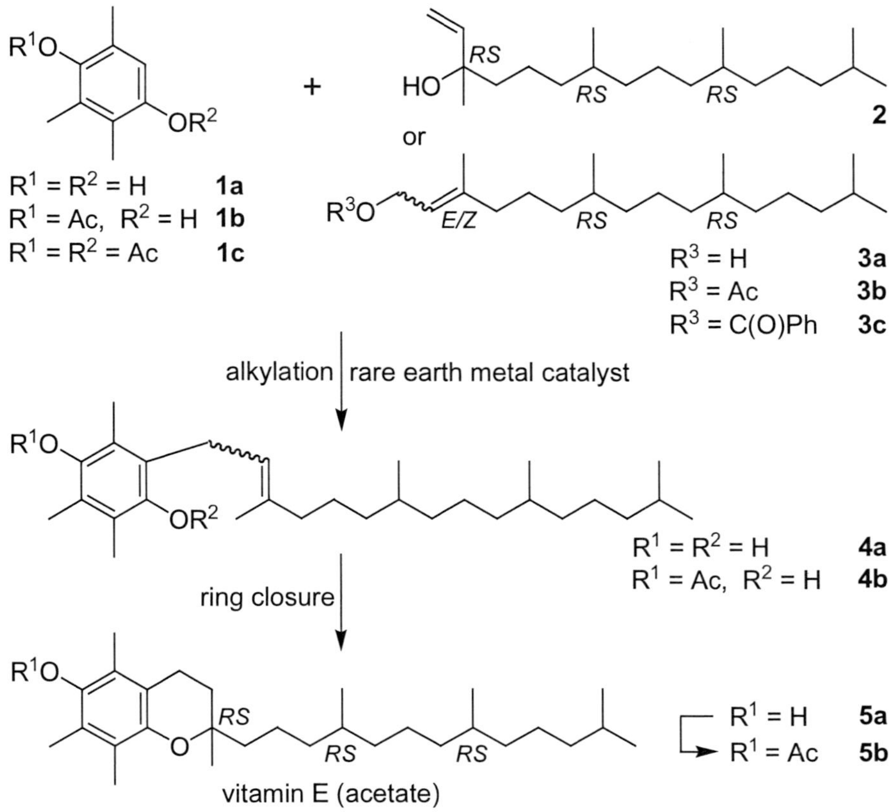

Scheme 1. General reaction sequence for the synthesis of tocopheryl acetate.

Applications to Vitamin E Chemistry

(all-*rac*)- α-Tocopheryl acetate (*5b*) represents the most important commercial product from the group of vitamin E compounds, which play a crucial role as fat-soluble antioxidants. Ester *5b* is predominantly applied in animal nutrition, and compounds *5a*/*5b* are produced on an industrial scale of more than 35'000 t/a [9]. The industrial processes for the preparation of this synthetic vitamin E include the Friedel-Crafts alkylation of trimethylhydroquinone (TMHQ, *1a*) with isophytol (IP, *2*), phytol (*3a*), or phytyl derivatives (e.g. *3b*, *3c*) followed by an acylation reaction (Scheme 1) [9].

Conventionally the alkylation reaction is catalyzed by a combination of a Lewis acid, e. g. zinc chloride, and a Brønsted acid, e. g. hydrochloric acid. Main disadvantages are the formation of by-products, waste, and corrosion problems [19,20]. Several approaches to by-pass these problems are based on new catalyst types and new procedures. A summary of recent accomplishments is given in [19]. A broad variety of lanthanide triflates were tested in the condensation of TMHQ (*1a*) and isophytol (*2*) by the group of Yamamoto, and $Sc(OTf)_3$ was found to be the most suitable catalyst [21,22]. Also the imide $Sc(NTf_2)_3$ and other rare earth salts [23,24], as well as microencapsulated scandium triflate [25] were used successfully and pointed to the potential of rare earth catalysts for this type of reaction.

Results from our laboratories showed that TMHQ and isophytol or suitable derivatives can be transferred into tocopherol with high selectivity and conversion by using gadolinium triflate as a catalyst (Table 1). Yields of 88-95% were obtained when performing the condensation reaction in a two-phase solvent system (e.g. ethylene carbonate/ heptane). In one-phase systems, good results could also be obtained by using polar aprotic solvents (γ-butyrolactone; $Sc(OTf)_3$, 76-79% yield). As side chain components, isophytol (*2*) and also esters like acetate *3b* and benzoate *3c* delivered good results (yields up to 95%) with up to 1 mol% of catalyst [26].

For comparison, it should be noted that metal triflimides of group 3-12 catalyze the reaction of *1a* and *2* in up to 93% yield [24]. Catalysts based on Hf, Bi or In salts, e.g. the triflates, are also efficient in this reaction. The application of Hf, Bi or In triflates or the corresponding tris[(perfluoroalkyl/aryl)sulfonyl]methides results in 92-98% yield of tocopherol under similar conditions [24,29,31]. Imide salts containing CF_3, C_2F_5, C_4F_9, C_8F_{17} or C_6F_5 groups, or mixtures thereof, are also useful catalysts for the condensation of *1a* and *2*. Yields of up to 94% could be observed. Especially catalysts based on Ag (93%), Ni (91%) and Tm (90%) are very selective [30].

Another group of catalysts based on methide salts of type $Met[C(SO_2C_nF_{2n+1})_3]_m$ (Met = metal, n = 1-8, m = valence of metal, e.g. Ag, Cu, Zr) is also a highly active and selective one for the synthesis of *5a*. Tocopherol *5a* could be synthesized in up to 94% yield [32].

Table 1. Synthesis of tocopherol 5a from TMHQ (1a) and isophytol (2) or a phytyl ester (3b,c)

Catalyst	s/c[a)]	Side chain component	Solvent	Yield 5a (%)[b)]
$Gd(OTf)_3$	100	**2**[c)]	EC/PC/hep	91.4
$Gd(OTf)_3$	100	**3b**[c)]	EC/hep	93.6
$Gd(OTf)_3$	100	**3c**[c)]	EC/hep	95.0
$Sc(OTf)_3$	2'000	**2**[d)]	γ-but	78.3
$Sc(OTf)_3$	1'000	**2**[d)]	γ-but	78.8
$Sc(OTf)_3$	200	**2**[d)]	γ-but	76.5
$Sc(OTf)_3$	100	**2**[d)]	γ-but	75.5
$Gd(OTf)_3$	500	**2**[c)]	EC/PC/hep	87.6
$Gd(OTf)_3$	200	**2**[c)]	EC/PC/hep	89.8
$Ce(NTf_2)_4$	1'000	**2**[e)]	tol	90.2
$Nd(NTf_2)_3$	750	**2**[e)]	tol	89.5
$Pr(NTf_2)_3$	750	**2**[e)]	tol	92.8
$Eu(NTf_2)_3$	750	**2**[e)]	tol	82.7
$Dy(NTf_2)_3$	750	**2**[e)]	tol	87.5
$Yb(NTf_2)_3$	750	**2**[e)]	tol	88.6
$Sc(NTf_2)_3$	5'000	**2**[e)]	tol	86.6
$La(NTf_2)_3$	750	**2**[e)]	tol	91.1

[a)] Molar substrate-to-catalyst ratio, based on IP or derivative (*2,3b,3c*), [b)] isolated, [c)] ratio *1a:2/3b/3c* = 1.5 [26], [d)] performed according to example F in ref. [26], 1 h, 110°C, ratio *1a:2* = 1.5, [e)] ratio *1a:2* = 1.0 [24]; EC = ethylene carbonate, PC = propylene carbonate, hep = heptane, γ-but = γ-butyrolactone, tol = toluene; reaction conditions: reflux, reaction time and specific conditions see data in references [24,26] given; yields are not optimized.

Next we pointed our attention to a modified process for the direct preparation of α-tocopheryl acetate (*5b*) from trimethylhydroquinone-1-acetate (TMHQ-1-MA, *1b*), easily synthesized from the corresponding diacetate *1c* by a lipase catalyzed saponification [27], with isophytol (IP, *2*). The advantage of this approach is that the final acetylation reaction (*5a*→*5b*, cf. Scheme 1) can be avoided. The Friedel-Crafts reaction (alkylation and ring-closure through-process to *5b*) can be carried out with excellent results (up to 94% of *5b*, see Table 2) in a polar, especially bi-phasic solvent systems, e.g. ethylene carbonate and a hydrocarbon like heptane, in presence of a rare earth catalyst [26,28]. In less polar solvent systems the reaction proceeds less efficiently.

Substrate-to-catalyst ratios of up to 8'000 are suitable for this reaction. The final product *5b* can be obtained in high yields (up to 94%). Here it must be pointed out that the selectivity of the reaction is remarkable high. Under standard reaction conditions, the saponification of tocopheryl acetate *5b* to tocopherol *5a* takes place only to a very low extent (<3%).

Depending on the conditions of the reaction, it is even possible to prepare the open-chain intermediate *4b* selectively (Table 2). The most active catalyst $Sc(OTf)_3$ (0.01 mol% in EC/heptane) delivered alkylation product *4b* in 86% yield. When using Ho, La, Tm, or Eu triflate, yields of 88-92% *4b* were obtained with 0.1 mol% of catalyst under prolonged heating.

Additionally the ring closure reaction of phytyltrimethylhydroquinone-1-acetate was studied in detail. It was observed that the cyclization reaction can also be carried out with high selectivity (Table 3). Reaction temperatures of 120 to 140°C and low catalyst loadings result in >95% yield.

Table 2. Alkylation reaction of TMHQ-1-MA (1b) and IP (2) to 4b and 5b

Catalyst	s/c[a)]	Solvent	Reaction time (h)	Yield 4b (%)	Yield 5a (%)	Yield 5b (%)
$Sc(OTf)_3$	8'000	EC/hep[b)]	1.25	0.6	0.6	92.1
$Sc(OTf)_3$	2'000	EC/hep[b)]	0.5	0.0	2.5	92.3
$Gd(OTf)_3$	1'000	EC/hep[c)]	1			94.0
$Sc(OTf)_3$	1'000	EC/hep[c)]	1			92.3
$Gd(OTf)_3$	200	EC/hep[b)]	17	11.1	0.5	64.3
$Gd(OTf)_3$	500	EC/hep[b)]	22	40.2	0.0	35.2
$Sc(OTf)_3$	10'000	EC/hep[b)]	0.17	72.5	0.0	1.9
$Sc(OTf)_3$	10'000	EC/hep[b)]	0.17	85.9	0.0	2.6
$Ho(OTf)_3$	1'000	EC/hep[d)]	3	92.1	0.0	1.1
$La(OTf)_3$	1'000	EC/hep[d)]	19	88.5	0.0	0.3
$Tm(OTf)_3$	1'000	EC/hep[d)]	3	89.5	0.0	1.0
$Eu(OTf)_3$	1'000	EC/hep[d)]	4	88.3	0.0	0.2

All reactions were carried out with a TMHQ-1-MA (*1b*): IP (*2*) ratio of 1.5, [a)] molar substrate-to-catalyst ratio, based on IP (*2*), [b)] for detailed information see ref. [28], [c)] for detailed information see ref. [9], [d)] performed according to example 4 in ref. [28]; EC = ethylene carbonate, hep = heptane, reaction conditions: reflux, reaction time and specific conditions see data in references given; yields are not optimized.

Table 3. Ring closure reaction of phytyltrimethylhydroquinone-1-acetate (4b) to tocopheryl acetate 5b

Catalyst	s/c	Conversion (%)	Solvent	Yield 5a (%)	Yield 5b (%)[a]
$Sc(OTf)_3$	2'000	>99.5	EC	0.1	94.9
$Sc(OTf)_3$	1'000	>99.5	EC	0.0	>99.5
$Sc(OTf)_3$	2'000	98.0	γ-but	1.0	96.0
$Sc(OTf)_3$	2'000	96.5	γ-but		89.0
$Gd(OTf)_3$	100	>99.5	EC/hep		94.4
$Gd(OTf)_3$	100	>99.5	γ-but	0.0	>99.5[b]
$Ho(OTf)_3$	100	>99.5	γ-but	0.0	>99.5
$Ho(OTf)_3$	100	72.0	EC	0.0	68.9
$La(OTf)_3$	100	>99.5	γ-but	0.0	>99.5[b]
$Yb(OTf)_3$	100	>99.5	γ-but	0.0	>99.5
$Yb(OTf)_3$	100	>99.5	EC	6.0	94.0
$Eu(OTf)_3$	100	>99.5	γ-but	0.0	>99.5
$Eu(OTf)_3$	1'000	53.0	EC	0.0	46.6
$Tm(OTf)_3$	100	>99.5	γ-but	0.0	>99.5
$Tm(OTf)_3$	1'000	91.0	EC	0.0	85.5[c]
$Y(OTf)_3$	100	>99.5	γ-but	0.0	>99.5
$Y(OTf)_3$	1'000	>99.5	EC	0.0	>99.5[b]
$Y(OTf)_3$	100	62.5	EC	0.0	62.5

Reaction time 1 h at 120-140°C, [a] isolated, [b] reaction time 5 h, [c] reaction time 8 h; EC = ethylene carbonate, γ-but = γ-butyrolactone, s/c = molar substrate-to-catalyst ratio; reaction conditions are not optimized.

In conclusion, it has been shown that various rare earth metal triflate and triflimide salts are efficient mediators of both Friedel-Crafts alkylation and cyclization reactions for the preparation of the vitamin E compounds (all-*rac*)- α-tocopherol (*5a*) and (all-*rac*)- α-tocopheryl acetate (*5b*), as well as the open-chain intermediate phytyltrimethylhydroquinone-1-acetate (*4a*). High selectivities were found for these reactions. When starting from the mono-ester trimethylhydroquinone-1-acetate (*1b*), both alkylation and ring-closure reaction were catalyzed with remarkable chemoselectivity without considerable competing ester hydrolysis. The catalysts were effective in amounts of 1 mol% and below.

References

[1] S. Kobayashi, K. Manabe, *Acc. Chem. Res.* 2002, 35, 209-217.

[2] S. Kobayashi, *Synlett* 1994, 689-701.

[3] G. A. Molander, *Chem. Rev.* 1992, 92, 29-68.

[4] S. Kobayashi in *Handbook on the Physics and Chemistry of Rare Earths* , K. A. Gscheidner Jr., L. Eyring (eds.), Elesevier Science B. V., 2000, vol. 29, chapter 186, p. 315-374.

[5] C. F. Baes, R. E. Mesmer, *The Hydrolysis of Cations*, Wiley, New York, 1976.

[6] J. H. Forsberg, V. T. Spaziano, T. M. Balasubramanian, G. K. Liu, S. A. Kinsley, C. A. Duckworth, J. J. Poteruca, P. S. Brown, J. L. Miller, *J. Org. Chem.* 1987, 52, 1017-1021.
[7] A. Kawada, S. Mitamura, J.-i. Matsuo, T. Tsuchiya, S. Kobayashi, *Bull. Chem. Soc. Jpn.* 2000, 73, 2325-2333.
[8] K. Rück-Braun, *Nachr. Chem. Tech. Lab.* 1998, 46, 1073-1076.
[9] W. Bonrath, C. Dittel, L. Giraudi, T. Netscher, T. Pabst, *Catal. Today* 2007, 121, 65–70.
[10] L. A. Sloan, D. J. Procter, *Chem. Soc. Rev.* 2006, 35, 1221-1229.
[11] G. Dujardin, S. Leconte, L. Coutable, E. Brown, *Tetrahedron Lett.* 2001, 42, 8849-8852.
[12] B. A. Kulkarni, A. Ganesan, *J. Comb. Chem.* 1999, 1, 373-378.
[13] Y. Wang, S. R. Wilson, *Tetrahedron Lett.* 1997, 38, 4021-4024.
[14] W. Zhang, W. Xie, J. Fang, P. G. Wang, *Tetrahedron Lett.* 1999, 40, 7929-7933.
[15] A. Graven, M. Grøtli, M. Meldal, *J. Chem. Soc., Perkin Trans.* 1 2000, 955-962.
[16] B. Se Lee, S. Mahajan, K. D. Janda, *Tetrahedron Lett.* 2005, 46, 807-810.
[17] D. Jayaprakash, Y. Kobayashi, S. Watanabe, T. Arai, H. Sasai, *Tetrahedron: Asymmetry* 2003, *14*, 1587-1592.
[18] S. Luo, L. Zhu, A. Talukdar, G. Zhang, X. Mi, *Mini-Reviews in Org. Chem.* 2005, 2, 177-202.
[19] W. Bonrath, T. Netscher, *Appl. Catal. A: General* 2005, 280, 55-73.
[20] K.-U. Baldenius, L.v.d. Bussche-Hünnefeld, E. Hilgemann, P. Hoppe, R. Stürmer, in *Ullmann's Encyclopedia of Industrial Chemistry*, vol. A 27, VCH Weinheim, 1996, p. 478-488.
[21] M. Matsui, N. Karibe, K. Hayashi, H. Yamamoto, *Bull. Chem. Soc. Jpn.* 1995, 68, 3569-3571.
[22] M. Matsui, H. Yamamoto, EP 658,552, 1995.
[23] K. Ishihara, M. Kubota, H. Yamamoto, *Synlett* 1996, 1045-1046.
[24] M. Baak, W. Bonrath, H. Pauling, WO 98/21197, 1998.
[25] F. Schager, W. Bonrath, *Appl. Catal. A: General* 2004, 202, 117-120.
[26] L. Giraudi, W. Bonrath, US 2005/0038269, 2005.
[27] W. Bonrath, R. Karge, T. Netscher, *J. Mol. Catal. B: Enzymatic* 2002, 19-20, 67-72; W. Bonrath, D. Eisenkrätzer, V. Enjolras, R. Karge, T. Netscher, M. Schneider, EP 1239045, 2002.
[28] W. Bonrath, C. Dittel, T. Netscher, T. Pabst, L. Giraudi, WO 2004/063182, 2004.
[29] A. Wildermann, Y. Foricher, T. Netscher, W. Bonrath, *Pure Appl. Chem.* 2007, 79, 1839-1846.
[30] W. Bonrath, A. Haas, E. Hoppmann, H. Pauling, EP 1180517, 2002.
[31] W. Bonrath, L. Giraudi, WO 2005/005407, 2005.
[32] W. Bonrath, A. Haas, E. Hoppmann, T. Netscher, H. Pauling, EP 1134218, 2001.

In: Research Progress in Materials Science
Editors: W. Olsson and F. Lindberg, pp. 173-208

ISBN: 978-1-60456-629-1

Chapter 8

PHOTOLUMINESCENCE AND ELECTROLUMINESCENCE STUDIES ON THE MORPHOLOGY OF ORGANIC MOLECULES IN THIN SOLID FILMS

Taiju Tsuboi*
Faculty of Engineering, Kyoto Sangyo University,
Kamigamo, Kita-ku, Kyoto 603-8555, Japan

Abstract

Photoluminescence, excitation and absorption spectra are investigated for thin films lightly and heavily doped with TPD, $Ir(ppy)_3$, Alq_3, PtOEP and mCP and for these neat films at various temperatures between 10 and 290 K. The photoluminescence spectra are compared with the electroluminescence spectra of organic light emitting diodes with emitting layers of $Ir(ppy)_3$ and PtOEP. Aggregates such as dimer and trimer are observed in the neat films of $Ir(ppy)_3$ and Alq_3. On the other hand, dimer and trimer are formed in TPD, PtOEP- and mCP-doped films and these neat films, together with monomer. In the case of PtOEP, large size aggregates which consist of many PtOEP molecules are also formed even in lightly doped films. Such a difference between the group of $Ir(ppy)_3$ and Alq_3 and the group of TPD, PtOEP and mCP is caused by difference of competition and balance among solid state solvation effect, dipole-dipole interactions between dopant molecules, and intermolecular covalent bonding. It is suggested that molecular geometrical structure gives rise to the difference.

1. Introduction

Organic light emitting diode (OLED) is used for creating flexible, full-color and thin flat panel display. Comparing with the currently popular liquid-crystal display, the most peculiar advantage of the OLED display is the possibility of roll-up display on the curved surface. The

* E-mail address: tsuboi@cc.kyoto-su.ac.jp Fax: +81-75-705-1914

OLED display is considered as a near future display with high performance including high brightness, fast response, light weight, and low voltage operation.

The display is an application of electroluminescence (EL) generated from OLEDs. The OLEDs have the multi-thin-layer structure that consists of metal cathode, electron injection layer, transport layer, hole blocking layer, emitting layer, hole transport layer, hole injection layer, and transparent indium-tin-oxide (ITO) anode. The materials of these layers are organic semiconductors with low carrier mobility of 10^{-3}-10^{-7} cm^2/Vs. Applied voltage to OLEDs induces the recombination of the injected electrons and holes in the emitting layer, giving rise to EL from the emitting materials.

Thin layers in OLED devices are fabricated by thermal evaporation in vacuum, spin-coating, or ink-jet printing. The optical properties of OLEDs have been explained using the properties of isolated organic molecules so far although each molecule is expected to have interaction with not only neighboring molecules in the confined layer but also molecules of neighboring layers. It has not been confirmed whether each molecule is present independently from neighboring molecules without formation of aggregate (e.g. dimer grain and island-like structure) in amorphous film and whether the constituent molecules form a lattice that repeats periodically in space. The electrical and optical properties of OLEDs depend on not only the location and distribution of each molecule but also the structure and geometry of aggregated molecules in thin film because they change the structure and alignment of the electronic energy levels of the constituent molecules. Therefore it is necessary to clarify the morphology of molecules in thin amorphous films, i.e. to clarify whether each molecule is distributed randomly without strong interaction between the neighboring molecules or several molecules form aggregates. No detailed spectroscopic investigation, however, has been made for the morphology of organic thin films except film of tris(8-hydroxyquinoline) aluminum (Alq_3) doped in N,N'-bis(3-methylphenyl)-N,N'-bis(phenyl)-benzidine (TPD) and of [2-methyl-6-[2-(2,3,6,7 -tetrahydro-1H, 5H-benzo [ij] quinolizin-9-yl) ethenyl]-4H-pyran-4-ylidene] propane-dinitrile (DCM2) doped in Alq_3 [1] and neat film of Alq_3 [2].

Baldo et al have suggested that the polar Alq_3 and DCM2 molecules form ordered domains by the dipole-dipole interaction among the neighboring molecules in less polar molecular host [1]. Using a model of locally ordered domains, they explained the peak shifts of the Alq_3 and DCM2 dopant emission bands which were observed with increasing the concentration of dopant. On the other hand, Brinkmann et al have suggested that there is a large distribution of molecular packings by π-π overlaps between facing ligands in amorphous Alq_3 films [2].

Thin film of Alq_3 shows large Stokes shift (about 130 nm) and nearly Gaussian-shape emission band without any vibronic structure even at low temperature like 10 K as shown in Fig. 1. Such a spectrum is quite different from those of the other organic molecules that show much smaller Stokes shift and asymmetric line shape with sharp drop at shorter wavelength side and well-resolved vibronic structure at the longer wavelength side. Therefore it is not clear whether the model of domain or packing formation for Alq_3 is applicable to the other organic molecules used for OLED devices.

In this Chapter we study the morphology of various organic thin films from the photoluminescence (PL), EL and absorption spectra at not only room temperature but also low temperature like 10 K.

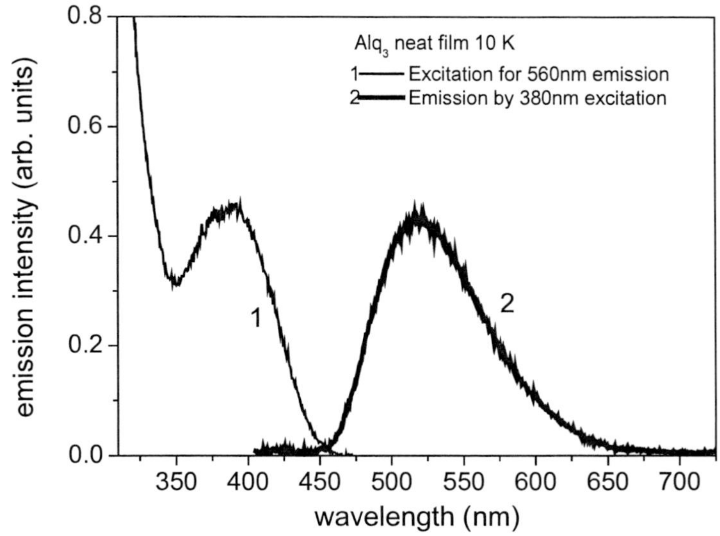

Figure 1. Photoluminescence and excitation spectra of a 10-nm thick Alq_3 neat film at 10 K. Excitation was made at 380 nm and excitation spectrum was obtained for 560 nm emission.

2. General Characteristics of Emission Spectra of Organic Molecules

Most organic molecules in solution show the absorption and PL spectra with vibronic structure and vibrational progressions. The emission and absorption spectra are nearly mirror images of each other. These characteristics are explained using the adiabatic potential-energy parabola curves with vibrational quantum states for the ground and excited states and using the electronic transitions between the two states under the Franck-Condon principle [3, 4].

The vibronic structure of the absorption band is due to the electronic transitions from the vibrational quantum number n=0 in the ground state to the vibrational quantum numbers m =0, 1, 2, 3, , , in the excited state. The intense line is usually observed for the 0-0 transition (i.e. zero-phonon transition). The vibronic structure of the emission band is due to the electronic transitions from the vibrational quantum number m=0 in the excited state to the vibrational quantum numbers n=0, 1, 2, 3, , , , in the ground state. The intense emission line is usually the 0-0 line and the 0-1, 0-2, 0-3 lines locate at the lower energy side of the 0-0 line.

Such an emission line shape is also observed in organic solid films as seen in Fig. 2 where the absorption and fluorescence spectra due to 4,4'-N,N'-dicarbazole-biphenyl (CBP) in film are presented. The vibronic structure of 0-0, 0-1 and 0-2 lines is clearly observed in the two spectra together with usual structure of 0-0 line higher than 0-1 line. Mirror image is found between the absorption and emission spectra, which are plotted against the photon energy, except a region of low-energy tail of the absorption spectrum. The long low-energy tail arises from the weak absorption bands due to triplet state. Observation of the mirror image indicates that the absorption bands of CBP observed at 325-370 nm arise from the same S_1 singlet state. The mirror symmetry also indicates that only one center (i.e. CBP monomer) is responsible for the spectra and additional several centers such as dimer and aggregate are not.

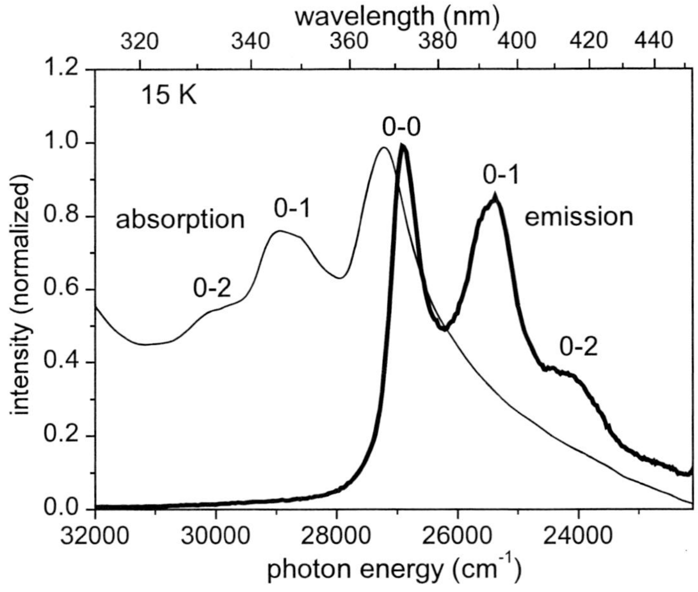

Figure 2. Absorption and photoluminescence (PL) spectra of CBP at 15 K. The PL spectrum was obtained by excitation at 300 nm. The vibrational lines are indicated as 0-0, 0-1 and 0-2 (see text).

When we measure the PL and EL spectra of various organic thin films, however, we realized that such an idealized emission spectrum is not observed for all organic materials. In the following sections we show several examples.

3. Emission Spectra of TPD

Figure 3 shows the PL and excitation spectra of 100 % TPD film (i.e. neat film) of a thickness of 91 nm. The film was spin-coated on the glass plate. The PL spectrum at 10K shows the usual vibronic structure where the peak intensity decreases as going from the 0-0 line to 0-1 line and to 0-2 line. The 0-0 line peak locates at 404 nm and never changes the peak position for variation of temperature, while the 0-1 line shifts to short wavelength with increasing temperature from 10 K to 290 K, i.e. the peak is at 426 and 417 nm at 10 and 290 K, respectively. The peak-height ratio of the 0-1 line to the 0-0 line increases considerably with increasing temperature from 10 K to 290 K. At 290 K, the 0-1 line height is nearly the same as the 0-0 line height. The ratio of the 0-1 line to the 0-2 line, however, changes little. The absorption cross-section spectrum at 290 K is shown in curve 9 [5]. From the absorption spectrum, it is suggested that such a rapid decrease of the 0-0 line is caused by the self-absorption as discussed later.

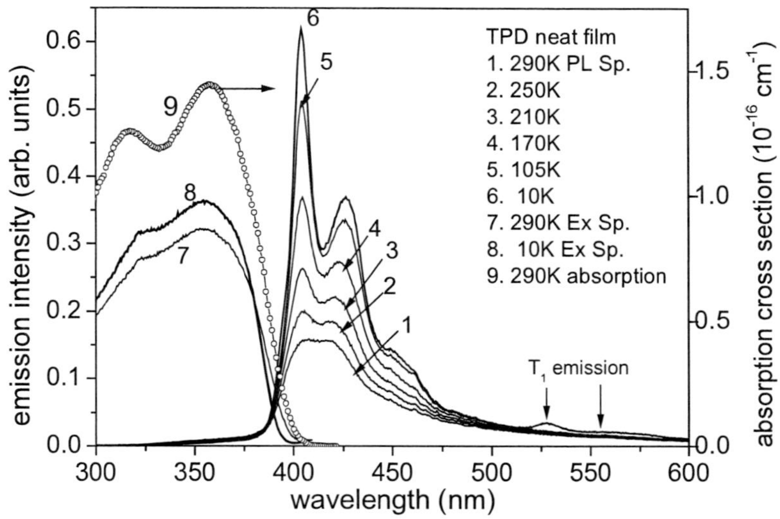

Figure 3. Photoluminescence spectra of a neat TPD film at various temperatures (curves 1-6), which was excited at 230 nm, together with the excitation spectra for 430 nm emission at 290 and 10 K (curves 7 and 8, respectively) and the absorption cross section spectrum (curve 9, right scale). Vertical arrows indicate the emission from the triplet state T_1.

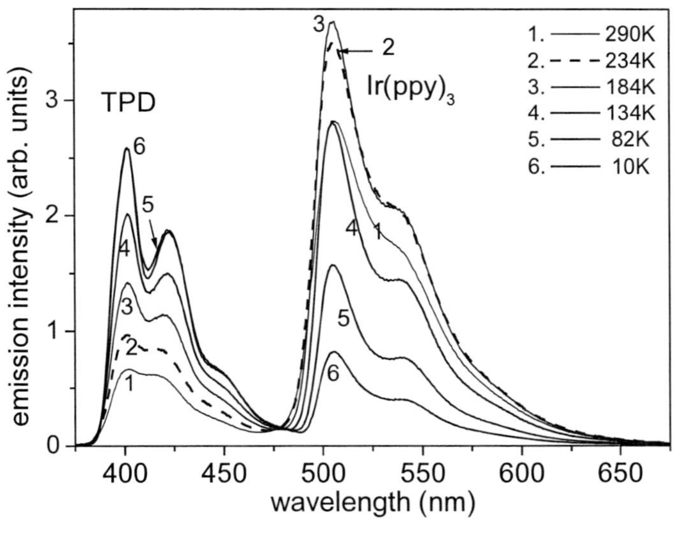

Figure 4. Photoluminescence spectra of a 60-nm thick film of TPD doped with 5wt% $Ir(ppy)_3$ at various temperatures, which was excited at 360 nm.

Emission bands appear at 520-600 nm at low temperatures with decreasing temperature from 150 K. The intensity increases with decreasing temperature as shown in Fig.3. These weak bands are attributed to the phosphorescence due to the transition from the triplet state T_1 to the ground state.

Figure 4 shows the PL spectra of a spin-coated film of 5wt% *fac* tris(2-phenylpyridine) iridium [$Ir(ppy)_3$] doped in TPD at various temperatures between 10 and 290 K. The emission band due to $Ir(ppy)_3$ is observed at 500-650 nm [6] besides the TPD band. Almost the same

TPD emission spectra as Fig.3 are obtained, except (1) the intensity ratio of the 0-0 and 0-1 line at 290 K, i.e. the 0-0 line intensity is higher than the 0-1 line and (2) the peak position of the 0-0 line is at 402 nm that shifts to the shorter wavelength from 404 nm of the neat TPD film. Unlike the case of TPD, the intensity ratio between the 0-0 and 0-1 lines does not change in the $Ir(ppy)_3$ emission for the variation of temperature.

In Figs. 3 and 4 are shown the PL spectra of TPD which is doped as host. We compare them with the spectra of TPD doped as guest. Figure 5 shows the PL spectra of a spin-coated film of 5wt% TPD doped in polystyrene (PS). The emission intensity increases with increasing temperature from 10 K to 290 K. The intensity ratio of 0-1 line to the 0-0 line change with increasing temperature from 10 K, but, unlike the case of 100% TPD film, the 0-0 line is still higher at 290 K than the 0-1 line.

The peak of the 0-0 line in Fig.5 is at 397 nm. It is noted that this peak wavelength was largely blue-shifted from the 404 nm 0-0 line of the 100% TPD film. The same is true for the 0-1 line, i.e. the 0-1 line peak is at 417 nm at 10 K for the dopant TPD, while at 426 nm at 10 K for the 100% TPD. There is peak shift but no change in the whole spectral line shape between the two films.

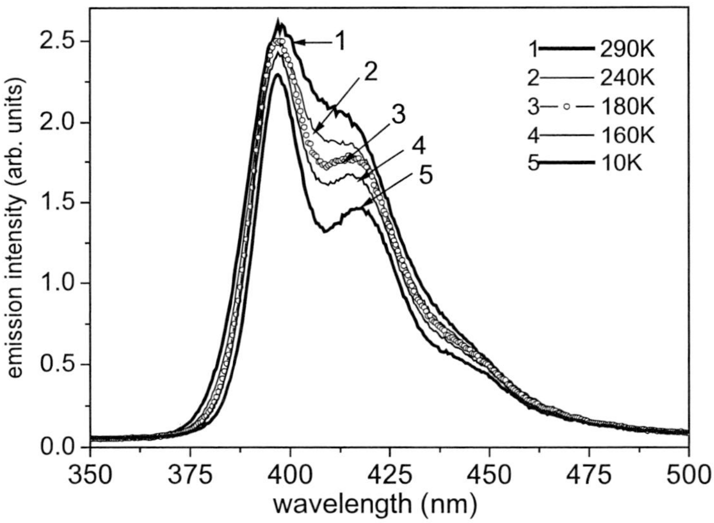

Figure 5. Photoluminescence spectra of TPD in a 90-nm thick film of polystyrene (PS) doped with 5wt% TPD at various temperatures, which was excited at 360 nm.

Figure 6 summarizes the emission spectra of the three films of 5wt% TPD doped in PS (called film TPD-A), TPD doped with 5wt% $Ir(ppy)_3$ (film TPD-B) and 100% TPD at 10 K. It is clearly observed by excitation at 360 nm that the intense peak wavelength is changed among the three films, i.e. 397, 402 and 404 nm for films TPD-A, TPD-B and neat film, respectively, and that the whole spectrum shows a red-shift with increasing the TPD concentration. It is found that both TPD-B and neat films have a shoulder at 397 nm which is the same position as the 0-0 line observed in the film TPD-A. The 397 nm band is attributed to monomer. Discussion is given in Sec.17 on the assignment of the emission bands of TPD-B and neat film.

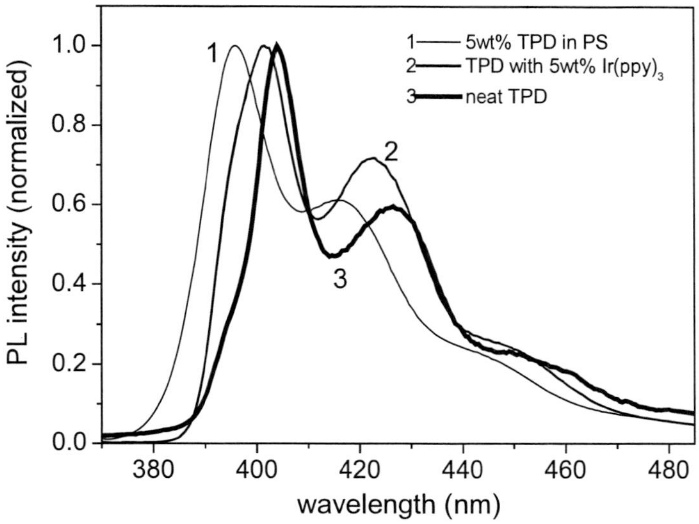

Figure 6. Photoluminescence (PL) spectra of the spin-coated films of 5wt% TPD doped in polystyrene (PS), TPD doped with 5wt% $Ir(ppy)_3$ and 100% TPD at 10 K, which were excited at 360 nm.

4. Self-absorption or Appearance of Dimer ?

As shown in Fig.3, the 0-0 line has the almost same peak height as 0-1 line at 290 K. This is quite different from the cases of low temperatures. The absorption band tail of TPD extends to longer wavelength (up to about 430 nm) with increasing temperature from 10 K to room temperature as seen in curves 8 and 9 of Fig.3. Therefore self-absorption is conceivable as one of the reasons for this unusual line shape. To check this possibility, we calculate the PL line shape corrected with the self-absorption.

The PL intensity $A(\lambda)$ (λ : emission wavelength) emitted from the molecules in film is changed to the PL intensity $B(\lambda)$ after the absorption $k(\lambda)$ in the film of thickness d. The observed spectrum is $B(\lambda)$, which is expressed by

$$B(\lambda) = A(\lambda)[1-\exp(-k(\lambda)d)]/k(\lambda)d.$$

From this equation we can derive the $A(\lambda)$ line shape before self-absorption.

Figure 7 shows the room temperature absorption and PL spectra (curves 1 and 2, respectively) of TPD neat film spin-coated on glass substrate. Curve 3 shows the PL spectrum $A(\lambda)$, i.e. line shape corrected with absorption. The main 0-0 line peak of the derived PL spectrum is at 402 nm, which is shifted from the observed peak wavelength at about 407 nm. This line shape is similar to the PL spectrum observed at 220 K. Thus quite reasonable PL spectrum is derived after the correction.

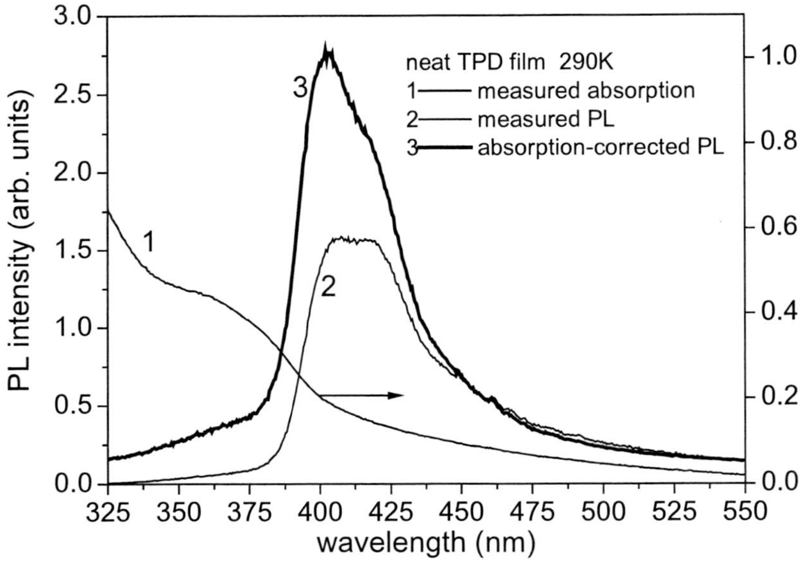

Figure 7. Absorption spectrum of neat TPD film at 290 K (curve 1), and the observed and expected emission spectra of TPD which are obtained from the same film at 290 K (curves 2 and 3, respectively) Curve 3 was derived after correction of self-absorption. The absorption spectrum contains the absorption by glass substrate.

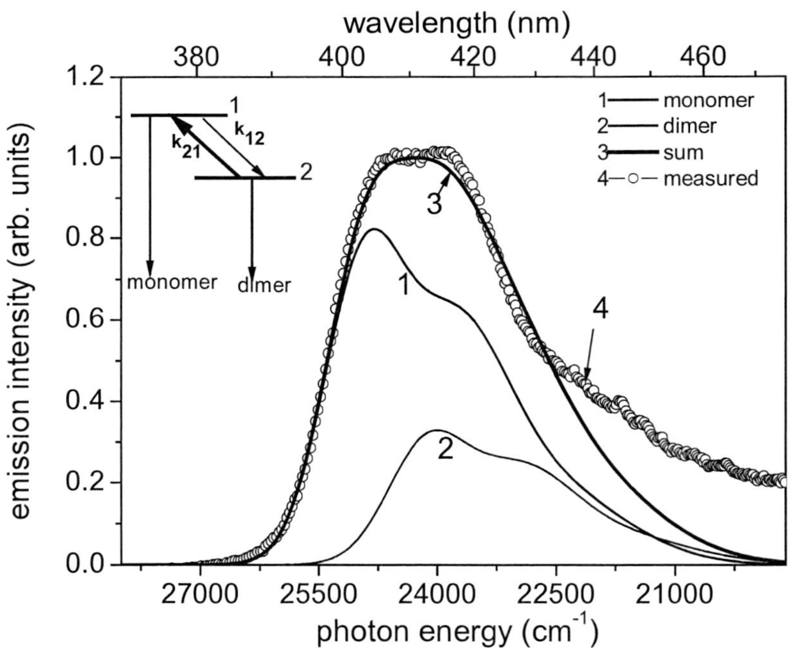

Figure 8. The photoluminescence (PL) spectra of monomer and dimer of TPD (curves 1 and 2, respectively) and the sum of the two spectra (curve3), compared with the PL spectrum of TPD neat film observed at 290 K (curve 4) (see text). Energy transfer between the emitting states of the monomer and dimmer is shown in inset. The parameter k_{12} is the transfer rates from the monomer to dimmer, and k_{21} is the reversed rate.

Another reason might be possible for the unusual line shape of TPD observed at 290 K in neat film. It is formation of dimer and aggregate of TPD molecules. We assumed that monomer and dimer coexist in the neat film, and their PL band line shapes are the same but the band of dimer is red-shifted by 800 cm^{-1} from the monomer band whose 0-0 line is at 24950 cm^{-1} (corresponding to 400.8 nm). The line shape due to the monomer was estimated from the PL spectrum at 290 K of Fig.5. A good fit is obtained between the calculation and observation as shown in Fig.8 where we choose a value of 0.4:1 as the intensity ratio of the dimer band to the monomer band.

This dimer model also gives a good fitting of the PL spectrum at 10 K by changing the intensity ratio to 4:1. Unlike the case of 290 K, it is estimated that the dimer-band intensity is much larger than the monomer-band intensity at 10 K. This is reasonable because the excited molecules are relaxed to the lower emitting state of dimer at low temperature, while the endothermic energy transfer occurs from the dimer level to the upper monomer level with increasing temperature (see inset of Fig.8), resulting in the decrease of dimer emission at 290 K. However, this model gives a deviation from the observed spectrum at the low energy side as seen in Fig.8. Although additional conformation such as trimer would be necessary to obtain better fitting, this model neglects the self-absorption effect. It is needed to calculate the line shape taking into account both self-absorption and aggregates.

5. Photoluminescence Spectra of Ir(ppy)$_3$

The PL spectra of Ir(ppy)$_3$ doped in TPD host are shown in Fig.4. Figure 9 shows the PL and excitation spectra of a spin-coated neat Ir(ppy)$_3$ film. As the case of the neat film of TPD, the

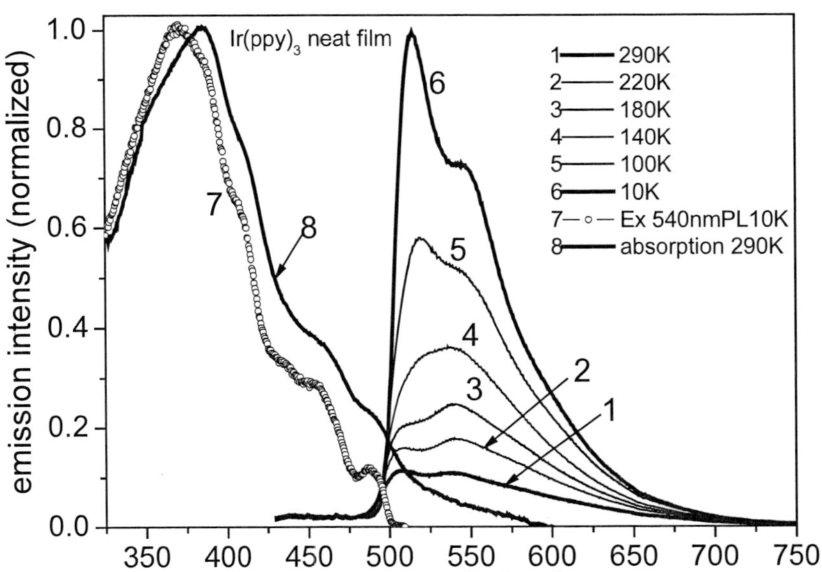

Figure 9. Photoluminescence (PL) spectra of a 60-nm thick film of Ir(ppy)$_3$ neat film at various temperatures (curves 1-6), which was excited at 360 nm, together with excitation (Ex) spectrum for 540 nm emission at 10 K (curve 7) and absorption spectrum of the neat film at 290 K (curve 8).

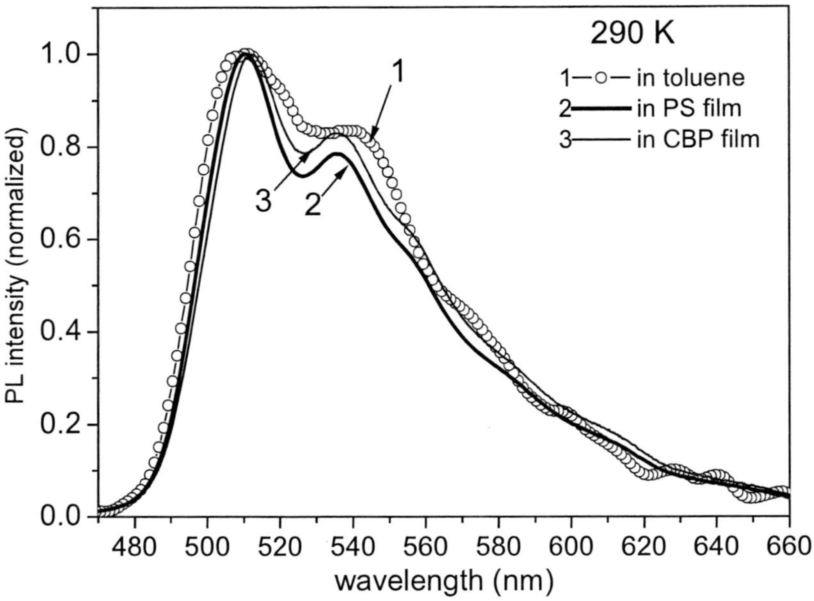

Figure 10. Photoluminescence spectra of Ir(ppy)$_3$ in toluene solvent (curve 1), in spin-coated polystyrene (PS) film doped with 4 wt% Ir(ppy)$_3$ (curves 2) and CBP film doped with 7.9 wt% Ir(ppy)$_3$ (curves 3), which were measured at room temperature.

emission intensity decreases with increasing temperature from 10 K to 290 K. Vibronic structure is also observed, but the intensity ratio of the 0-0 line to the 0-1 line changes drastically for the variation of temperature. The 0-1 line is higher at 140-220 K than the 0-0 line. The peak height of the 0-0 line is a little higher than that of the 0-1 line at 290 K, but almost comparable to each other. This behavior is different from the case of Ir(ppy)$_3$ doped in TPD as seen in Fig.4.

The excitation spectrum for the Ir(ppy)$_3$ emission at 10 K is shown in curve 7 in Fig.9, while the absorption spectrum at 290 K is shown in curve 8. From these spectra, it is suggested that the decrease of the 0-0 line intensity, which is observed at high temperatures, is caused by the self-absorption because of strong overlap of the absorption tail with the emission band at high temperatures. In the case of Ir(ppy)$_3$ doped as guest, the 0-0 line never becomes lower than the 0-1 line by changing temperature as shown in Fig.4. This is understood as follows. The absorption intensity of Ir(ppy)$_3$ is too weak in the lightly doped film to give a strong effect of self-absorption on the PL intensity.

Figure 10 shows the PL spectra of Ir(ppy)$_3$ doped in PS and CBP and in toluene solution at 290 K. Unlike the case of the neat film at 290 K, these doped Ir(ppy)$_3$ molecules show the usual vibronic structure, i.e. the 0-1 line intensity is smaller than the 0-0 line one, which is consistent with the case of Ir(ppy)$_3$ doped in TPD (Fig.4). These vibronic structures are similar to each other in the three samples although the intensity ratio of the 0-0 line to the 0-1 line is a little different among them. Only the 0-0 and 0-1 emission lines are observed for Ir(ppy)$_3$ doped in TPD (spin-coated film, at 290 K, see Fig.4), while not only the 0-0 and 0-1 lines but also 0-2, 0-3 and 0-4 lines are observed in Fig.10. This suggests that the resolution in vibronic structure of Ir(ppy)$_3$ depends on the host molecules.

6. Electroluminescence Spectra of IR(PPY)$_3$

The EL spectra of Ir(ppy)$_3$ were measured using two kinds of OLEDs #Ir1 and #Ir4 with emitting layer of 2.9 and 5.8 wt% Ir(ppy)$_3$-doped CBP, respectively. These devices have the following multi-layer structures.

#Ir1: ITO/α-NPD/2.9wt%Ir(ppy)$_3$:CBP (35nm)/BCP/Alq$_3$ /Al
#Ir4: ITO/α-NPD/ 5.8wt%Ir(ppy)$_3$:CBP(35nm)/BCP/Alq$_3$ /Al.

The layer thickness of the emitting layer is indicated in the parenthesis. Layers of 2,9-dimethyl-4,7-diphenyl -1,10-phenanthroline (BCP) and Alq$_3$ are used as hole-blocking and electron-transport layers, respectively. The devices are sealed by the glass plate to avoid water vapor and oxgen. In Fig.11 the EL and PL spectra of #Ir4 are shown. Similar spectra are also obtained for #Ir1, but the 0-1 intensity relative to the 0-0 line is higher in #Ir4 than in #Ir1. This is understood as follows. The more strong self-absorption effect occurs in #Ir4 than in #Ir1 because the concentration of #Ir4 is higher by two times than that of #Ir1.

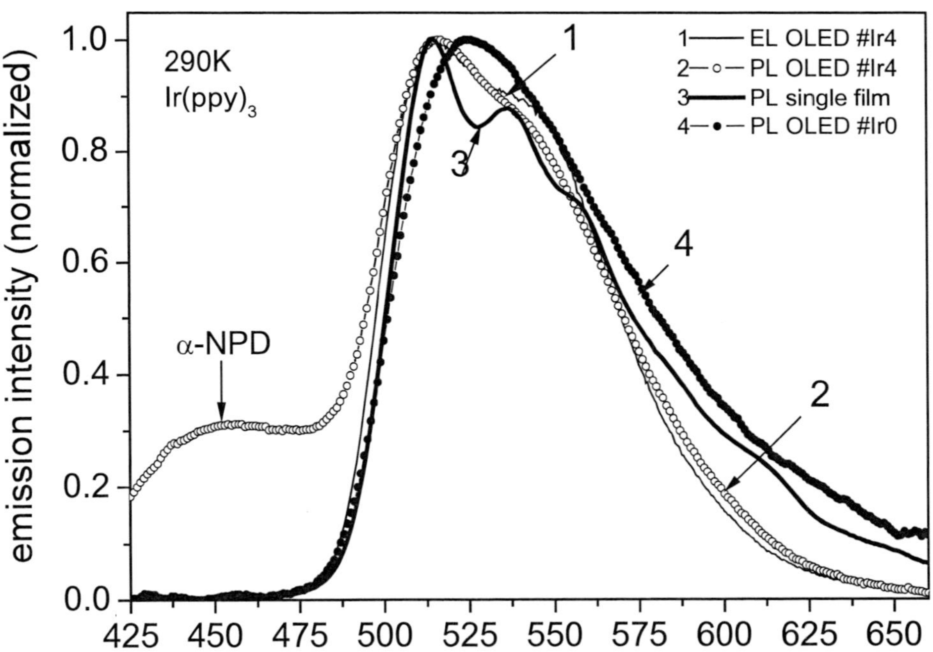

Figure 11. Electroluminescence (EL, curve 1) and photoluminescence (PL, curve 2) spectra of the OLED #Ir4 with emitting layer of CBP doped with 5.8wt% Ir(ppy)$_3$ at 290 K, compared with PL (curve 3) of a single layer film of CBP doped with 7.9wt% Ir(ppy)$_3$, which was spin-coated on galss plate, at 290 K. Curve 4 shows EL of OLED #Ir0 with emitting layer of 100% Ir(ppy)$_3$. The EL was obtained at 9 V. The PL of OLED and single layer was obtained by excitation at 365 nm. The 450 nm emission band in curve 2 is due to the emission from hole transport layer of α-NPD.

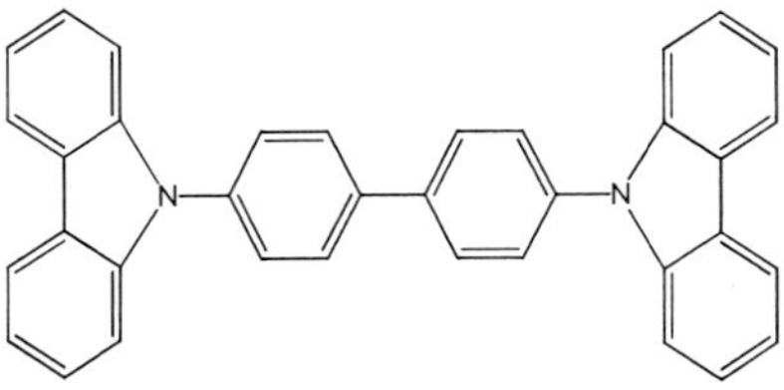

CBP

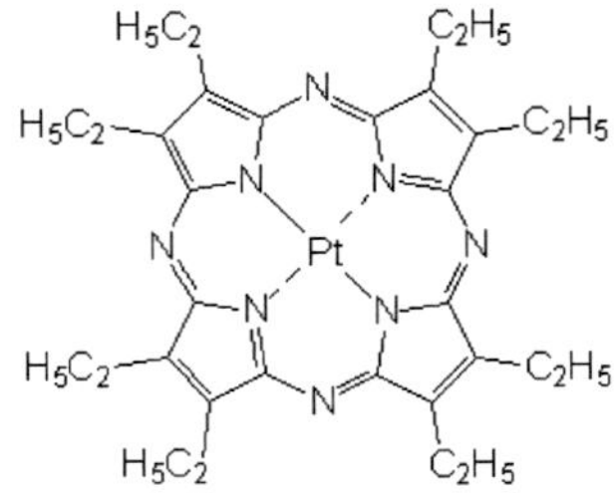

PtOEP

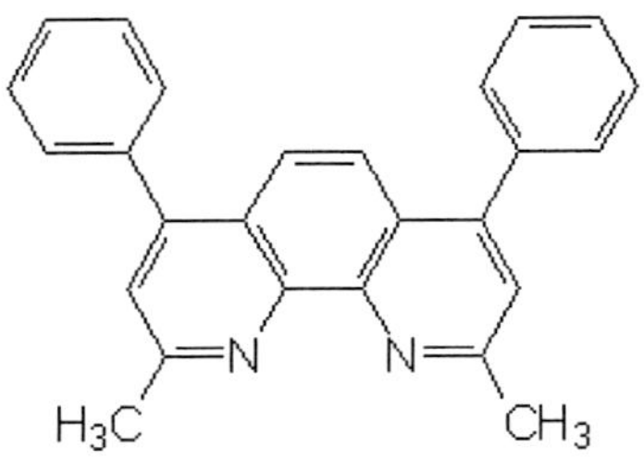

BCP

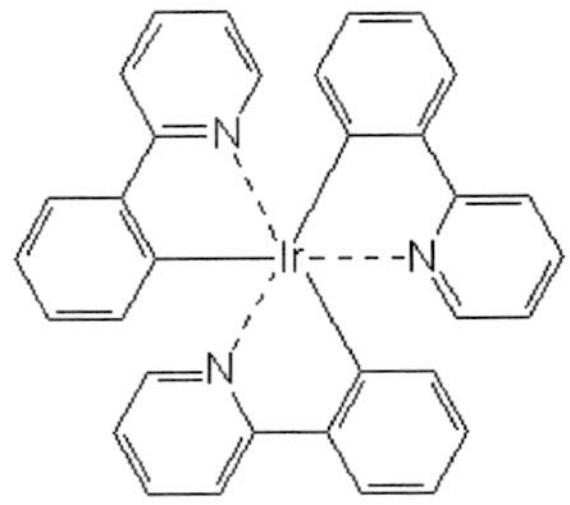

Ir(ppy)3

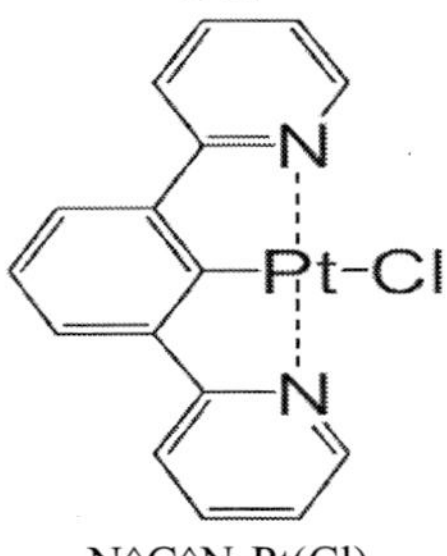

N^C^N-Pt(Cl)

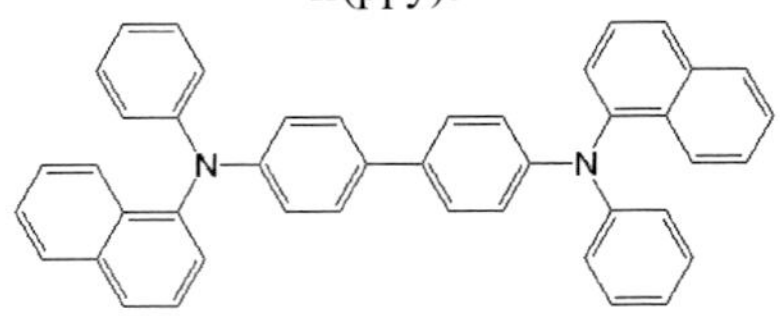

α-NPD

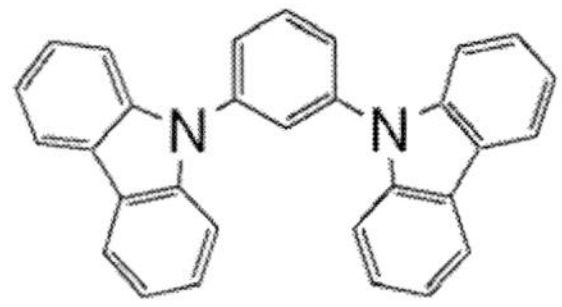

mCP

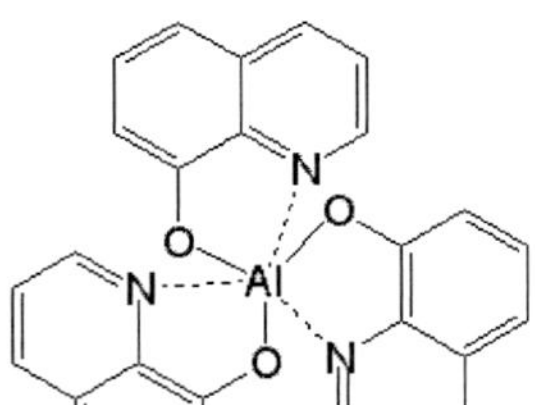

Alq$_3$

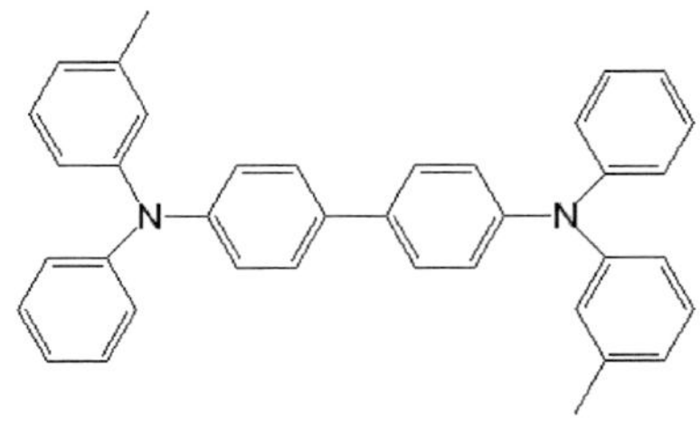

TPD

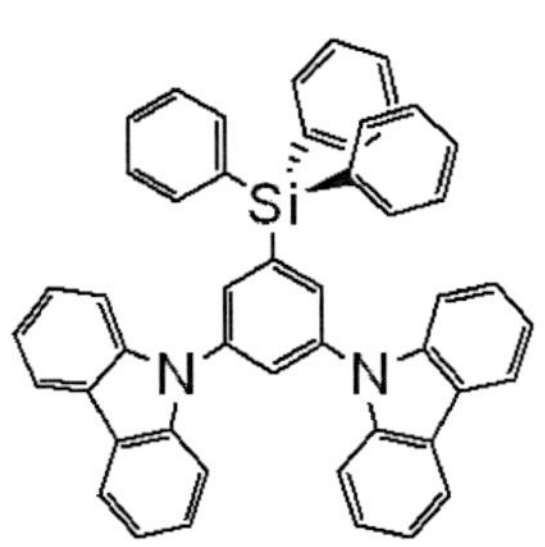

SimCP

Figure 12. Chemical structures of organic molecules.

When these devices are excited with 380 nm light which was introduced through the ITO transparent anode, it is observed that the PL spectra of $Ir(ppy)_3$ are almost the same as EL spectra. Additionally a very weak and broad PL band with at peak about 450 nm is observed. This is due to N, N'-bis(1-naphtyl)-N, N'-diphenyl-1,1'-biphenyl-4,4'-diamine (α-NPD) used as hole transport layer, whose chemical structure is shown in Fig.12, together with the other organic molecules used in this Chapter. In the EL spectrum of #Ir1 and #Ir4 the α-NPD emission band is also observed although its intensity is much weaker than the case of PL.

The room temperature PL spectrum of a single layer film of CBP doped with 7.9 wt% $Ir(ppy)_3$ is also shown in curve 3 of Fig.11 [7]. This film was spin-coated on glass substrate. It is noted that the well-resolved vibronic structure appears in this single layer $Ir(ppy)_3$:CBP film than in the emitting $Ir(ppy)_3$:CBP layer of the multi-layer structured OLEDs (curves 1 and 2 of Fig.11). Regarding this difference between the same $Ir(ppy)_3$-doped CBP films, we discuss in Sec.15.

Curve 4 of Fig.11 shows the EL spectrum of OLED device with emitting layer of 100% $Ir(ppy)_3$ (called #Ir0). The #Ir0 device structure is

#Ir0 : ITO/ $Ir(ppy)_3$(50nm)/LiF/Al.

This device has neither electron transport layer nor hole transport layer. It is noted that EL spectrum of #Ir0 has a single structureless band and its band peak is at 528 nm which is largely shifted from the emission peak of #Ir4. The same 528 nm emission band was observed in #Ir0 by photoexcitation at 380 nm. It is suggested that this red-shifted band is caused by (1) strong self-absorption from neat $Ir(ppy)_3$ film in #Ir0 because the absorption is much intense in 100% $Ir(ppy)_3$ film than in the 5.8wt% $Ir(ppy)_3$ doped film of #Ir4, and (2) solvation effect due to different environments of $Ir(ppy)_3$ molecule between #Ir0 and #Ir4. Regarding the solid state solvation effect, we discuss in Sec.13.

7. Emission Spectra of PtOEP

Figure 13 shows the PL spectra of platinum octaethyl porphyrin (PtOEP) in film of PS doped with 5wt% PtOEP, neat film evaporated on glass plate, and multi-layer structure OLED #Pt3 with emitting layer of CBP doped with 5 wt% PtOEP, together with EL spectra of #Pt0 with a single layer of 100% PtOEP and multi-layer structure OLED #Pt5 with emitting layer of 100% PtOEP. The #Pt0, #Pt3 and #Pt5 OLEDs have the following layer structures.

#Pt0: ITO/PtOEP(50nm)/LiF/Al
#Pt3: ITO/α-NPD/5wt%PtOEP: CBP(33nm)/BAlq/Alq_3 / LiF/Al
#Pt5: ITO/α-NPD/100%PtOEP(45nm)/BAlq/Alq_3 / LiF/Al

In #Pt3 and #Pt5, bis(2-methyl-8-quinolinolato) mono (4-phenylphenolato) aluminum (BAlq) is used as hole blocking layer.

The PL spectrum was obtained by excitation at 360 nm, while the EL spectrum was obtained under the applied voltage of 10 V. The EL spectrum of #Pt3 was the same as the PL spectrum of #Pt3. PtOEP molecules doped in PS and CBP give the same PL band (with a peak at about 645 nm) as the case of film of 5wt% PtOEP doped in TPD. The EL of #Pt3 with CBP doped with 5wt% PtOEP shows the same 645 nm band. Taking into account that the 644.5 nm emission band is observed in PtOEP dissolved in dichloromethane (CH_2Cl_2) solvent [10], the 645 nm emission band is attributable to PtOEP monomer.

The neat film and #Pt0 OLED show additional PL band at 654 nm besides the 645 nm band (curves 3 and 4 of Fig.13). Same emission band is also observed in EL of #Pt0. This indicates that the neat film has an additional emission center (e.g. dimer and trimer) which is different from monomer responsible for the 645 nm emission band. Two EL bands are also obtained from the OLED #Pt5. One band is at 645 nm which is the same as the cases of #Pt0 and doped film. The other is at 651 nm (curve 5) which is different from the cases of #Pt0 and neat film although the EL is generated from layer of 100% PtOEP as #Pt0 and neat film. The 651 nm band peak is shifted from the 654 nm band peak of #Pt0. This peak shift is caused by the different concentration ratio of dimer to monomer between #Pt0 and #Pt5.

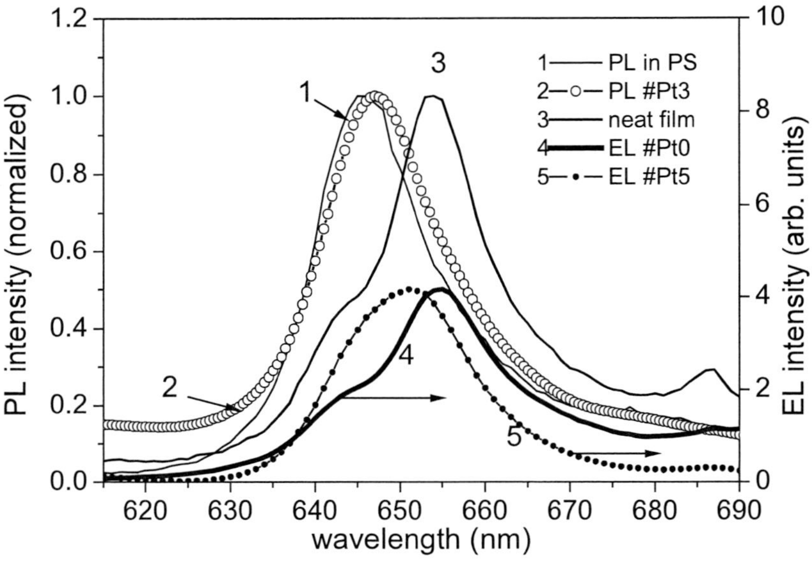

Figure 13. Photoluminescence (PL) spectra of PtOEP in film of polystyrene (PS) doped with 5wt% PtOEP (curve 1), in the emitting CBP layer doped with 5.8wt% PtOEP of OLED #Pt3 (curve 2), and in PtOEP neat film evaporated on quartz plate (curve 3) at 290 K, together with electroluminescence (EL) spectrum from single 100% PtOEP layer structure OLED #Pt0 (curve 4) and multi-layer structure OLED #Pt5 with 100% PtOEP layer (curve 5).

As mentioned in Sec.6, organic molecules of emitting layer in multi-layer structure OLEDs interact with not only the molecules within the same layer but also molecules in the neighboring layers. Therefore it is reasonable to consider that the environment of 100% PtOEP layer is quite different between #Pt0 and #Pt5, therefore different concentration ratio of dimer to monomer is obtained in the two OLEDs. Of the two OLEDs, the ratio is higher in #Pt0 than in #Pt5.

Figure 14 shows the absorption spectra of 60-nm thick neat PtOEP film (curve 1) and diluted PtOEP in toluene solution (curve 2 [8]), together with the excitation spectra for the 645 nm PtOEP emission from film of PS doped with 5wt% PtOEP (curve 3) and from OLED

#Pt3 (curve 4) at 290 K. Kalinowski et al measured the absorption spectra using 70-nm thick PtOEP neat films, film of polycarbonate (PC) doped with 6% PtOEP and PtOEP in toluene solution at room temperature [9]. Their results are consistent with Fig.14, e.g. additional absorption bands were observed at the low energy side of the 645 nm band in the neat films. From Fig.14, it is suggested that PtOEP monomer, which is present in the films doped as guest, has absorption bands at about 535 and 502 nm, while the neat film has additional intense aggregate bands at the longer wavelength side (i.e. at 548 and 510 nm).

Figure 15 shows the low temperature spectra of the neat film and 2wt% PtOEP doped in TPD. The absorption spectrum of PtOEP in CH_2Cl_2 solvent at 290 K is also shown for comparison. The solvent spectrum was obtained from [10]. A well-resolved spectrum is obtained in the neat film at 13 K. Absorption bands are observed at 547.7, 537.5, 517, 508 and 496 nm. Additionally a weak band is also observed at about 532 nm. In the lightly doped film three absorption bands are observed at 532.3, 500 and 495 nm, which is similar to the absorption spectrum of diluted PtOEP in CH_2Cl_2 solvent at room temperature. Red shift is observed for the films. This is due to the solid state solvation effect. Intense PtOEP absorption bands are observed at 529 and 496 nm in CH_2Cl_2 (curve 3 of Fig.15), while at 534.5 and about 500 nm in toluene (curve 2 of Fig.14). The whole spectrum of CH_2Cl_2 is shifted by about 5 nm from the spectrum in toluene. Taking into account that the dielectric constant of CH_2Cl_2 is 9.1 which is largely different from 2.4 for toluene, the peak shift between the two solvents is understood by the solvent effect. When two molecules are in close proximity and form dimer, a low-energy photoluminescence band which is red-shifted from the band of monomer is observed. The same is true for trimer, i.e. the trimer band is red-shifted from the dimer band. There are many reports for the observation of dimer in Pt^{2+}-compounds [11,12,13].

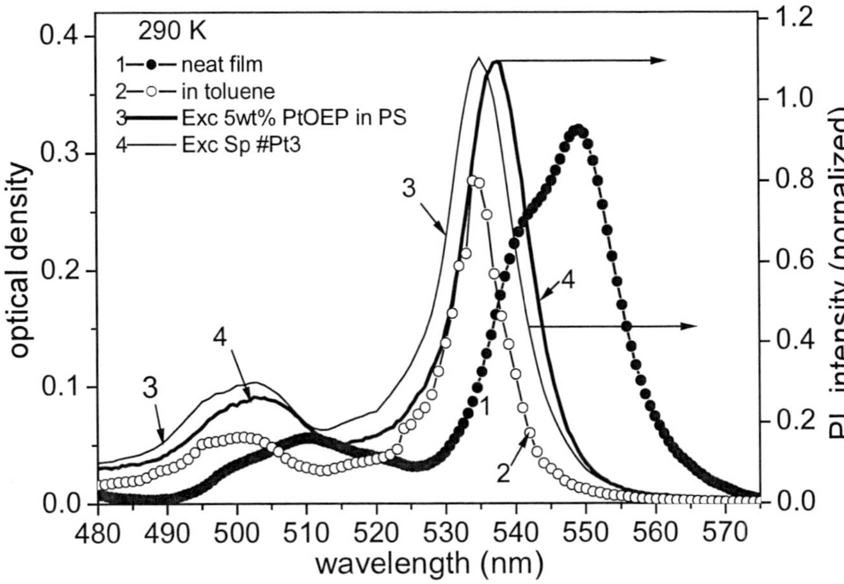

Figure 14. Absorption spectra of 60-nm thick PtOEP neat films (curve 1) and of PtOEP in toluene solution (curve 2) at 290 K, together with excitation spectra for 645 nm emission in film of polystyrene (PS) doped with 5wt% PtOEP (curve 4)and in the CBP layer with 5wt% PtOEP of OLED #Pt3 (curve 4) at 290 K.

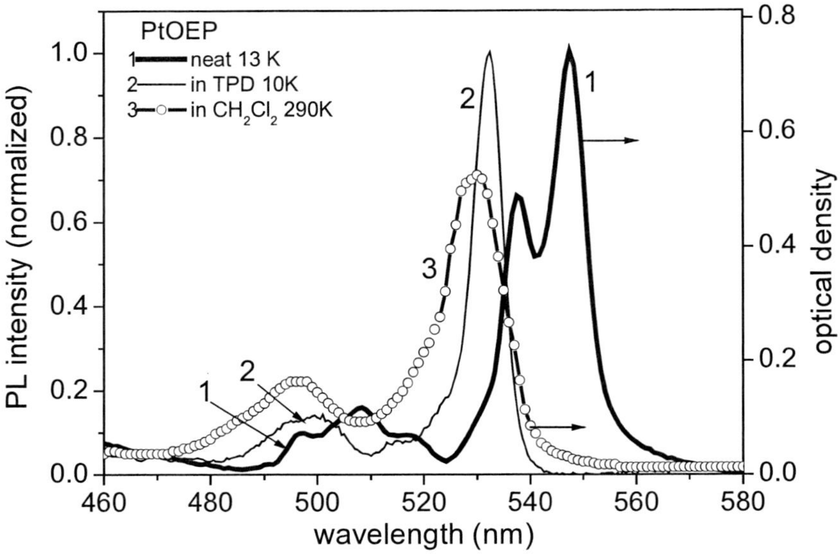

Figure 15. Absorption spectrum of neat PTOEP film at 13 K (curve 1) and excitation spectrum for 640 nm emission of PtOEP in a film of TPD doped with 2 wt% PtOEP at 10 K (curve 2), together with the absorption spectrum of PtOEP in CH_2Cl_2 solvent at 290 K (curve 3).

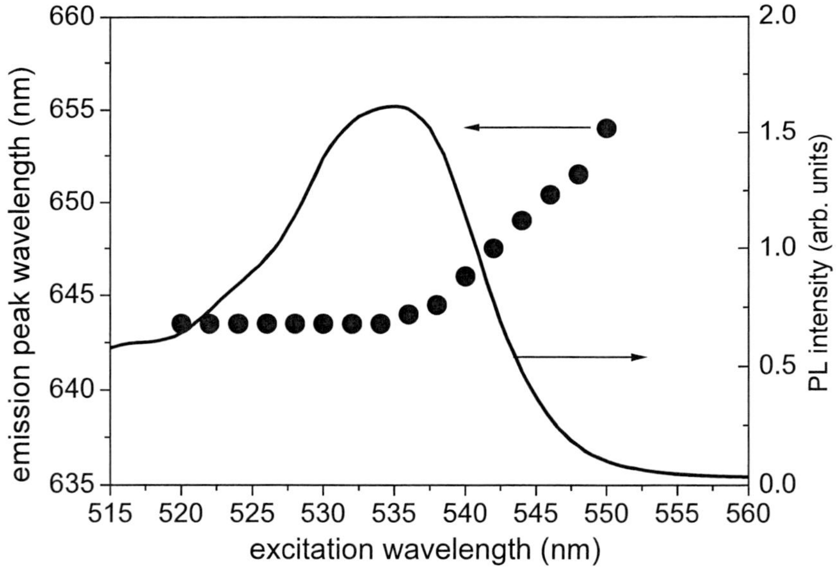

Figure 16. Excitation spectrum (solid line)for 645 nm emission in film of TPD doped with 5 wt% PtOEP at 10 K. Dependence of the peak wavelength of the intense PtOEP emission band around 645 nm on the excitation wavelength at 10 K (closed circle, left scale).

From comparison of the spectra among the neat film at 13 K, lightly doped film at 10 K and solution at room temperature, it is suggested that the neat film contains dimer and trimer (responsible for the 537.5 and 547.5 nm absorption bands, respectively) besides the monomer (responsible for 532 nm absorption band). The 645 nm emission band observed in lightly doped films (Fig.13) is attributed to monomer. Presence of dimer is confirmed by the PL spectra at low temperature. In Fig.16 is shown the excitation spectrum for 645 nm emission of the same film at 10 K. A broad band is observed at about 537 nm. When the film of 5wt% PtOEP doped in TPD is excited with lights of various wavelengths at 10 K, the peak wavelength of the intense PL band changes depending on the excitation wavelength as shown in Fig.17. The excitation wavelength was changed from 520 nm to 550 nm to excite the singlet state of PtOEP directly. No change was observed between 520 and 535 nm where the PL shows a peak at 643 nm. The emission peak shifts to longer wavelength by changing the excitation wavelength from 535 nm to 550 nm as plotted in Fig.16.

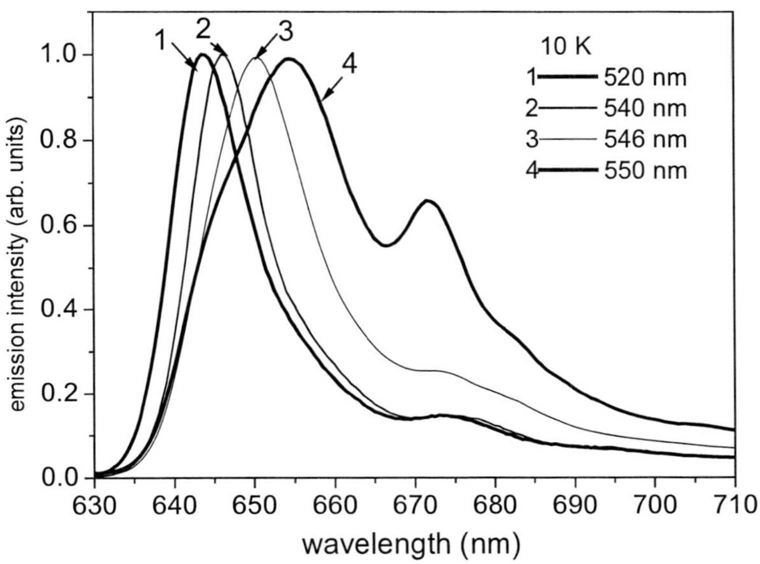

Figure 17. Photoluminescence spectra of PtOEP in a 60-nm thick film of TPD doped with 5wt% PtOEP, excited at 520, 540, 546 and 550 nm (curves 1, 2, 3 and 4, respectively) at 10 K.

The absorption band of the 5wt% PtOEP film does not show a structured shape at 535-550 nm clearly. However, observation of different emission bands by different excitation wavelengths indicates that the absorption band consists of at least three bands of different origins, one locates at 520-535 nm, the others at 535-550 nm. The first is monomer which is responsible for the emission band with a peak at 643 nm, the second and third are attributed to dimer and trimer which are responsible for the emission bands with peaks at 654 and 672 nm, respectively (see Fig.17).

When the film of 5wt% PtOEP doped in TPD is excited with lights of wavelengths longer than 550 nm at 10 K, it is observed that additional emission bands appear at 750-800 nm as shown in Fig.18. The main emission bands have peaks at 751, 760 and 770 nm. The appearance of these emission bands depends on the excitation wavelength. Figure 19 shows the excitation spectra for 751 and 645 nm emissions at 290 K. The excitation spectra for 650 nm emission in of neat film and for 645 nm emission of 5 wt% PtOEP film are consistent with the absorption spectra shown in Fig.14. The 751 nm emission band is excited at the same absorption bands at 536 and 504 nm as the 645 nm emission band.

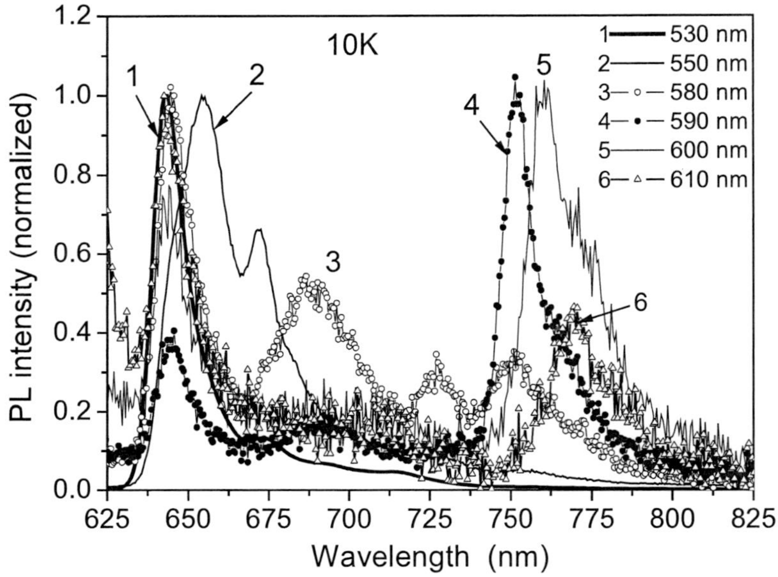

Figure 18. Photoluminescence spectra of PtOEP in film of TPD doped with 5wt% PtOEP at 10 K, which was excited with lights of various wavelengths.

Therefore these emission bands are attributed to aggregate, which consists of many PtOEP molecules, because the 750-800 nm emission bands are also observed 100% PtOEP neat films [8,9]. Figures 18 and 19 indicate that such aggregates are present in not only the 100 % neat film but also a lightly doped film like 5 wt % concentration.

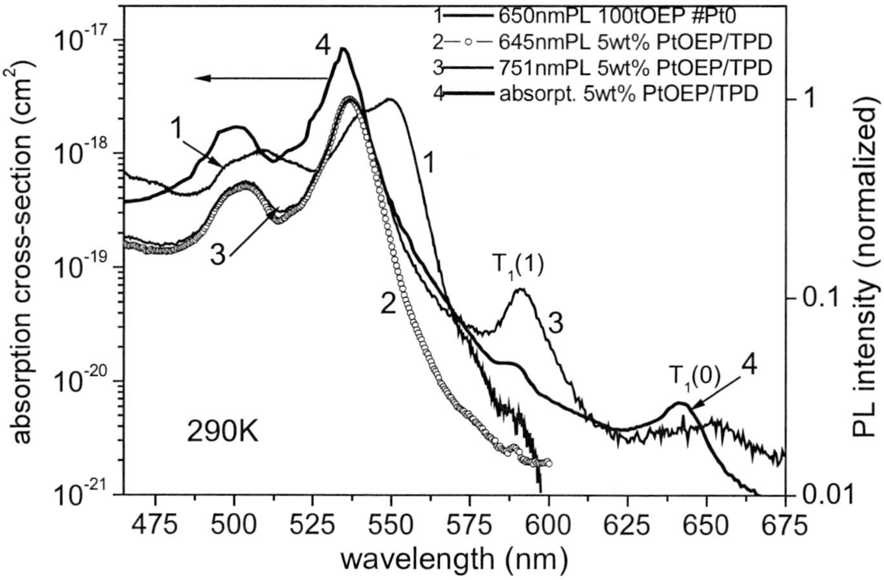

Figure 19. Excitation spectra for 650 nm emission from 50-nm thick 100% PtOEP layer in #Pt0 OLED device (curve 1), and for 645 and 751 nm emission (curves 2 and 3, respectively) and absorption cross-section spectrum (curve 4, left scale) from 61-nm thick film of TPD doped with 5 wt% PtOEP at 290 K.

Figure 20 shows the PL spectra of a 60-nm thick film of TPD doped with PtOEP which was spin-coated on glass plate. The PtOEP concentration is very low, about 1 wt%. The PL spectra were obtained by excitation with light of various wavelengths at 10 K. The very lightly doped film exhibits a sharp excitation band at 532.5 nm for 640 nm emission. When excited with 520-560 nm light, no peak shift was observed with excitation between 520 and 532 nm light, while the emission band shifts to longer wavelength with increasing the excitation wavelength. The emission band peak is at 640, 643, 645, 650 nm by the 530, 536, 538 and 542 nm excitations, respectively (Fig.20). This indicates the dimer-like center is present even in the very low concentration PtOEP film although there are very few. No emission, however, was observed at 700-850 nm region when excited with light of various wavelengths of 580-650 nm, suggesting that the large aggregate is not created in the very lightly doped film.

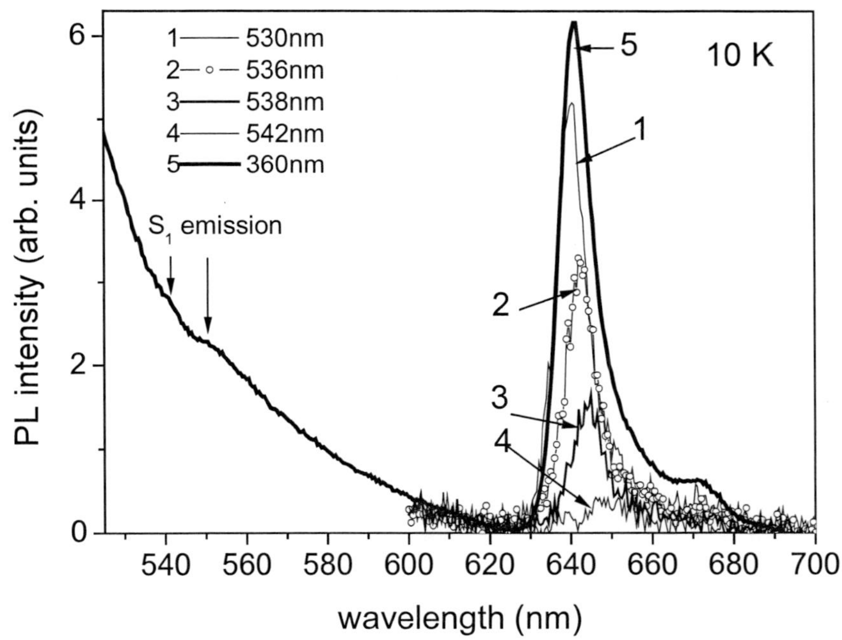

Figure 20. Photoluminescence (PL) spectra of PtOEP in film of TPD doped with 1wt% PtOEP at 10 K, which was excited with lights of various wavelengths. The vertical arrows indicate the emission from the singlet S_1 state.

8. The S_1 Emission and T_1 Absorption Bands of PtOEP

The emission spectrum by 360 nm excitation is shown in curve 5 of Fig.20. The peak position of the emission band is the same as the band excited with 530 nm. Very weak emission bands are observed at 540 and 552 nm which are superimposed on the tail of the TPD emission band. The peak height of the 552 nm band is about 0.015 of that of the 640 nm intense band. These emission band are attributed to the emission from the singlet state S_1 of PtOEP.

The absorption and PL spectra of PtOEP dissolved in CH_2Cl_2 solution have been studied by Singh and Eftekhari [10]. The absorption spectrum is shown in curve 3 of Fig.15. The absorption bands are observed at 531, 496, 376 and 288 nm, and the intense emission band is observed at 644.5 nm [10]. The 531 and 496 nm absorption bands are close to the 535 and 502 nm absorption bands that are observed in the PtOEP films doped as guest (Fig.19). Besides the 644.5 nm emission band, a weak and broad emission band is observed in the solution at room temperature, which is located at 538-590 nm region. This corresponds to the 540 and 552 nm S_1 emission bands of Fig.20. The peak height of 538-590 nm band is about 0.1 of the 644.5 nm emission band, which is much higher ratio than the case of the film. It is noted that the singlet emission in solution is much enhanced in solution than in solid film.

The absorption band at about 535 nm is attributed to the singlet state of PtOEP monomer as mentioned above. Unlike the case of $Ir(ppy)_3$, the absorption bands due to the triplet state of PtOEP are so weak that it is difficult to observe them by the usual absorption spectroscopic measurement. A very weak excitation band is observed at 588 nm in the excitation spectrum for 650 nm emission in #Pt0 (with a emitting layer of 100% PtOEP) as shown in curve 1 of Fig.19. The lowest-energy absorption band due to the T_1 state is expected to locate near 650 nm from the Stokes shift. The T_1 absorption band was clearly found at 642 nm using the film of 5wt% PtOEP doped in TPD by a highly-resolved absorption measurement as shown by curve 4 of Fig.19 [5]. The 642 and 588 nm bands are attributed to the 0-0 and 0-1 lines associated with the T_1 state due to metal-ligand charge transfer triplet state (^{3}MLCT), respectively. The intensity of the 642 nm band is about three order smaller than the singlet 535 nm absorption band. This means that PtOEP has much weaker spin-orbit interaction than $Ir(ppy)_3$. In the case of $Ir(ppy)_3$, the absorption bands due to ^{3}MLCT are easily observed at 450-500 nm (Fig. 9) [6, 14, 15]. Later we discuss on the reason why the T_1 absorption is much weaker in PtOEP than in $Ir(ppy)_3$.

9. Self-absorption in Emission Spectra of CBP

Figure 21 shows the PL and excitation spectra of CBP in a 30 nm thick neat film of CBP at various temperatures. The emission spectrum at 290 K consists of three intense bands at 372, 388 and 407 nm and a very weak band at about 430 nm. The middle 388 nm band is the most intense of the four bands. The intensity increases for all the bands with decreasing temperature from 290 K to 10 K. At 10 K the shortest-wavelength emission band at 371 nm is the most intense of the four bands. Therefore the 371, 388 and 430 nm bands observed at 290 K are attributed to the 0-0, 0-1 and 0-2 line, respectively. It is noted that the 0-0 band is much lower than the middle 0-1 at 290 K. Like the case of the TPD neat film (Figs. 3 and 7), this unusual vibronic structure is understood by the self-absorption because the absorption band strongly overlaps with the 0-0 line at 290 K (see curves 5 and 6 of Fig.21).

Similar unusual vibronic spectrum is observed in a film of CBP doped with 7 wt% iridium bis [(4,6-difluoropheny)-pyridinato-5-(pyridin-2-yl)-1H-tetrazolate (FIrN4) at 290 K as shown in Fig.22. From the absorption spectra monitored by the excitation spectra (curves 6 and 7 of Fig.22), it is also suggested that the absorption gives rise to depression of the 0-0 emission line at high temperatures.

In conclusion, the unusual PL spectra with lower height 0-0 line at 290 K is caused by the self- absorption and small Stokes shift.

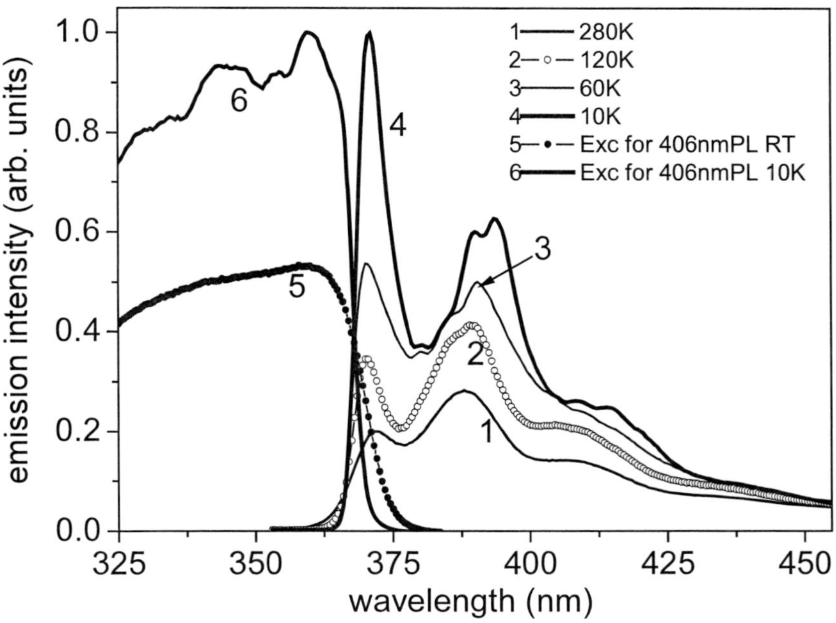

Figure 21. Temperature dependence of photoluminescence (PL) spectra of CBP in 30-nm thick neat film excited with 300 nm light (curves 1-4), together with the excitation spectra for 406 nm emission at 290 and 10 K (curves 5 and 6, respectively).

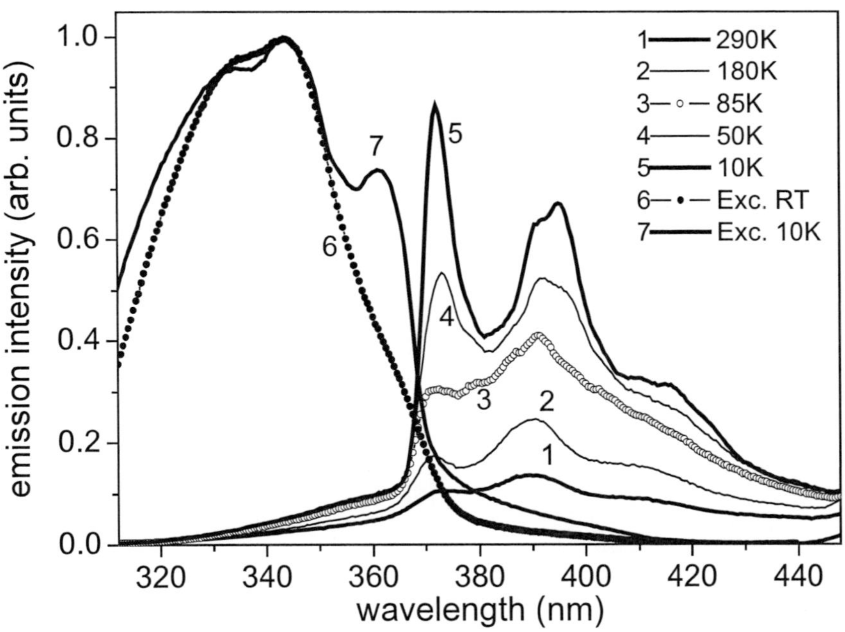

Figure 22. Temperature dependence of photoluminescence spectra of CBP in 30-nm thick film of CBP doped with 7wt% FirN4, excited with 300 nm light (curves 1-5), together with the excitation spectra for 460 nm FirN4 emission at 290 and 10 K (curves 6 and 7, respectively).

10. Emission Spectra of mCP and SimCP

Figure 23 shows the PL and excitation spectra of 1,3-bis (9-carbazolyl) benzene (mCP) in three films of polystyrene (PS) doped with 1, 5 and 20 wt% mCP and in an mCP neat film at 10 K. The 0-0 line peak position of the PL spectrum shows red-shift due to self-absorption. The amount of shift becomes large with increasing the dopant concentration from 1 to 5 to 20 wt%. No change is observed for the 0-1 line peak position between the 1 and 5 wt% films, but the 0-1 line of 20 wt% film shows red shift, i.e. peak at about 365 nm in 20 wt% film, while at 360 nm in the 1 wt% film. This shift is understood by solid state solvation effect which is more strong in highly doped film than in lightly doped films. This is confirmed from the peak position of the corresponding 0-0 absorption line which is monitored by the excitation spectra of curves 1', 2' and 3' of Fig.23.

In the neat film are observed three triple sharp PL lines with vibronic progression. Their peaks are at 347.5 nm (called a0 line, hereafter, see Fig.23), 351.9 nm (b0), 355 nm (c0), 364.2 nm (a1), 368.6 nm (b1), 373.4 nm (c1), 382.8 nm (a2), 387.5 nm (b2) and 392 nm (c2). The excitation spectrum of the neat film has additional bands at the longer wavelength side in each of the absorption bands due to monomer which are observed in the lightly doped 1 wt% film. Two bands are observed at 342 and 347 nm (called B and C bands, respectively, see Fig.23) at the long wavelength side of the intense band at 340.6 nm (A band). The A absorption band is the 0-0 absorption line of the mCP monomer. Its 0-1 and 0-2 vibrational absorption lines are observed at 324 and 307 nm, respectively.

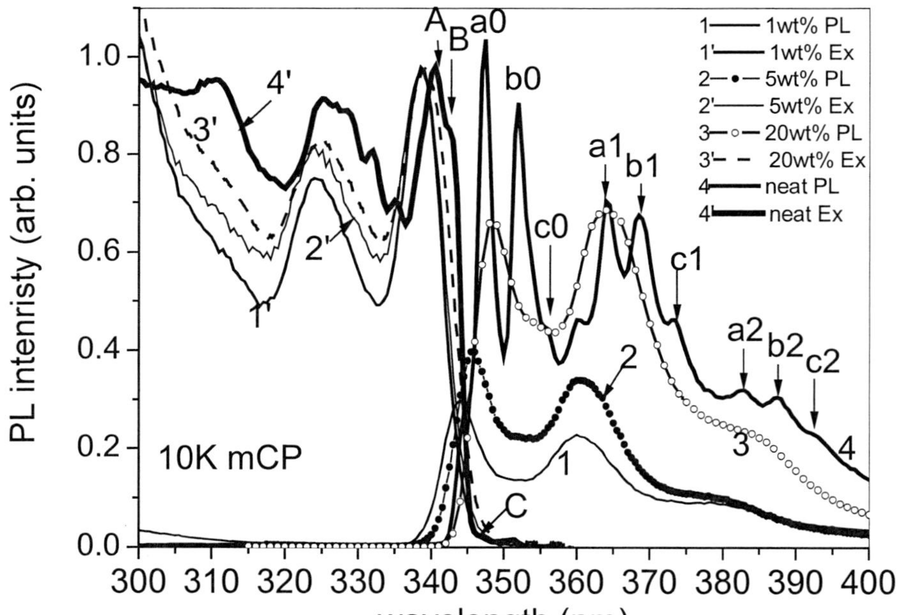

Figure 23. Photoluminescence (PL) and excitation (Ex) spectra of mCP in films of polystyrene doped with 1, 5 and 20 wt% mCP (curves 1(1'), 2(2') and 3(3'), respectively) and in mCP neat film (curve 4 (4')) at 10 K. The PL and Ex spectra were obtained by 305 nm excitation and for 375 nm emission, respectively.

From characteristics of the peaks observed in Fig.23 (e.g. the a_0-a_1 separation is the same as the b_0-b_1 separation and also same as the c_0-c_1 separation) and from comparison with lightly doped films, it is suggested that there are at least three kinds of conformations in the neat film. One is the PtOEP monomer responsible for the a0, a1 and a2 emission lines and the A absorption band, another is dimer responsible for the b0, b1 and b2 emission lines and the B absorption band, and the other is trimer responsible for the c0, c1 and c2 emission lines and the C band. The a0, a1 and a2 emission lines are attributable to the 0-0, 0-1 and 0-2 vibrational lines associated with the singlet state S_1 of monomer, respectively, the b0, b1 and b2 emission lines are attributable to the 0-0, 0-1 and 0-2 lines associated with the S_1 state of dimer, and the c0, c1 and c2 emission lines are attributable to the 0-0, 0-1 and 0-2 lines associated with the S_1 state of trimer.

Similarly, three triple peaks with vibronic progression are observed at longer wavelength of 440-520 nm in the mCP neat film at 10 K as shown in Fig.24 although they are much weaker than the bands at 345-395 nm. The observed peaks are at 443.2 nm (called m0), 471.7 nm (m1) and 504 nm (m2) which are attributable to the 0-0, 0-1 and 0-2 vibrational lines, respectively, associated with the T_1 state of monomer, at 448.7 nm (d0), 477.0 nm (d1) and 510.5 nm (d2) which are attributable to the 0-0, 0-1 and 0-2 lines associated with the T_1 state of the dimer, and at 455 nm (t0), 482 nm (t1) and 516 nm (t2) which are attributable to the 0-0, 0-1 and 0-2 lines associated with the T_1 state of the trimer.

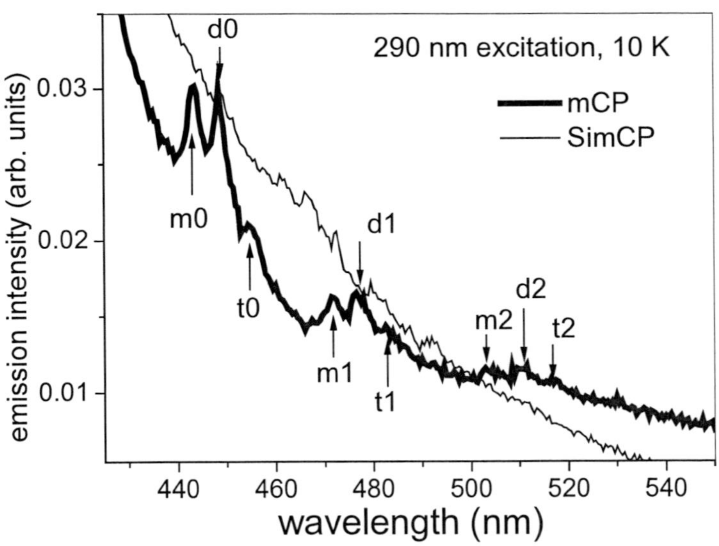

Figure 24. Photoluminescence spectra of mCP and SimCP neat films in the 425-550 nm spectral range, which are excited at 290 nm at 10 K.

Figure 25 shows the PL and excitation spectra of 3,5-bis (9-carbazolyl) tetraphenylsilane (SimCP) in three films of PS doped with 3, 10 and 20 wt% mCP and in SimCP neat film at 10 K. SimCP is a derivative of mCP molecule [16], whose chemical structures shown in Fig.12. No big change is observed in the line shapes among three doped films as the case of mCP. Additional sharp lines are observed in the neat film, indicating the presence of dimer, but not so well resolved from the lines due to monomer. This indicates that, unlike mCP, SimCP has a tendency of avoiding the formation of dimer. This will be understood from the difference of chemical structure between two molecules. It is suggested that Si-ligand in SimCP avoids intermolecular covalent bonding because of the localization of electrons near Si atom. Such an electron distribution induces stronger intramolecular covalent bonding, resulting in more stable chemical structure than mCP without central Si atom. This is confirmed that the glass transition temperature Tg is higher for SimCP (Tg=101 ^{0}C) than for mCP (Tg=55 ^{0}C) [16].

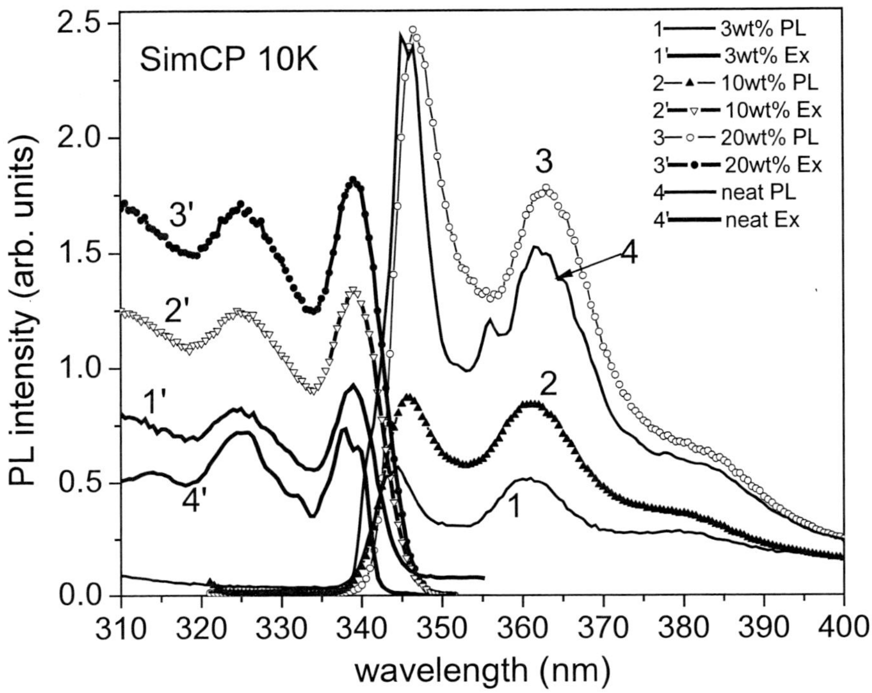

Figure 25. Photoluminescence (PL) and excitation (Ex) spectra of SimCP in films of polystyrene (PS) doped with 3, 10 and 20 wt% SimCP (curves 1(1'), 2(2') and 3(3'), respectively) and in SimCP neat film (curve 4 (4')) at 10 K. The PL and Ex spectra were obtained by 305 nm excitation and for 360 nm emission, respectively.

11. Why PtOEP Shows Much Weaker Luminancein OLEDs Than Ir(ppy)$_3$?

It is known that OLEDs with PtOEP red emitter show much weaker luminance than those with Ir(ppy)$_3$ green emitter. This is confirmed by comparison of luminance of PtOEP OLED with that of Ir(ppy)$_3$ OLED which has the same multi-layer structure as the PtOEP OLED. For example, we obtained the luminance of 7500 cd/m^2 at applied voltage of 10V in #Ir1, while 40 cd/m^2 in #Pt2 at the same 10 V. The #Pt2 shows lower luminance efficiency by one order than #Ir1 as seen in Fig.26. The organic phosphors PtOEP and Ir(ppy)$_3$ are doped in the same CBP in the emitting layer in the two OLEDs, where no energy transfer from the lowest triplet state T_1 of PtOEP and Ir(ppy)$_3$ to the CBP host because the T_1 state locates below the T_1 state of CBP at 2.56 eV. Why such a large difference occurs between PtOEP and Ir(ppy)$_3$?

The luminescence intensity P is given by

$$P = N\, k_r\, \Phi\ ,$$

where N is the population at the emitting state of the excited molecules, k_r is the radiative transition probability which is the inverse of radiative lifetime τ_r, Φ is the quantum efficiency which is given by $\Phi=k_r/(k_r + k_{nr})$, k_{nr} is non-radiative transition probability from the emitting state. Here the emitting state means the lowest triplet state T_1 in the case of phosphorescent materials.

The photoluminescence quantum efficiency (i.e. internal quantum efficiency in OLED device) is nearly 100% for $Ir(ppy)_3$ doped in CBP at room temperature [17, 18], while 40 % for PtOEP doped in CBP [19] and 41.5 % for PtOEP in both de-aerated toluene and chloroform [8]. The radiative lifetime of the T_1 state is estimated as approximately 2 μs at room temperature for $Ir(ppy)_3$ [7, 20], while about 120 μs for PtOEP at room temperature [8]. This leads to about 0.004 as the luminescence intensity ratio of PtOEP to $Ir(ppy)_3$, i.e. a much weaker PtOEP emission than the $Ir(ppy)_3$ emission. Therefore we understand that the radiative transition probability of the T_1 state is strongly responsible for the difference of emission intensity between the two phosphors. The probability depends on the spin-orbit coupling, which becomes strong with increasing the atomic number of the central metal ion in the organic compounds (called heavy atom effect). The atomic number of Pt^{2+} and Ir^{3+} ions is 78 and 77, respectively, almost the same for the two ions. This means that, besides the heavy atom effect, another factor is responsible for difference of spin-orbit coupling between PtOEP and $Ir(ppy)_3$.

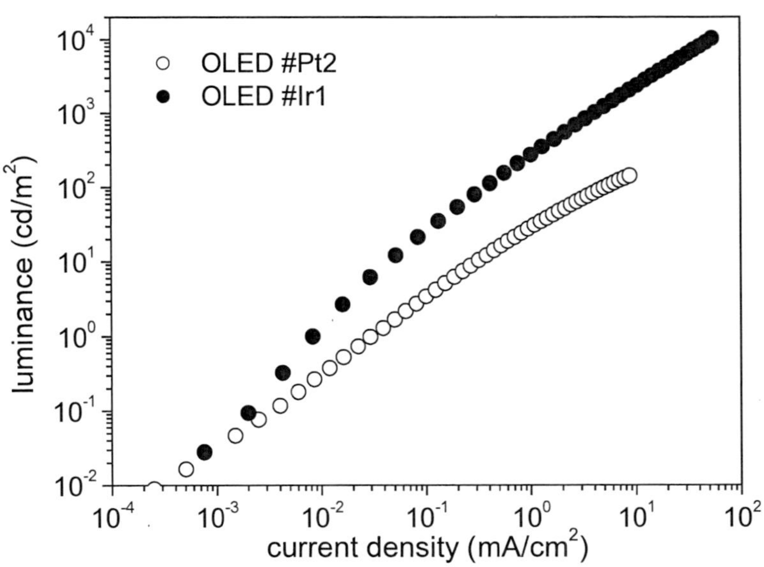

Figure 26. Luminance of the #Pt2 and #Ir1 OLEDs plotted against the current density.

Figure 27 shows the absorption and EL spectra of PtOEP, which are compared with those of $Ir(ppy)_3$ at 290 K. These spectra are plotted against photon energy. It is noted that the emission band width is much wider in $Ir(ppy)_3$ than in PtOEP. The same is true for the low-energy absorption bands. This suggests that the mixing of the electronic states of central Ir^{3+} metal ion with the ligand states is much higher than the case of PtOEP, i.e. the orbital overlap between the metal ion and ligand molecules is much larger in $Ir(ppy)_3$ than in PtOEP. Strong overlap gives rise to intense spin-orbit coupling. The next question is why PtOEP has much weaker orbital overlap.

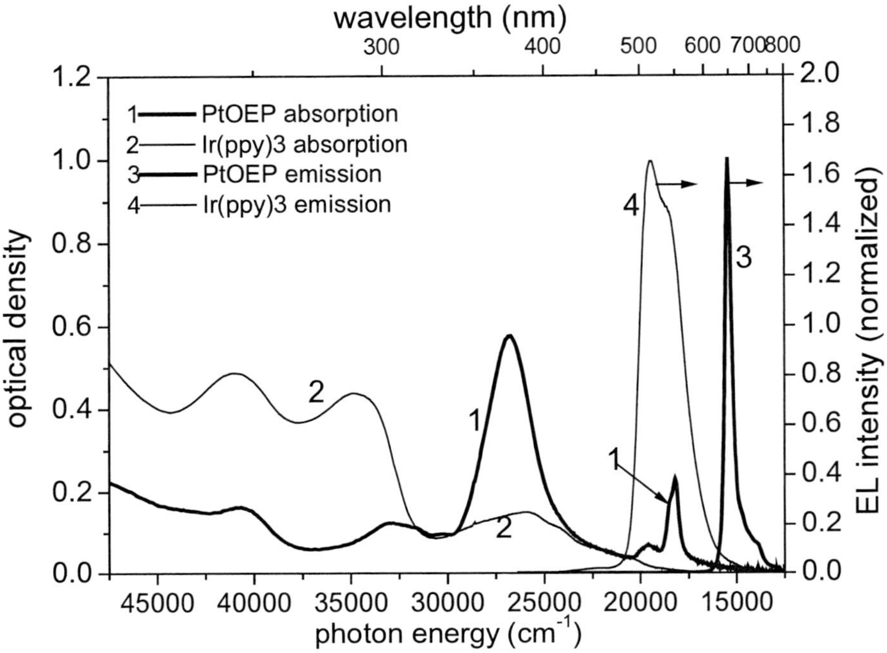

Figure 27. Absorption spectra of 60 nm-thick films of PtOEP and $Ir(ppy)_3$ at 290 K and electroluminescence (EL) spectra of #Pt2 and #Ir1 OLEDs, which are plotted against photon energy.

The electronic states of the highest occupied molecular orbital (HOMO) and the lowest unoccupied molecular orbital (LUMO) have been clarified for $Ir(ppy)_3$ using the time-dependent density function theory TDDFT [14,21]. According to the TDDFT calculation, the electronic population of d-electrons of Ir^{3+} with $(5d)^6$ outer electron configuration is about 50 % in the HOMO level and the rest is π electrons of ligand phenyl pyridine (ppy) molecule.

This means that an efficient electron overlapping is achieved in $Ir(ppy)_3$, leading to delocalization of the excited triplet state over the ligands and Ir^{3+} metal and to formation of the metal-ligand charge transfer (MLCT) state. It is suggested that such an overlapping is possible by the following reason. The d orbital of Ir^{3+} splits to two levels of six-fold degenerate t_{2g} and four-fold degenerate e_g in octahedral symmetry. The t_{2g} level fully occupied by six d-electrons is located close to the π levels of ligand. Therefore the d- and π electrons are mixed to each other easily, leading to the strong spin-orbit interaction.

The symmetry of Pt^{2+} ion in PtOEP is D_{4h}. In this symmetry the lower-energy t_{2g} level splits into two fold degenerate e_g and non-degenerate b_{2g}, while the higher-energy e_g level splits into two non-degenerate a_{1g} and b_{1g}, where the a_{1g} level is lower than the b_{1g} level and also the b_{2g} level. Pt^{2+} ion has $(5d)^8$ outer electron configuration. The HOMO level is formed by the fully occupied b_{2g} level. The π electrons of porphyrin ligand surrounding the central

Pt^{2+} ion has localized in porphyrin itself, and very weak overlapping with Pt^{2+} ion. This suggests that the π level is energetically far from the b_{2g} level of Pt^{2+}. As the result, the spin-orbit interaction becomes much weaker in PtOEP than in $Ir(ppy)_3$.

The radiative lifetime of the T_1 state is two order longer in PtOEP than in $Ir(ppy)_3$. The long lifetime causes the energy dissipation between the neighboring PtOEP molecules. In the case of photoluminescence of organic molecules excited with high power laser, triplet-triplet annihilation occurs, leading to enhancement of the non-radiative probability [3,7,8]. Such a nonlinear phenomenon, however, does not occur when molecules are excited with non-coherent light like radiation from Xe-lamp. In the case of non-coherent light excitation, PtOEP with long radiative lifetime of 120 μs gives rise to the energy dissipation from the excited PtOEP to unexcited neighboring PtOEP molecule by overcome of potential barrier between two neighboring molecules by thermal activation energy [22] (see Sec.12). This is confirmed from the observation that the PtOEP emission intensity decreases with increasing temperature from 10 K to room temperature [5], because the energy dissipation probability increases with increasing temperature.

There is additional reason for the very weak luminance in PtOEP OLEDs. Dimer of PtOEP are easily formed even in lightly PtOEP-doped films. The concentration of PtOEP is usually 3-6 wt % in OLEDs, where small aggregates like dimer are present as mentioned in Sec.7. The non-radiative processes from the monomer PtOEP to the dime occur because the T_1 state of dime is lower than that of monomer, giving rise to weak EL luminance intensity. This process is not negligible in the actual OLED devices.

12. What Pt^{2+}-Compounds Emit Intense Phosphorescence ?

If the b_{2g} level of Pt^{2+} locates energetically close to the π level of ligand, one can expect strong spin-orbit interaction. Such a situation is produced by the strong ligand field for the central Pt^{2+} ion. There is a Pt^{2+}-compound, terdentate N^C^N-coordinating 1,3-di(pyridyl) benzene Pt(Cl) with Cl at one of the four coordinate of Pt, which is called N^C^N-Pt(Cl) [23] or Pt(dpb)Cl [24]. The chemical structure is shown in Fig.12. The symmetry of Pt^{2+} is C_{2v}, which is lower than that of PtOEP. Therefore the ligand field splitting of d electrons of Pt^{2+} is wider than the case of PtOEP. It is expected that, unlike the case of PtOEP, the b_{2g} level of Pt^{2+} is located close to the π level of ligand, leading to the mixing of the two orbitals. In fact the strong mixing of the d orbital and π orbital is found by the TDDFT calculation for N^C^N-Pt(Cl) [24]. As a result we can expect high phosphorescence intensity for this compound.

Intense EL luminance is certainly obtained in the OLED with N^C^N-Pt(Cl) doped in CBP, e.g. about 17000 cd/m^2 at applied voltage of 10 V. The phosphorescence lifetime of N^C^N-Pt(Cl) is about 6.8 μs and the photoluminescence quantum efficiency Φ is 60 % at room temperature [23]. These values gives 11 μs as the radiative lifetime τ_r, which is one order shorter than τ_r of PtOEP, indicating that the spin-orbit interaction of the emitting triplet T_1 state with the singlet state is much stronger in N^C^N-Pt(Cl) than in PtOEP. In this way we understand the reason why N^C^N-Pt(Cl) gives much higher EL luminance than PtOEP doped in CBP (which gives 40 cd/m^2 for at the same 10 V).

In conclusion, it is suggested that the energy levels of d-electrons of Pt^{2+} ion relative to the energy levels of ligand orbitals determine the spin-orbit coupling of the molecules. The relative positions of energy levels are determined by the local symmetry of Pt^{2+}. The lower the symmetry of metal ion, the stronger the spin-orbit coupling, leading to high phosphorescence intensity.

The integrated PL intensity of N^C^N-Pt(Cl) decreases with increasing temperature from 8 K, e.g. the intensity at 300 K reduces to 68 % of the intensity at 8 K in thin film of N^C^N-Pt(Cl)-doped polycarbonate (PC) [23]. This ratio is close to the quantum efficiency of 60 % which was measured using dichloromethane solution at room temperature. Such a reduction is explained by dissipation (i.e. diffusion) of excited energy from the T_1 state to the neighboring N^C^N-Pt(Cl) molecule. The dissipation rate k_D from the T_1 state of the excited guest molecule M_1 to the T_1 state of unexcited neighboring guest molecule M_2 is given by

$$k_D = K_D[A_T + \exp(-E_D/kT)],$$

where K_D is coupling parameter between two guest molecules, A_T is parameter due to tunneling of the excited energy from the excited to unexcited molecules, and E_D is thermal activation energy to overcome the potential barrier between the neighboring molecules [18]. Back transfer from the M_2 molecule to M_1 might be possible at high temperatures like room temperature. However it is negligible because the M_2 molecule has many processes after excitation, e.g. energy dissipation and diffusion to neighboring molecules. The probability of back transfer to the original M_1 molecule is very small.

The parameter K_D depends on the radiative lifetime of the T_1 state, i.e. the radiative lifetime is important factor to make the dissipation possible. If the lifetime is short, the rate of radiative transition from the T_1 state to the ground state S_0 becomes much higher than the rate of energy transfer to the neighboring molecule, resulting in depression of dissipation process. The τ_r=11 μs of N^C^N-Pt(Cl) is rather long, therefore the dissipation process occurs simultaneously with the radiative process to the S_0 state, giving rise to decrease of PL intensity with increasing temperature. If τ_r is more shorter than 11 μs, we would obtain higher quantum efficiency than 60 %. To shorten the radiative lifetime of N^C^N-Pt(Cl), one has to modify the chemical structure so as to reduce the local symmetry around Pt^{2+} ion.

The parameters K_D and A_T are temperature independent, therefore the temperature dependence of PL intensity is determined by the thermal activation. Since the quantum efficiency (0.60) at room temperature is close to the intensity ratio (0.68) of the RT intensity of PL to the 8K intensity [23], we predict the PL quantum efficiency at 8 K is nearly 100 % for N^C^N-Pt(Cl). The following three reasons are conceivable for the 100% efficiency. (1) The non-radiative process of dissipation to neighboring molecule is negligible at such a low temperature because the thermal activation does not occur and because the tunneling process is negligible in very lightly doped PC film. (2) Energy transfer from the T_1 state of N^C^N-Pt(Cl) to the PC host is impossible because the T_1 state of host locates far above that of guest. (3) The non-radiative multi-phonon transition from the T_1 state to the ground state does not occur by the energy gap law because of wide energy gap (2.48 eV) between the T_1 and S_0 states although the higher-energy singlet states including the S_1 state mix with the T_1 state by the spin-orbit interaction.

13. Solid State Solvation Effect

Absorption and PL bands of organic solute molecule in liquid solutions change the peak wavelength without changing their line shapes, depending on molecule of solvent. The shift becomes large in the solvent with higher dielectric constant because the permittivity of solvent changes the local electric field around the solute and then the energy states of the solute is changed by the interaction of dipole of solute molecule with the electric field due to solvent. This is called solvent effect or solvation effect [25].

This idea of solvation effect was applied to the case of solid films since the similar phenomena were observed for dye in films [26, 27, 28]. For example, when Alq_3 is doped in TPD film, it was observed that the green PL band of Alq_3 shows red shift from 498 to 522 nm, without change in spectral shape and peak width, as the concentration of Alq_3 is increased from 2.5 to 100 % [1,26]. This is understood as follows. When the DCM2 concentration is increased in the relatively less-polar Alq_3 host (dipole moment: 5.5 D for Alq_3 and 11.3 D for DCM2 [27]), the distance between nearest-neighbor DCM2 molecules decreases, resulting in increase of the local polarization field and then leading to red shift with increasing DCM2 concentration [28]. A theory to explain this sold state solvation effect was suggested by Baldo et al [1]. They considered that the polar Alq_3 molecules form ordered domain in TPD host, which consists of many Alq_3 molecules, by the dipole-dipole interaction among the neighboring molecules in less polar molecular TPD host (dipole moment: 1.5 D [1]). The energy gap between the excited and ground states of the central molecule in the domain decreases with increasing the domain size. It was assumed that the observed emission is caused by the central molecule in the maximum-size domain although domains with various sizes are present.

The solvation effect is also observed for TPD, $Ir(ppy)_3$ and PtOEP when we measure the absorption and PL spectra by changing solvent molecule in solution [7, 8] and by changing dopant concentration and host molecule of film variously (see e.g. Figs. 3 and 5 for TPD). The peak shift in the emission band is observed not only between films but also between liquid solution and sold film as seen in e.g. Figs. 10 and 15 for PtOEP. The amount of the relative peak shift is mainly understood by the relative values of dielectric constant and polarizability between host and guest molecules.

14. Unusual ALQ_3 Emission Spectrum

As mentioned in Sec.1, the emission spectrum of Alq_3 is quite different from the other organic molecules. For example, large Stokes shift and structureless broad emission band are pointed out. We shall examine the reasons.

There are many reports on the EL spectra of Alq_3 from 100% Alq_3 neat layer since the first report of EL band with a peak at 550 nm with a 60-nm thick film by Tang and VanSlyke [29]. We notice that the EL peak wavelength is different among the reports although a Gaussian-shape emission band is observed. For example, the peak wavelength is about 532 nm [30], 528 nm [31], 525 nm [32], 520 nm [33], and 510 nm [34]. Similarly the PL spectrum of neat film at room temperature is different among the investigators although the difference is much smaller than the case of EL. For example, the PL peak is at 530 nm in 50-nm thick film [34], about 530 nm in 100-nm thick film [30], 525 nm in 60-nm thick film [32], 522 nm in 200-nm thick film [1]. We obtained a peak at 522 nm at room temperature in 10-nm thick film (Fig.1). Observation of different peak positions in neat films indicates that the morphology of neat film is different among the films fabricated. One of these reasons is a different mixture ratio of two isomers of *mer*- and *fac*-Alq_3 among the films [2,35].

The PL spectrum of Alq_3 in solution shows a single Gaussian-shape band, i.e. the peak at 540 nm in N,N-dimethylformamide (DMF) solution at 298 K, at about 510 nm in the solution of the same DMF at room temperature [30] and at 475 nm in ethanol at 77 K [36]. No vibronic structure is observed even for isolated single Alq_3 molecule in liquid solution. The same is true for the PL spectrum from a film of lightly doped TPD (2.5% Alq_3) at room temperature [1]. Why Alq_3 molecule does not show the vibronic structure?

The large Stokes shift means a large Huang-Rhys parameter (or called Huang-Rhys factor) S [4,37], indicating a large Franck-Condon shift. The Franck-Condon shift is caused by difference of the bottom of adiabatic potential parabola (i.e. equilibrium position in the configurational coordinate) between the excited and ground states. The spectral distribution of emission band involving 0-n (n=0, 1,2,3, , ,) vibrational transitions is proportional to the Franck-Condon factor F_{0n} :

$$F_{0n}= \exp(-S)S^n/n!$$

Therefore the S factor is approximately estimated from the intensity ratio of the 0-1 line to the 0-0 line. Since no vibronic structure appears in Alq_3 thin films even at 10 K (Fig.1), we cannot estimate the S factor for Alq_3. Brinkmann et al found the vibronic structure with progression of 525 cm^{-1} in polycrystalline α- and β-Alq_3 at 4.2 K [2], then estimated S=2.6. Usually the S factor is less than 1 for most organic molecules. Although we cannot estimate the S factor in thin film of Alq_3, it is suggested from the large Stokes shift that the S factor is more than 2.6 in thin film.

According to calculation with TDDFT, the electron density in HOMO is localized on different ligand from that in LUMO [30, 38]. Different charge distribution between HOMO and LUMO means the different charge distribution (and different orientation of dipole moment) between the ground state S_0 and the emitting singlet state S_1, resulting in the large shift of the equilibrium coordinate between the ground and excited state adiabatic potential parabolas. In this way we understand the reason why large Stokes shift occurs in Alq_3.

Broad emission band is observed in Alq_3. This is also understood by the large Huang-Rhys factor S. Simulation of spectral distribution of emission band using various S values indicates that the much more vibrational states are involved in the emission spectrum with increasing the S number [37, 39]. For example, only four or five vibrational transitions are involved for S=0.4, while at least twenty vibrational transitions are involved for S=20. This means that Alq_3 with large S value gives a broad emission band.

15. Appearance of Resolved Vibronic Structure

Organic molecular complexes exhibit a more or less vibronic structure in the emission and absorption spectra. The resolution in the vibronic structure becomes high with decreasing temperature from room temperature to low temperature near 10 K as shown in Fig.5. There are three molecule groups regarding the PL line shape. The first group is molecules with a well resolved vibronic structure, the second is molecules with a less resolved structure, and the last is molecules with no structure even at 10 K. For example, CBP, TPD, PtOEP, mCP and SimCP belong to the first group, and Alq_3 to the last group. $Ir(ppy)_3$ in multi-layer OLED device belongs to the second group, while $Ir(ppy)_3$ doped in CBP film, which is spin-coated on glass plate, belongs to the first group (Fig.11).

It has been suggested that the highly resolved vibronic structure is due to the transition from the ligand-centered (LC) $\pi-\pi^*$ state, while poorly resolved one from the metal-ligand-charge-transfer (MLCT) state [15, 40, 41, 42]. From this suggestion, the less resolved green emission of $Ir(ppy)_3$ is attributed to the MLCT state, which is consistent with the TDDFT calculation [14, 21]. However, it should be reminded that the vibronic resolution in PL spectrum becomes high with decreasing temperature from room temperature (Figs. 4 and 5). This indicates that the appearance of resolved vibronic structure is caused by not only the electronic state (either LC or MLCT) but also the electron-vibration interaction.

The vibronic structure of the emission spectrum is caused by the vibrational transitions from the vibrational quantum number m=0 in the excited state to the vibrational quantum numbers n=0, 1, 2, 3, , , in the ground state as mentioned in Sec.2. The vibrational mode involved in the ground state determines the vibronic progression, i.e. the separation of the 0-0 and 0-1 lines, etc. If only one mode is involved, we expect to observe highly resolved vibronic structure. This means that if many vibrational modes with different vibrational energies are involved, we observe a less structured spectrum by overlapping among various vibrational lines. Similarly if there is a large distribution of molecular environments in thin film, inhomogeneous broadening occurs, resulting in the smearing out of vibronic structure in the emission spectra. Decrease of temperature induces to depress the distribution due to thermal movement, giving rise to well-resolved vibronic spectra.

The LC $\pi-\pi^*$ transition is responsible for the green emission of Alq_3 and then we expect a resolved spectrum. But no vibronic structure is observed for Alq_3. This is explained as follows. In Alq_3, no long-range order is present in thin films, which was clarified by the absence of X-ray diffraction peaks, while there is not only a large disordered distribution of molecular packings [2] but also mixture of two isomers of *mer*- and *fac*-Alq_3 in the amorphous films. As a result Alq_3 does not show any resolved spectrum.

In Pt^{2+}–compounds such as PtOEP with porphyrin-plane, intermolecular stacking interactions occurs between the planar cations, leading to formation of planar-parallel orientation in the constituent molecules, i.e. ordered stacking structure. As a result, unlike the case of Alq_3 with random distributions of various conformations, resolved vibronic spectrum is observed. In conclusion it is suggested that the spectral line shape of whether it is well resolved or not depends on the chemical structure of molecule and the distribution of conformations in amorphous thin films.

Regarding $Ir(ppy)_3$-doped CBP film, it shows different resolved spectrum depending on environment of the film. A well-resolved vibronic structure appears in this single layer $Ir(ppy)_3$:CBP film than in the emitting $Ir(ppy)_3$:CBP layer of the multi-layer structured OLEDs (see Sec.6). This difference is understood from the above-mentioned model. The $Ir(ppy)_3$ molecules in OLEDs interact with not only the molecules within the same layer but also molecules in the neighboring layers. Diffusion of the components from the neighboring layers and the interface is not negligible in the multi-layer structure OLEDs. Such various interactions induce a large distribution of environments and geometrical strutures for $Ir(ppy)_3$

molecules. As a result, the $Ir(ppy)_3$ film of OLED shows less-resolved vibronic structure and disappearance of weak 0-2 and 0-3 vibrational lines.

Watanabe et al observed EL spectra with different intensity ratio between the 0-0 and 0-1 vibronic lines in the EL spectra of $Ir(ppy)_3$ in 6.5 wt% $Ir(ppy)_3$-doped CBP layer in multi-layer OLEDs by changing the thickness of electron transport layer of Alq_3 variously between 20 and 60 nm [43]. The structure of OLED device is quite similar to that of #Ir4. When the Alq_3 layer becomes thick, the lower-height 0-1 line becomes close to the 0-0 line height. When the thickness of emitting layer was changed from 10 to 60 nm but the Alq_3 layer was fixed at 30 nm, no change was observed in the intensity ratio between the 0-0 and 0-1 lines [43]. These experimental results indicate that the conformation of $Ir(ppy)_3$ molecules in the emitting layer is changed by not only the nearest-neighboring layer but also next-neighboring layer, suggesting inter-layer interaction which is due to dipole-dipole interaction of molecules among the neighboring layers.

16. Formation of Aggregates in Thin Films

As mentioned in Sec.7, the PL and absorption spectra indicate that PtOEP molecules forms not only small aggregate such as dimer in neat film but also large aggregate besides monomer. PtOEP dimer was formed even in lightly PtOEP-doped TPD, polystyrene and polycabonate films. Dimer and trimer of mCP are also formed in mCP neat film. TPD dimer is also formed in the TPD neat film. On the other hand, dimer is not found in Alq_3 and $Ir(ppy)_3$ neat films, which show peak shift in the emission bands, by the solid state solvation effect, with increasing concentration of the dopant. Thus, there are two groups in the doped and neat films. The first group is that the constituent molecules form aggregates as the cases of PtOEP and mCP, the second group is that the constituent molecules do not form aggregate as the cases of Alq_3 and $Ir(ppy)_3$.

Formation of aggregate is due to the intermolecular covalent bonding. PtOEP has plane of porphyrin which has a strong tendency towards $\pi-\pi$ stacking as the case of square planar Pt^{2+}-compounds [11]. Therefore dimer is easily formed, but monomer PtOEP also exists together with the aggregates even in heavily doped films.

Alq_3 and $Ir(ppy)_3$ molecules do not have such a planar structure, resulting in poor stacking. The TDDFT calculation for Alq_3 shows absence of effective HOMO overlap between neighboring molecules [38]. This indicates that formation of stable dimer and trimer are difficult in Alq_3 film although each of Alq_3 molecules has dipole-dipole interaction with neighboring molecules.

Molecules confined in thin film take their stable positions and conformation finally after they are randomly distributed by evaporation or spin-coating. When the concentration of dopant is increased and the intermolecular distance becomes short, molecules like PtOEP and mCP form aggregates by strong intermolecular covalent bonding, while molecules like Alq_3 and $Ir(ppy)_3$ do not. Like this the strength of intermolecular bonding depends on the chemical structure of each molecule.

17. Microcrystals in Organic Thin Films

Atomic force microscopy (AFM) and friction force microscopy (FFM) were used to study the morphology of TPD and Alq_3 film surfaces [44]. The images of AFM and FFM indicate that partial crystallization appears in TPD single layer film when it is kept at room temperature in a week after evaporation on glass, while Alq_3 film is extremely stable even for a few years and such microcrystals are not formed even at high temperature of 80 °C [45]. This means that, unlike Alq_3, TPD molecules form microcrystals easily.

We observed the almost the same PL line shapes of TPD at 10 K between the TPD neat film and 5wt% TPD doped polystyrene (PS) film (Figs. 3 and 5) although the peak shift is observed in the two films by solvation effect. The spectra of Figs.3 and 5 were measured by photoexcitation in the absorption bands of TPD dopant and PS host several weeks after fabrication of films. Therefore it is confirmed that the amorphous molecules are still present for a long time.

As shown in Fig.6, TPD doped as host for $Ir(ppy)_3$ guest and TPD in the neat film give emission band at about 404 nm in addition to the monomer emission band at 397 nm. It is suggested the 404 nm band is due to not excimer but aggregate such as dimer, which consists of small numbers of TPD molecules, because it locates close to the monomer band. Where is the emission band due to microcrystals suggested from the AFM and FFM images ?

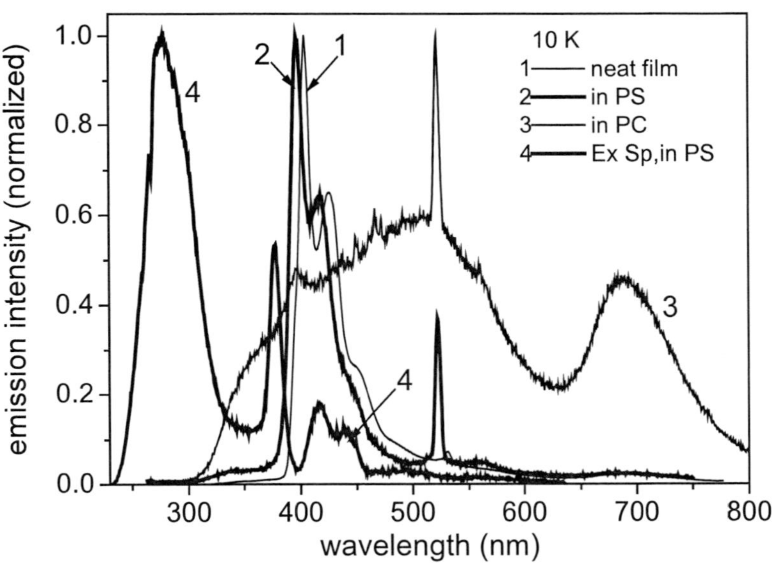

Figure 28. Photoluminescence spectra of TPD neat film (curve 1) excited at 265 nm and films of 5wt% TPD doped polystyrene (PS) and polycarbonate (PC) (curves 2 and 3, respectively), which are excited at 260 nm, together with excitation spectrum for 690 nm emission in the 5wt% TPD doped PS (curve 4). The measurements were made at 10 K. Sharp lines at 530 nm of curve 1 and at 530 nm of curves 2 and 3 are due to the harmonic of the excitation light.

When the neat film and 5 wt% TPD-doped PS film are excited at light of 260-295 nm, additional PL band is observed at 690 nm at 10 K as shown in curves 1 and 2 of Fig.28. The same PL band is also observed in 5wt% TPD-doped polycarbonate (PC) film (curve 3 of

Fig.28). The 690 nm emission band is excited by the absorption bands different from the bands of isolated TPD and PS molecules (curve 4 of Fig.28). This structureless Gaussian-shape 690 nm emission band was also observed in TPD film doped with 5wt% $Ir(ppy)_3$. From these results, it would be suggested that the 690 nm emission band is attributed to the microcrystals.

A very weak PL band is observed at 700 nm in CBP neat film besides the PL band of isolated CBP molecule. The 700 nm band is also observed in films of CBP doped with 7 wt% FIrpic and doped with 7 wt% FIrN4. The excitation spectrum for this emission is different from the excitation spectrum for emission of the isolated CBP molecule. The 700 nm band enhances with decreasing temperature to 10 K. This emission band is quite similar to the 690 nm emission band observed in films containing TPD. Therefore, like the case of TPD, the 700 nm emission band is also attributed to the microcrystals of CBP.

Presence of domain in films has been suggested for Alq_3 films [1]. Unlike the case of TPD neat film, however, the AFM image does not show such a formation in Alq_3 neat films [45]. The PL spectral line shape is the same between the neat film and lightly doped film. We could not find any additional emission band in the Alq_3 neat film clearly (see Fig.1). This is understood as follows. Each molecule of Alq_3 does not have strong interaction with neighboring molecules in the domain but weak dipole-dipole interaction that induces small energy shift from the isolated molecule by the perturbation from molecules in domain. This means that the domain is not practically present in film and not detected by physical measurements like AFM. We believe that the domain means presence of sphere with maximum radius where each molecule in film has intermolecular interaction.

Acknowledgments

We thank Profs. Alfons Penzkofer and Chin-Ti Chen for kindly supplying the spin-coated and evaporated organic films that are used in the present study. We also thank Mr. Hideyuki Murayama for assistance in photoluminescence measurements and Mr. Nadeer Aljaroudi for assistance in numerical calculation. The present work was partly supported by the Grant-in-Aid for the Scientific Research from the Japan Society for Science Promotion.

References

[1] M.A. Baldo, Z.G. Soos and S.R. Forrest, *Chem. Phys. Lett.* **347** (2001) 297.

[2] M. Brinkmann, G. Gadret, M. Muccini, C. Taliani, N. Masciocchi and A. Sironi, *J. Am. Chem. Soc.* **122** (2000) 5147.

[3] N.J. Turro, *Modern Molecular Photochemistry* (University Sci. Books, Sausalito, 1991).

[4] M. Pope and C.E. Swenberg, *Electronic Processes in Organic Crystals*, (Clarendon Press, Oxford, 1982).

[5] T. Tsuboi, H. Murayama and A. Penzkofer, *Thin Solid Films* 499(2006)306.

[6] T. Tsuboi and M. Tanigawa, *Thin Solid Films* **438-439** (2003) 301.

[7] W. Holzer, A. Penzkofer and T. Tsuboi, *Chem. Phys.* **308** (2005) 93.

[8] A.K. Banzal, W. Holzer, A. Penzkofer and T. Tsuboi, *Chem. Phys.* (2006), in press.

[9] J. Kalinowski, W. Stampor, J. Szmythowski, M. Cocchi, D. Vrgili, P. Di Marco, *J. Chem. Phys.* **122** (2005) 154710.
[10] J.J. Singh and A. Eftekhari, *NASA Technical Paper* **3619** (1996).
[11] S.W. Lai, M.C.W. Chan, T.C. Cheung, S.M. Peng and C.M. *Che, Inorg. Chem.* **38** (1999) 4046.,
[12] V.M. Miskowski and V.H. Houlding, *Inorg. Chem.* **28** (1989) 1529.
[13] T. Dienel, H. Proehl, T. Fritz and K. Leo, *J. Lumin.* **110**(2004) 253.
[14] P. J. Hay, *J. Phys. Chem.* **A 106** (2002) 1634.
[15] M.G. Colombo, T.C. Brunold, T. Riedener, H.U. Gudel, M. Fortsch and H.B. Burgi, *Inorg. Chem.* **33** (1994) 545.
[16] S.-J. Yeh, M.-F. Wu, C.-T. Chen, Y.-H. Song, Y. Chi, M.-H. Ho, S.-F. Hsu and C. H. Chen, *Advanced Mat.* **17**(2005) 285.
[17] K. Goushi, R. Kwong, J.J. Brown, H. Sasabe and C. Adachi, *J. Appl. Phys.* **95** (2004)7798.
[18] T. Tsuboi, *J. Lumin.* **119-120** (2006) 127.
[19] C. Adachi, M.A. Baldo, S.R. Forrest, S. Lamansky, M.E. Thompson and R.C. Kwong, *Appl. Phys. Lett.* **78** (2001) 1622.
[20] W.J. Finkenzeller and H. Yersin: *Chem. Phys. Lett.* **377** (2003) 299.
[21] N.G. Park, G.C. Choi, J.E. Lee and Y.S. Kim, *Current Appl. Phys.* **5** (2005) 79.
[22] T. Tsuboi and N. Aljaroudi, unpublished.
[23] T. Tsuzuki and S. Tokito, *Proc. Int. Super-Functionality Organic Devices*, *IPAP Conf. Series* **6** (2005) 99.
[24] W. Sotoyama, T. Satoh, H. Sato, A. Matsuura and N. Sawatari, *J. Phys. Chem.* **A 109** (2005) 9760.
[25] A.P. Wilde and R.J. Watts, *J. Phys. Chem.* 95 (1991) 622.
[26] V. Bulovic R. Deshpande, M.E. Thompson and S.R. Forrest, *Chem. Phys. Lett.* **308** (1999) 317.
[27] V. Bulovic, A. Shoustikov, M.A. Baldo, E. Bose, V. Kozlova, M.E. Thompson and S.R. Forrest, *Chem. Phys. Lett.* **287**(1998) 455.
[28] C.F. Madigan and V. Bulovic, *Phys. Rev. Lett.* **91** (2003)247403.
[29] C.W. Tang and S.A. VanSlyke, *Appl. Phys. Lett.* **51** (1987) 913.
[30] P.E. Burrows, Z. Shen, V. Bulovic, D.M. McCarty, S.R. Forrest, J.A. Cronin and M.E. Thompson, *J. Appl. Phys.* **79** (1996)7991.
[31] S.A. VanSlyke, C.H. Chen and C.W. Tang, *Appl. Phys. Lett.* **69** (1996) 2160.
[32] H.Q. Tang, H.X. Liao, K, Xu, Z.X. Zhou and L.H. Zhu, *J. Lumin.* **118** (2006) 39.
[33] J. Kalinowski, N. Camaioni, P.D. Marco, V. Fattori and G. Giro, *Int. J. Electron.* **81** (1996) 377.
[34] R.J. Curry and W.P. Gillin, *J. Appl. Phys.* **88** (2000) 781.
[35] M. Muccini, M.A. Loi, K. Kenevey, R. Zamboni, N. Masciocchi and A. Sironi, *Adv. Materials* **16** (2004) 861.
[36] R. Ballardini, G. Varani, M.T. Indelli and F. Scandola, *Inorg. Chem.* **25** (1986) 3858.
[37] B. Henderson and G.F. Imbusch, *Optical Spectroscopy of Inorganic Solids* (Oxford Sci. Publ., Oxford, 1989).
[38] B.C. Lin, C.P. Cheng, Z.Q. You and C.P. Hsu, *J. Am. Chem. Soc.* **127** (2005) 66.
[39] Y. Farge and M.P. Fontana, *Electronic and Vibrational Properties of Point Defects in Ionic Crystals* (Amsterdam, North-Holland, 1974)

[40] M. Maestri, D. Sandrini, V. Balzani, A. von Zelewsky and P. Jolliet, *Helv. Chim. Acta* **71** (1988) 1053.

[41] S. Lamansky, P. Djurovich, D. Murphy, F. Abdel-Razzaq, S.R. Forrest and M.E. Thompson, *J. Am. Chem. Soc*, **123** (2001) 4304.

[42] C.B. Blanton, Z. Murtaza, R.J. Shaver and D.P. Rillema, *Inorg. Chem.* **31** (1992) 3230.

[43] T. Watanabe, K. Nakamura, S. Kawami, Y. Fukuda, T. Tsuji, T. Wakimoto,. S. Miyaguchi, M. Yahiro, M. Yang and T. Tsutsui, *Synth. Met.* **122** (2001) 203.

[44] H. Mu, H. Shen and D. Klotzkin, *Solid-St. Electro.* **48** (2004) 2085.

[45] E.M. Han, L.M. Do, N. Yamamoto and M. Fujihira, *Thin Solid Films* **273** (1996) 202.

In: Research Progress in Materials Science
Editors: W. Olsson and F. Lindberg, pp. 209-252

ISBN: 978-1-60456-629-1

Chapter 9

FRACTAL PHYSICAL CHEMISTRY OF POLYMER SOLUTIONS

G.V. Kozlov*[1], *I.V. Dolbin*[2] *and G.E. Zaikov

[1]Researching Institute of Applied Mathematics and Automatization of Kabardino-Balkarian Scientific Center of Russian Academy of Science Nal'chik – 360000, Shortanov st., 89 "a", Russian Federation

[2]Institute of Biochemical Physics of Russian Academy of Science Moscow – 119991, Kosygin st., 4, Russian Federation

Abstract

The main aspects of the application of the fractal analysis for the description of the behavior of macromolecular coils in the diluted solutions are considered. The coil structure is quantitatively described with the help of its fractal (Hausdorff) dimension characterizing the distribution of coil elements in space. A number of the experimental methods of the determination this dimension is offered. The interrelation of the classical and fractal (structural) characteristics of macromolecular coils is shown. The fractal model is used for the description of such phenomena as film formation from polymer solutions, flocculation and self-diffusion of macromolecular coils. The example of the application of the theory of fractional derivatives for the description of macromolecular coil the main parameters of a in solution is give.

Key words: Polymer solution, macromolecular coil, structure, fractal analysis, polymer film, flocculation, self-diffusion.

1. Introduction

Developed in the physical chemistry of polymer solutions the basic ideas are the basis of our understanding of the peculiar properties of polymers. Therefore, this problem is always given much attention and a great number of the monograph summed up these studies exist. The principal feature of the offered review is the introduction of the idea of macromolecular coil structure within the framework of the fractal analysis. Such an approach allows to receive the direct correlations "structure-properties", that is the main task of any physical domain including the physical chemistry of polymer solutions.

2. Experimental Methods of Determintion of Fractal Dimension of Macromolecular Coils in Solution

For quantitative characterization of macromolecular coil structure are required estimation methods of its fractal dimension D_f. Most simply the value D_f is determined from its interrelation with known and easy defined properties [1] and number of such methods in present chapter will be considered.

Linear polymer macromolecules are known to occur in various conformational and/or phase states, depending on their molecular weight, the quality of solvent, temperature, concentration, and other are a random coil in an ideal (θ) solvent, an impermeable coil in a good solvent, and a permeable coil. In each of these states, a macromolecular coil in solution is a fractal, i.e., a self-similar object with dimension D_f, which is uncoincided to its topological dimension. The fractal dimension D_f characterizes the volume distribution of its constituent elements in space [3]. Determination of D_f is the first step of studying macromolecular coils within the framework of the fractal analysis, and D_f is usually estimated by finding the exponents in equations of the Mark-Kuhn-Houwink type, which relate the intrinsic viscosity [η], the translational diffusion coefficient D_0, or the rate sedimentation coefficient S_0 with the molecular weight M of polymers [4]:

$$[\eta] \sim M^{b_\eta}, \tag{1}$$

$$D_0 \sim M^{-b_D}, \tag{2}$$

$$S_0 \sim M^{b_S}. \tag{3}$$

The D_f value can be calculated according to the following relationship [4]:

$$D_{f_\eta} = \frac{3}{b_\eta + 1}, \tag{4}$$

$$D_{f_D} = \frac{1}{b_D}, \tag{5}$$

$$D_{f_S} = \frac{1}{1 - b_S}. \tag{6}$$

All considered methods require quite complex and laborious measurements [4-6]. The simplest of these methods, which requires no complicated instrumentation, is measurement of [η]. Therefore, in papers [7, 8] are proposed a simple rapid method of estimating D_f of macromolecular coils in solutions, which is based on the same principles as those applied in deriving the equations (4)-(6).

The coefficient of swelling α of a macromolecular coil is known to be defined as [9]:

$$\alpha = \left(\frac{\langle \bar{r}^2 \rangle}{\langle \bar{r}^2 \rangle_\theta} \right)^{1/2}, \tag{7}$$

where $\langle \bar{r}^2 \rangle^{1/2}$ and $\langle \bar{r}^2 \rangle_\theta^{1/2}$ are the root-meansquare distances between the ends of the macromolecule in an arbitrary solvent and an ideal (θ) solvent, respectively.

In its turn, the value of α is related with the intrinsic viscosities [η] and $[\eta]_\theta$ of the polymer in arbitrary and ideal solvents, respectively, so [9]:

$$\alpha^3 = \frac{[\eta]}{[\eta]_\theta}. \tag{8}$$

The parameter ε of the bulk interactions (which cause a deviation of the coil shape from the ideal, Gaussian shape) is found according to the relationship [9]:

$$\varepsilon = \frac{d \ln \alpha^2}{d \ln M} = \frac{\alpha^2 - 1}{5\alpha^2 - 3}. \tag{9}$$

In its turn, both ε and D_f depend on the exponent b_η in Mark-Kuhn-Houwink equation (1) [6]:

$$\varepsilon = \frac{2b_\eta - 1}{3}, \tag{10}$$

and the dependence of D_f on b_η is given by the equation (4).

Combining the equations (4) and (7)-(10), one can obtain the following relationship, which enables one to determine D_f only from the intrinsic viscosities [η] and $[\eta]_\theta$ [7]:

$$D_f = \frac{5([\eta]/[\eta]_\theta)^{2/3} - 3}{3([\eta]/[\eta]_\theta)^{2/3} - 2}. \tag{11}$$

Table 1 offers a comparison between experimental and calculated data for biopolymer of polysaccharide class (rodexman). The D_f calculation according to the equation(4) the value ~ 1,71 is give. Estimations of D_f according to the equation (11) are listed in Table 1 for eight fractions of rodexman with the molecular weights in the range (27-103)×10^3. Table 1 also presents the [η] values [5] and the $[\eta]_\theta$ values, calculated according to the equation (1) at condition b_η=0,5 [2]. As can be see from the data of this Table, the value of D_f varies within a very narrow range 1,68-1,70, which agrees well with the estimation according to the equation (4). There is a certain increase in D_f (a coil compactness increase) with decreasing M, which is also consistent with the known data [6].

The authors [10] for obtaining of D_f estimation method are used two-component model of solubility parameter, according to which [11]:

$$\vec{\delta}^{\,2} = \delta_f^2 + \delta_c^2, \tag{12}$$

where component δ_f of solubility parameter includes a dispersion interaction energy and interaction energy of dipole bonding and component δ_c – hydrogen bonding interaction energy and the interaction energy between an electron-deficient atom of one molecule (acceptor) and an electron-rich atom of another molecule (donor), which requires a certain orientation of these two molecules. In this treatment is no need to have separate component for description of interactions between polar molecules [11].

Table 1. Comparison between the fractal dimensions D_f calculated by two methods for a macromolecular coil of rodexman in aqueous solution [7]

M×10^3 [5]	[η]×10^3, m^3/kg [5]	$[\eta]_\theta$×10^3 m^3/kg, the equation (1)	D_f, the equation (4)	D_f, the equation (11)
103	135	7,48	-	1,68
93	115	7,11	-	1,69
90	104	6,99	-	1,69
72	98	6,25	1,71	1,69
67	82	6,03	-	1,69
45	62	4,94	-	1,69
32	51	4,17	-	1,69
27	37	3,83	-	1,70

Within the framework of this model the following equation was obtained [10]:

$$D_f = 1{,}50 + 0{,}2(\Delta\delta_f)^{2/(1+\delta_c)}, \tag{13}$$

where $\Delta\delta_f$ is the difference of values δ_f for polymer and solvent.

The equation (13) gives a good correspondence with calculation according to the equation (4) for selection including 7 polymers and 9 solvents (the average dispersion ~ 4 %) [10].

As it is known [12], the value of D_f is determined by two interactions groups: an interactions polymer-solvent and interactions amongs coil elements. Can be supposed, that second group is determined by molecular characteristics of polymer, for example, its characteristic ratio C_∞ (the characteristic of chain statistical flexibility [13]) and cross-sectional area of macromolecule S. To remove of influence of interactions of first group the authors [14] were obtained the equation for D_f calculation in good solvent, i.e., the minimal value D_f [2], which has form:

$$D_f = 2 - 4\left(\frac{2}{C_\infty^2 S}\right)^{1/2}, \tag{14}$$

where S gives in Å.

The equation (14) gives also good correspondence with calculation according to the equation (4) for 9 pairs of polymer-solvent (the average dispersion ~ 4,3 %) [14].

3. The Interrelation of the Classical and Fractal Characteristics of Polymer Solutions

As noted above, the value of the fractal dimension of a macromolecular coil in solution is determined by two groups of factors: interactions polymer-solvent and interactions among coil elements [12]. The dependence of the reduced viscosity η_{red} on the concentration c for the dilute polymeric solutions is usually described by the equation of Huggins [13]:

$$\eta_{red} = [\eta] + k_x[\eta]^2 c + \ldots, \tag{15}$$

where k_x is a Huggins constant, characterizing the level of polymer-solvent interactions.

In Fig. 1 there is the dependence of value D_f, calculated according to the equation (4), on the parameter k_x^{-2} (such a form of the dependence is chosen with the purpose of the obtaining of the linear correlation) for polymers having $M=5\times10^5$. As we can see there is, the good linear correlation for 30 various pairs of polymer-solvent (the correlation 0,930), which allows to predict quite simply the value of D_f. It is expected that for the other M values the correlation $D_f(k_x^{-2})$ will have an analogous form, but another slope. This dependence allows to make some important conclusions.

First, it seems that this correlation describes the dependence of D_f only on the interactions polymer-solvent, characterized by a Huggins parameter k_x, and not take into consideration the interactions among coil elements. However, the linearity itself of this correlation supposes that it includes the factors of the second group, mentioned above. For example, it is well known [16] that the increase of the chain rigidity results to the increase of exponent b_η in Mark-Kuhn-Houwink equation and, consequently, to D_f reduction (the equation (4)). At the same time the increase of the chain rigidity results to $[\eta]$ the growth at the other equal conditions. Thus, both the increase of the chain rigidity and the improvement of the solvent thermodynamic quality give the same effect – [η] the increase and, correspondingly, the k_x growth according to the equation (15). This, in its turn, results to k_x^{-2} decrease and D_f reduction that is expected. This supposition is confirmed experimentally – in paper [17] the increase of Kuhn statistical segment length A, characterizing the chain thermodynamical rigidity, at the improvement of the solvent thermodynamical quality for two polyarylates, is shown.

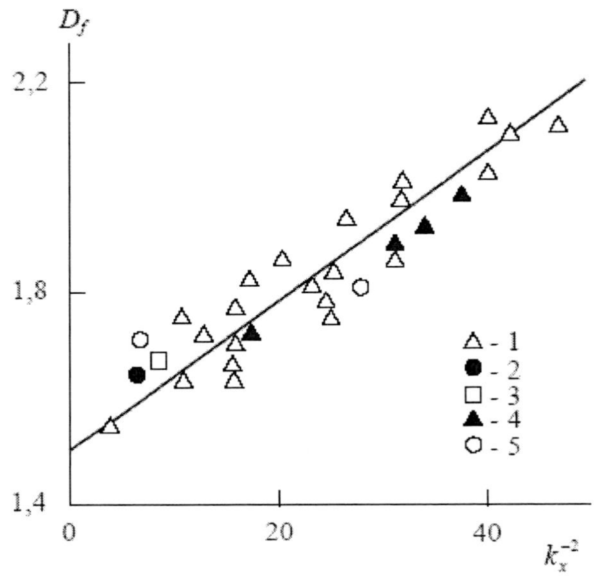

Figure 1. The relationship between a macromolecular coil fractal dimension D_f and parameter k_x^{-2} for polyarylate (1), polymethylmethacrylate (2), polyvynil chloride (3), polysulfone (4) and polycarbonate (5) [15].

Second, as it is shown in paper [15], the minimal value k_x=0,14 (or maximal value k_x^{-2} ≈ 51) reaches at [η]=0. From the graph of Fig. 1 is follows that $D_f \approx 2{,}25$ is corresponds to this value k_x. As it is known [12], the screening of the interactions of the exluded volume results to D_f increase and at the complete screening (the compensation of the pointed effects) the value D_f^c corresponds to so called compensated state. Within the framework of Flory's theory, when the compensation is realized by the interactions with the other coils, D_f^c=2,5 (for the three-dimensional Euclidean space). The other method of the decrease of repulsive interactions among coil elements is the introduction of the attractive interactions. For this case, corresponding to an isolated coil (a dilute solution) can be written [12]:

$$D_f^c = \frac{4(d+1)}{7}, \qquad (16)$$

where d is dimension of an Euclidean space, in which the fractal is considered. It is obvious, in our case d=3 and then D_f^c=2,286, that practically exactly corresponds to the limiting D_f value, obtained according to the graph of Fig. 1.

Another limiting case, corresponding only to the repulsive interactions, corresponds to the condition k_x^{-2}=0. In this case the extrapolation of the graph of Fig. 1 to k_x^{-2}=0 gives D_f=1,50, that corresponds to the dimension of a permeable coil [2]. Thus, obtained from the graph of Fig. 1 the limiting values of D_f completely correspond to the theoretical conclusions.

Third, as noted above, the decrease of exponent b_η at M growth is observed. The authors [15] obtained the following relationship:

$$D_f = \frac{3\ln M}{\ln M + \ln(7{,}14k_x - 1) - \ln K_\eta - \ln k_x}, \tag{17}$$

where K_η is a coefficient of the proportionality in Mark-Kuhn-Houwink equation (1).

The equation (17) gives the analytical interrelation between D_f and M and, in essence, is one more method of D_f estimation [15].

The common aspects of the dependence of D_f on a parameter of the solvent solubility δ_s (or, more exactly, on the difference δ_s and a parameter δ_p of the polymer solubility, $\Delta\delta=\delta_s-\delta_p$) were considered in papers [18-20]. They allow to obtain the following tendencies of the change of the fractal dimension of a macromolecular coil in solution as the function of a number of factors [19]:

1. a poorness of the thermodynamic quality of a solvent in respect of polymer (an increase $|\Delta\delta|$) always results to increase D_f;
2. the bigger the flexibility of a polymer main chain is, the weaker the tendency of D_f growth is at $|\Delta\delta|$ increase;
3. in addition to point 2) we note that the introduction of the rigid para-links instead of more flexible metha-links in the main chain results to the sharp decrease D_f at the other equal conditions;
4. the polymers with the bulk side groups and a flexible main chain will have D_f values, which are bigger on approximately constant value $\Delta D_f \approx 0{,}17$ (representing itself the difference of coil dimensions in a good solvent and a permeable coil) in comparison with the polymers, which doesn't have the noted features;
5. the existence of the hydrogen bonding polymer-solvent results to much faster growth D_f at $|\Delta\delta|$ increase (approximately twice) in comparison with the solvent, which is not capable to the formation of the mentioned bonds.

These common rules allow to choose purposely the solvent for the synthesis of any polymer. As it was shown before [21, 22], the D_f decrease results to the essential acceleration of the reaction at the other equal conditions.

As it is noted above, between the length of Kuhn segment A, characterizing statistical rigidity of chain, and D_f the certain correlation is observed, analytically expressed in such a way [23, 24]:

$$D_f = 2{,}0 - 1{,}32 \times 10^{-2} A, \tag{18}$$

for the flexible chain polymers and

$$D_f = 1{,}667 - 4{,}45 \times 10^{-4} A, \tag{19}$$

for the rigid chain polymers. In the equations (18) and (19) the value A is expressed in Å.

The equations (18) and (19) have the expected from the most general ideas form: the increase A, i.e., the increase of chain rigidity, results to D_f decrease, i.e., the decrease of the compactness degree of a macromolecular coil in solution [23]. As it is known [13], for Gaussian chain in an ideal solvent the condition is carried out:

$$\left\langle \bar{r}^2 \right\rangle_\theta^{1/2} = NA, \tag{20}$$

where N is a number of the statistical segments of a chain.

For the nonideal solvents the condition (20) is broken and the conformation of a macromolecular coil in solution differs from the Gaussian one. This results to the variation of A as a function of the thermodynamical quality of a solvent in respect to the polymer. Let's draw attention to the principal difference of the equations (20), on the one hand, and (18), (19) on the other hand. The equation (20) doesn't allow the condition of the infinitely flexible chain, i.e., A=0, then the equations (18) and (19) give such condition at $D_f > 2{,}0$ (i.e., in the ideal θ-solvent) and $D_f = 1{,}667$, respectively. This supposes that the equations (18) and (19) give the dependence of A on two interactions groups: polymer-solvent and coil elements between themselves, i.e., those groups of the interactions, which define D_f value [12]. It is possible that more exact dependences $D_f(A)$ will be extrapolated for the flexible chain polymers at $D_f = 2{,}0$ to A value for θ-solvent, for which the macromolecular coil has the Gaussian conformation. However, the equation (18) and (19) allow to make the estimation of value A even in terms of the obtained approximations in papers [23, 24].

In Fig. 2 and 3 the comparison of dependences $A(D_f)$, calculated according to the equations (18) and (19) for the flexible and rigid chain polymers, respectively, and cited in papers [25, 26] is shown. As we can see, in both cases the expected tendency of $A(D_f)$ variation is obtained and for the most part of polymer-solvent pairs the good quantitative correspondence is obtained. The substantial dispersion of A values for two used methods of the estimation in the case 5 out of 19 shown in Fig. 2 and 3 polymer-solvent pairs is due to the corresponding dispersion of the literary values A, which can be fourfold [27] for the rigid chain polymers. It is specific that the authors [28] made the correction of A value for three rigid chain polymers in the terms of the fractal analysis that allows to obtain the most probable values of this parameter for the noted polymers.

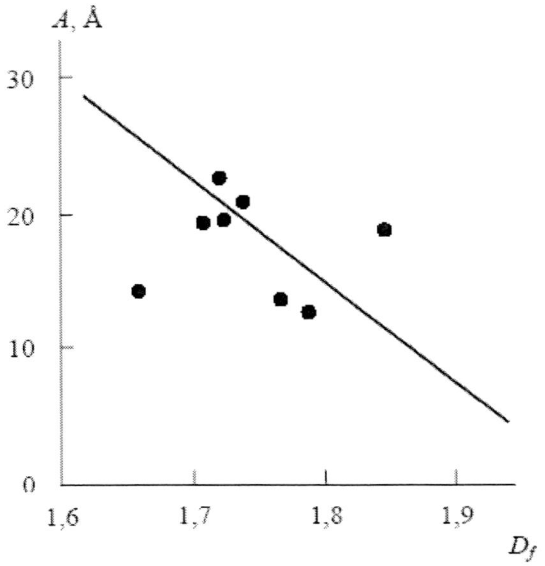

Figure 2. The dependence of Kuhn segment length A on a macromolecular coil fractal dimension D_f for flexible chain polymers.

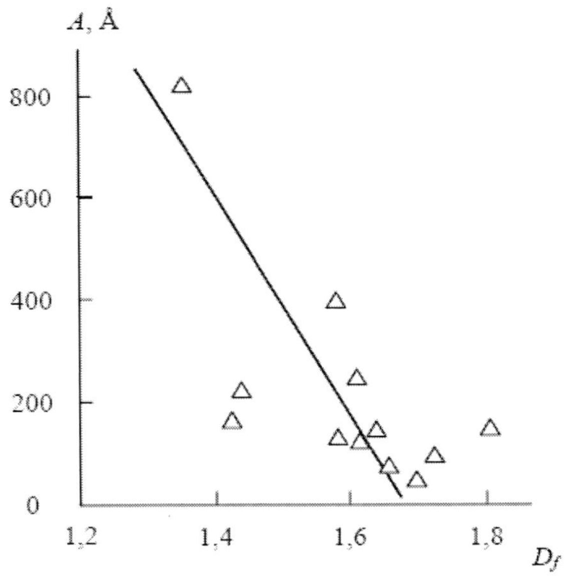

Figure 3. The dependence of Kuhn segment length A on a macromolecular coil fractal dimension D_f for rigid chain polymers. The straight line – the calculation according to the equation (19), points – data of paper [26] [24].

Let's note some more important aspect. The comparison of the equations (18) and (19) allow to give one more definition of the flexible and rigid chain polymers. The mentioned equations suppose that the maximal dimensions D_f of a macromolecular coil of the linear polymers are equal to 2,0 (i.e., coil in ideal θ-solvent [2]) for the flexible polymers and 1,667 (i.e., coil in good solvent [2]) – for the rigid polymers. Therefore, to the rigid chain polymers it is necessary to refer such polymers, the rigidity of chain of which (intramolecular interactions) allows to compensate it by the interactions polymer-solvent (intermolecular interactions) only to the certain degree, i.e., to $D_f = 1,667$. The other criterion follows from the equation (18): at minimal D_f value 1,0 we obtain the value $A \approx 75$Å. Differently speaking, the polymers with $A \leq 75$ Å are the flexible chain (for them it can be reached $D_f = 2,0$ by the screening of the intramolecular interactions), and with $A > 75$ Å are the rigid chain polymers.

As the value of D_f of a macromolecular coil in solution is determined by two noted above groups of the interactions, then such definition allows to the authors [29] to connect D_f and the parameter χ_1 of Flory-Huggins interaction, which is defined as follows [13]:

$$\chi_1 = \frac{E_{vap}}{RT}\left(1 - \frac{\delta_p}{\delta_s}\right)^2 + \chi_s, \tag{21}$$

where E_{vap} is molar energy of solvent vaporization, R is a universal gas constant, T is temperature, χ_s is an empirical parameter.

Sometimes the value χ_1 is defined so [13]:

$$\chi_1 = \chi_H + \chi_s, \tag{22}$$

where χ_H and χ_s are enthalpic and entropic components of Flory-Huggins parameter, respectively.

The equation (21) assumes that with the help of parameter δ_p it is possible to determine the interactions of coil elements among themselves, and with the help of the ratio δ_p/δ_s – the interactions polymer-solvent. Besides, the entropic component χ_s takes into account the effects of ordering of the interacting elements of system.

The set of the experimental data for the various polymers has allowed to receive an empirical correlation $D_f(\chi_1)$ of the following form [29]:

$$D_f = 1,50 + 0,45\chi_1. \tag{23}$$

The theoretical estimations suppose that the rise of temperature makes the fractal (macromolecular coil) less compact, i.e., that leads to the D_f decrease. However, the experimental confirmation of this supposition is absent in the present time. The authors [30] considered the D_f change with T variation on the example of the solutions of polyarylate on the basis of phenolphtalein and isophthalic acid dichloroanhydride (F-1) [16].

As it is known [31], the reason of changes of the macromolecular coil structure in the diluted solution with varying the temperature can be the effects of long-range interaction (effects of exluded volume) as well as the effects of close-range interaction leading to the specific influence of the solvent on the undisturbed dimensions of a coil [16]. For the semirigid chain polymers to which itis necessary to refer polyarilates the macromolecules in the solution are characterized by the significant transparency, and therefore the effects of the long-range interaction can be neglected. For these polymers it is assumed [32], that the temperature coefficient of the viscosity for them $d \ln [\eta]/dT$ is negative according to the sign, and according to the absolute value it excesses the one for the flexible chain polymers of the same M. The temperature dependences of $[\eta]$ for the solution of F-1 in three solvents are given in Fig. 4. These dependences have the expected character, i.e., the decrease of $[\eta]$ with the increase of T is observed. As it has been already noted, it supposes that the macromolecular coil structure is mainly controlled by the close-range interaction effects, that is the interaction polymer-solvent. The value $d \ln [\eta]/dT$ is equal about to – 0,0053, what agrees with the corresponding parameter for the more rigid chain polyamides [31].

The dependences η_{red}/c (where η_{red} is reduced viscosity, c is polymer's concentration) on c for the solutions F-1 correspond to the Huggins's equation (15), that allows to obtain the values k_x=0,397, 0,314 and 0,294 for T=293, 323 and 353 K, respectively. It means that the interaction polymer-solvent decreases with the temperature increase and the graph of Fig. 4 allows to estimate D_f from ~ 1,60 to 1,68 due by this increase.

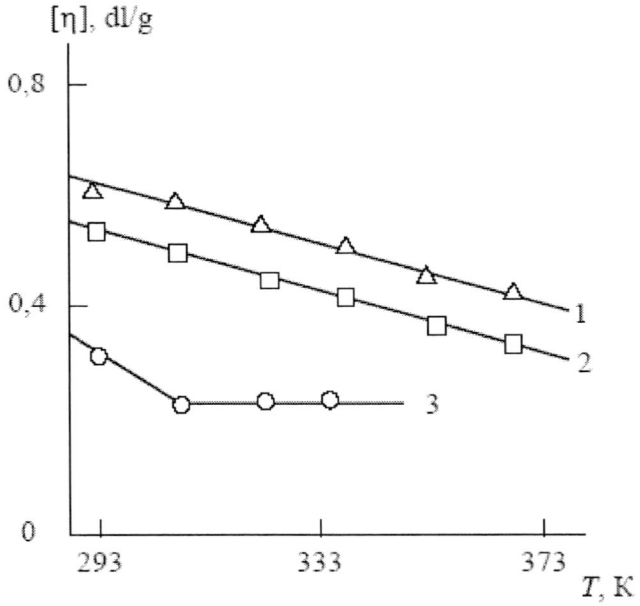

Figure 4. The dependence of the intrinsic viscosity $[\eta]$ on temperature T for the solutions of polyarylate F-1 in and tetrahydrofurane (3) [30].

Fig. 5 shows the dependences $D_f(T)$ for solutions F-1 in tetrachloroethane and N,N-dimethylformamide (c=0,5 weight %), where the value D_f is determined according to the equation (11). As we can see, the monotonous increase of D_f with T rise is observed, that is the increase of the compactness of a macromolecular coil [12]. The extrapolation of the graphs in Fig. 5 to D_f=2,0 [2] allows to estimate the values of θ-temperature for the indicated solvents. It should be noted, that the achievement of θ-conditions doesn't mean any critical state of the coil. Such state for the coil in the diluted solutions (practically the isolated macromolecule) can be achieved at the critical value of the fractal dimension D_f^c, estimated according to the equation (16). We should note, that about at $D_f = D_f^c = 2{,}285$ the dependence $[\eta](T)$ for F-1 in tetrahydrofurane (Fig. 4) becomes parallel to abscissa axis, i.e., this D_f value is critical [30].

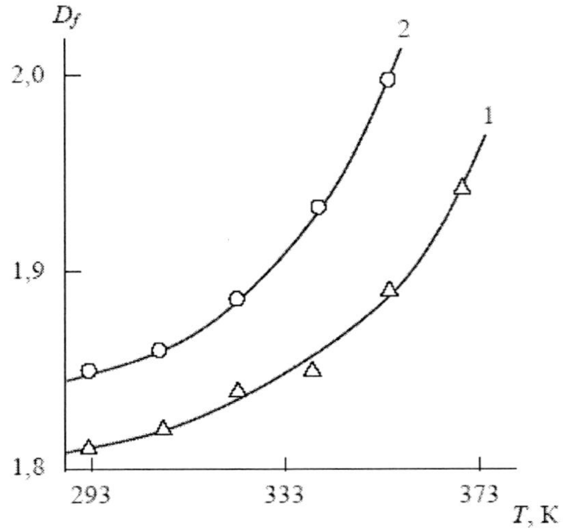

Figure 5. The dependence of the fractal dimension D_f of a macromolecular coil on temperature T for the solutions of polyarylate F-1 in symm-tetrachloroethane (1) and N,N-dimethylformamide (2) [30].

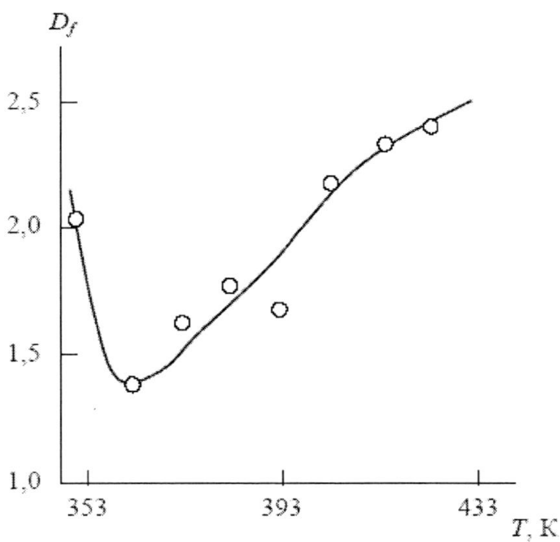

Figure 6. The dependence of the fractal dimension D_f of a macromolecular coil on temperature T for the solutions of triacetate cellulose in benzyle spirite according to the data of paper [33] [30].

Generally the dependence $D_f(T)$ is expected to be rather complicated. So, for the flexible chain polymers one can observe at first the increase of $[\eta]$ with the T rise and then its decrease [13]. This assumes at first the decrease D_f with the T rise for these polymers and then its increase. In Fig. 6 the dependence $D_f(T)$ for the solution of the triacetate cellulose in benzyle spirite [33] is shown, where the D_f value is calculated according to the equation (4). As one can see at first the D_f decrease is observed and in the interval T=363-423 K the D_f increase takes place. Such change of the dependence $D_f(T)$ for the triacetate cellulose can be explained by the complex interaction polymer-solvent: the D_f decrease at $T \leq 363$ K is explained by the desolvation process due to "squeezing" of the solvent out of the macromolecular coil at the pairwise interactions increase, and the D_f increase at $T > 363$ K is connected with the destruction of the solvent structure [33]. For the diacetate cellulose in the various solvents the decrease of $[\eta]$ with the T rise is observed [34], that supposes the corresponding D_f increase according to the equation (11).

Thus, the cited above results suppose, that the dependence $D_f(T)$ for the diluted solutions has rather complex character and for the semirigid and rigid chain polymers the D_f increase with T rise is observed in general [30].

A branching degree (and/or the existence of bulk side substituents) is the important property of a polymer chain, which defined in many things the polymer behaviour in solutions. It is shown [16, 35], that the increase of the molecular weight side substituents and of a branching degree results to the systematic decrease of exponent b_η in Mark-Kuhn-Houwink equation, that at the other equal conditions it means the decrease of a coil gyration radius, i.e., its rise of compactness. For the quantitative estimation of polymer branching degree the branching factor g is usually used, defined as follows [16]:

$$g^{2-b_\eta} = \frac{[\eta]_b}{[\eta]_l}, \tag{24}$$

where subseripts "b" and "l" mean the branched and linear macromolecules, respectively.

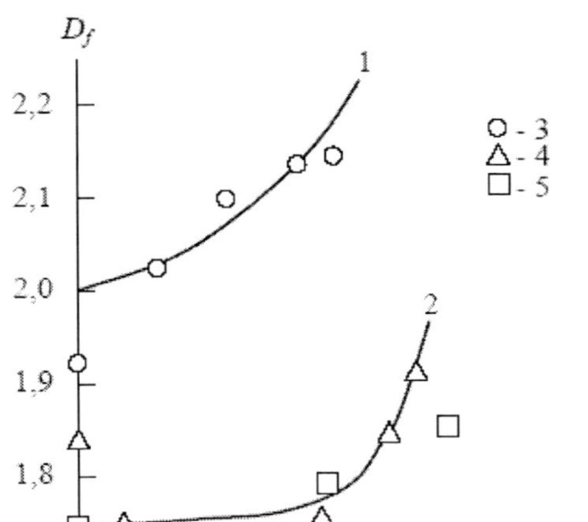

Figure 7. The dependences of the fractal dimension D_f of a macromolecular coil on branching factor g for PPQX (1-4) and B-PES (5). The solvents: N-methylpyrrolydone (1, 3) and chloroform (2, 4, 5). 1, 2 – the calculation according to the equation (25), 3-5 – experimental data [38].

In Fig. 7 there are the dependences D_f on g (points) for polyphenylquinoxaline (PPQX, solvents-chloroform and N-methylpyrrolydone) [35] and bromine-containing aromatic polyethersulfones (B-PES, solvent-chloroform) [36]. As one can see in any case the increase of branching degree, characterized the g decrease [16, 35], results to D_f rise, i.e., the increase of a coil compactness degree [37, 38]. Another factor influencing on the D_f value is the solvent thermodynamical quality in respect to polymer. As follows from the data of Fig. 7, a poor solvent for PPQX (N-methylpyrrolydone) results to much greater values D_f in comparison with a good solvent (chloroform) for both considered polymers. We should note, that the data for the linear analogues of these polymers (at g=1) do not lie on the dependences for the branched polymers, therewith the law for the relationship of the values D_f for the branched and linear analogues is absent [37].

The graphs of Fig. 7 allow to make two important conclusions. First, as it should be expected [12], the value D_f of the branched polymers is controlled by two factors: polymer-solvent interactions and the interactions among the coil elements. Second, the dominant parameter in the determination of the second factor is a branching degree g, as it is not observed any correspondence with the linear analogues.

In order to obtain the analytical dependence D_f on g the equations (4) and (24) can be used. Their combination allows to obtain the following relationship [38]:

$$D_f = \frac{3\ln g}{3\ln g - \left(\ln[\eta]_b - \ln[\eta]_l\right)}. \tag{25}$$

In Fig. 7 the comparison between the dependences $D_f(g)$, obtained according to the equations (4) and (24) (points) and the equation (25) (full curves) is shown. From this comparison the good correspondence of dependences $D_f(g)$, obtained by both noted methods [37, 38] follows.

As it is known [13], widely used now the traditional methods of the determination of macromolecular coil structural features of the branched polymers are connected with the necessity of the comparison of their hydrodynamic properties with the properties of the linear analogue of the same molecular weight. Theories used for this purpose are complicated, often nonunique and sometimes they do not allow the quantitative estimations of the branched polymers structure. Therefore the authors [39] used for these purposes the methods of a fractal analysis.

As it is known [40], the phantom (which does not account for the interactions of the exluded volume) fractal dimension of a macromolecular coil d_f is given by

$$d_f = \frac{2d_s}{2-d_s}, \tag{26}$$

and the swollen (which accounts for the exluded volume effects) fractal dimension of the same coil D_f is determined as follows [40]:

$$D_f = \frac{d_s(d+2)}{d_s+2}. \tag{27}$$

In the equations (26) and (27) d_s there is a spectral (fracton) dimension characterizing the connectivity of the object [41].

The equation (27) gives the value D_f in solvents consisting of the point (zero-dimensional) molecules, which have the fractal dimension δ_f=0 [40]. Generally, the value δ_f can be estimated according to the equation [42]:

$$\delta_f = 1{,}58\left(\delta_s^{1/2} - 2{,}83\right), \tag{28}$$

where δ_s is in $(cal/cm^3)^{1/2}$. The discussion of the equation (28) will be giveh in chapter 4.

From the equation (28) it follows, that the condition δ_f=0 is achieved for $\delta_s \leq 8{,}0$ $(cal/cm^3)^{1/2}$. As the values δ_f for the used solvents are higher than this value, for them $\delta_f > 0$. According to the equation (28) it can be estimated that the value δ_f is equal to ~ 0,19 for chloroform and ~ 0,91 for N-methylpyrrolydone. In this case the value D_f is determined according to the equation [40]:

$$D_f = \frac{d+2}{2-\left(\delta_f - 2\right)/d_s}. \tag{29}$$

In Fig. 8 the dependences D_f and d_f on d_s for all considered above cases are shown. The dependences $D_f(d_s)$ for the real polymers should be settled down between the curves $d_f(d_s)$ and $D_f(d_s)$, where the value D_f is determined according to the equation (27) (curves 1 and 2, respectively). For the macromolecular coils PPQX in N-methylpyrrolydone this condition is really carried out and what's more the experimental values D_f (the equation (4)) are excellently agreed to the calculation D_f according to the equation (29), where δ_f=0,91, and the value d_s is calculated according to the equation (27). At the same time the data for the macromolecular coils PPQX in chloroform estimated according to the equation (4), are lower than the dependence $D_f(d_s)$ for the swollen coils, though the equation (27) assumes the minimal value D_f for them (δ_f=0). The dependence $D_f(d_s)$ for the swollen coils (curve 2 in Fig. 8) is not agreed to the experimental results. For the explanation of this disagreement, let's consider the definition of the equation (29), cited in [40]. It is assumed that the fractal molecules of the solvent screen the interactions of the exluded volume between the macromolecule elements. Therefore, the increase δ_f leads to D_f rise. Then the value α, which is equal to [40]:

$$\alpha = \frac{\delta_f}{d_f}, \tag{30}$$

is incorporated. In the equation (30) δ_f is a phantom fractal dimension of the solvent molecules.

The final result of this treatment is the equation (29) or, if the spectral dimension d_s is used, the following equation is obtained [40]:

$$D_f = \frac{d_s(d+2)}{(1-\alpha)d_s + 2}. \tag{31}$$

It is easy to see that in this treatment the possibility of screening of the interactions of the attraction between the macromolecule elements is not allowed. This effect should give the decrease of a coil compactness and the reduction of D_f value that is observed for PPQX coils in chloroform (Fig 8). The simplest method of the incorporation of this effect is the replacement of the sign "plus" before δ_f in the equation (29) by "minus" or the replacement of the sign "minus" before α in the equation (31) by "plus". The calculation D_f according to the modified by the mentioned method equations (29) and (31) gave the excellent correspondence with the experiment (curve 5, Fig. 8). This allows to assume that the chloroform molecules screen the interactions of the attraction between the branches of PPQX macromolecule, but they do not screen them for the main chain of polymer (the experimental value D_f for linear PPQX corresponds well to the calculation according to the equation (29) or the equation (31) at d_s=1,0 [41], Fig. 8) [39].

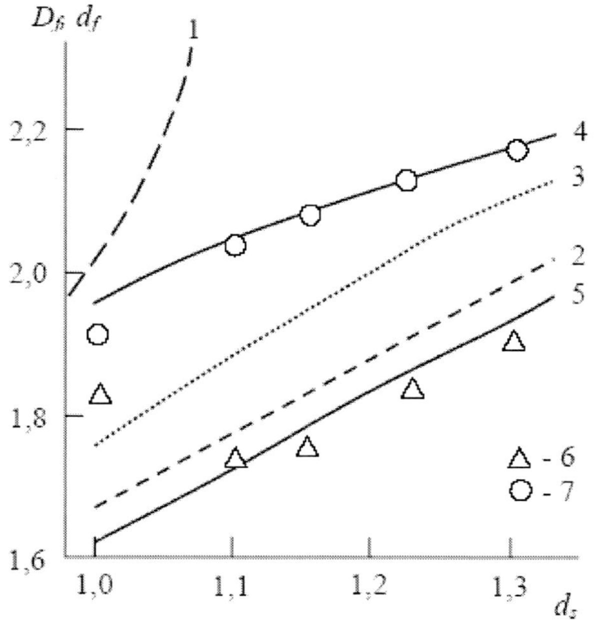

Figure 8. The dependences of swollen D_f and phantom d_f fractal dimensions of a macromolecular coil PPQX on spectral dimension d_s. The calculation according to: the equations (26) (1) and (27) (2), the equation (29) for chloroform (3) and N-methylpyrrolydone (4); the modified equation (29) for chloroform (5); the equation (4) for chloroform (6) and N-methylpyrrolydone (7) [39].

The offered above data suppose, the substantial role of the dimension d_s for the characteristic of the macromolecular coil structure of branched polymers (the equations (26), (27), (29), (31)). Therefore, the authors [43] considered the dependence d_s on a branching degree of PPQX chain more strictly. For the estimation of a branching degree the number of the nodes of branching m were used [13]. The value m can be determined by two methods. First of them is given in paper [35]:

$$m = \frac{N\gamma}{(2+\gamma)}, \tag{32}$$

where N is a polymerization degree, γ is a fraction of a functional branching component.

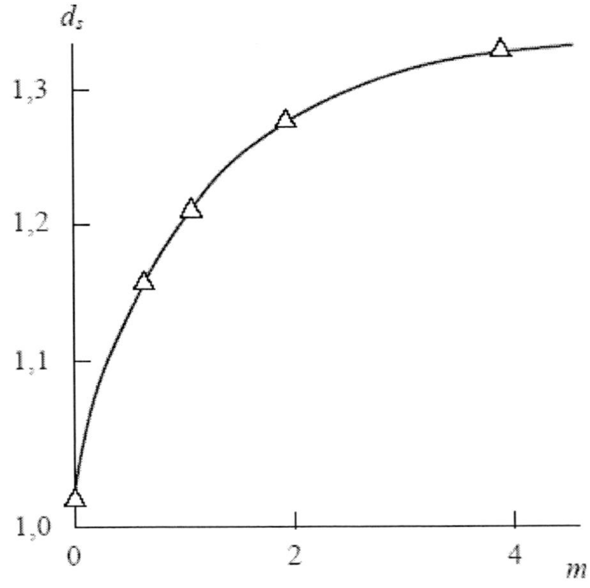

Figure 9. The dependence of the spectral dimension d_s, calculated according to the equation (31), on number branching nodes m for PPQX [43].

The second method uses the formula for the trifunctional branching nodes [13, 35]:

$$g = \left[\left(1+\frac{m}{7}\right)^{0,5} + \frac{4m}{9\pi}\right]^{-0,5}. \tag{33}$$

In Fig. 9 the dependence $d_s(m)$ for PPQX is shown. As it should be expected from the most common considerations, the value d_s monotonously rises at m increase, approaching to the asymptotic magnitude $d_s \approx 1,33$ [41] at $m \approx 4$. This dependence is approximately quadric and it can be linearized d_s with written as a function of $m^{1/2}$. Then the analytical form of the dependence $d_s(m)$ will have the form [43]:

$$d_s = 1 + 0{,}17m^{0,5}, \quad m \leq 4,$$
$$d_s = 1{,}33, \quad m > 4. \tag{34}$$

The equation (34) assumes that the degree of a chain branching reaches the saturation and the farther increase of γ (or m) does not change the macromolecular coil structure, i.e., the dimensions d_s and D_f [43].

Thus, the considered in chapters 2 and 3 results showed the interrelation of the macromolecular coil structure in solution, characterized by the dimension D_f, with practically all basic parameters of the classical physical chemistry of polymer solutions. This fact can be treated as the defining influence of the coil structure on these parameters. Besides, these correlations give a number of the simple and convenient methods for D_f estimation.

4. A Fractal Variant of the Mark-Kuhn- Houwink Equation

The Mark-Kuhn-Houwink equation derived from the analysis of a great body of experimental data is widely used for determining M_η from the measured intrinsic viscosity $[\eta]$ of polymer solutions (the equation (1)) [13].

Analysis of experimental data revealed [44] that the K_η and b_η constants involved in equation (1) correlate with one another for many flexible chain polymers dissolved in good solvents. Thus, it was found that

$$K_\eta \approx \frac{21}{m_0}\left(\frac{4\times10^{-4}M}{m_0}\right)^{b_\eta}, \tag{35}$$

where m_0 is the molecular mass of a monomer unit.

In some cases, the use of empirical relationships such as equation (35) enables molecular weight evaluation to be performed without detailed study of the samples. A more exact relationship for K_η was proposed by Budtov [13]:

$$K_\eta = \frac{0{,}46}{m_0}\cdot\frac{A^2 l\times10^{24}}{(7{,}3sm_0)^{b_\eta}} \tag{36}$$

where l is the length of a monomer unit, and s is the number of monomer units per segment.

For polymers with the same chain rigidity, the relationship (36) is very close to empirical formula [35]. Relationships (1), (35) and (36) were noted [13] to be very useful in the analysis of solitary polymer samples when they have not been characterized in a relevant solvent but their structural parameters are known.

For the derivation of the fractal variant of the Mark-Kuhn-Houwink equation authors [45, 46] started from the following well-known formula [13]:

$$[\eta]=6^{1/2}\Phi(\alpha)\frac{\langle R_g^3\rangle}{M}, \tag{5}$$

where $\Phi(\alpha)$ is the form factor of a macromolecular coil under rotational friction and R_g is coil gyration radius.

For the sake of convenience, α is adopted to be equal to the viscous swelling ratio defined according to the equation (8). The simplest relationship between Φ and α is [13]:

$$\Phi(\alpha)=\Phi_\theta\left(0{,}753+\frac{0{,}247}{\alpha^3}\right), \tag{38}$$

where Φ_θ is Φ for a equal to unity.

The relationship between R_g and M can be expressed as an following approximate [45]:

$$R_g\approx 37{,}5N^{1/D_f}=37{,}5\left(\frac{M}{m_0}\right)^{1/D_f}, \tag{39}$$

where R_g is give in Å.

The combination of the equations (8), (37)-(39) allows to obtain a relationship between $[\eta]$ and M, which is similar to the Mark-Kuhn-Houwink equation [45, 46]:

$$[\eta]=c(\alpha)\frac{M^{(3-D_f)/D_f}}{m_0^{3/D_f}}, \tag{40}$$

where c is some constant depending on α, because it contains the parameter $\Phi(\alpha)$. This circumstance makes it possible to evaluate the coefficient $c(\alpha)$ at α variation in accordance with the equation (37). The value α itself can be evaluated by the equations (1), (8) and (35), if $[\eta]_\theta$ is determined at b_η=0.5 (or D_f=2,0) [2]. We should to note one important aspect. As it is known [47], the ratio ($[\eta]/[\eta]_\theta$) is a single-valued function of D_f, as follows from the equation (11).

Thus, the equation (40) suggests that the proportionality coefficient K_η depends only on two parameters: D_f and m_0. The final equation for calculating the K_η value takes on the form [45, 46]:

$$K_\eta=\frac{8{,}1\left(0{,}753+0{,}247/\alpha^3\right)}{m_0^{3/D_f}}. \tag{41}$$

Fig. 10 compares theoretical K_η values K_η^T calculated according to the equation (41) with corresponding experimental values K_η^e [16] for polyarylates. This plot demonstrates good agreement between K_η^T and K_η^e. Such agreement between K_η^T and K_η^e is obtained in paper [48] for series of cationic polyelectrolytes-copolymers of acrylamide with trimethyl ammonium methylmethacryate.

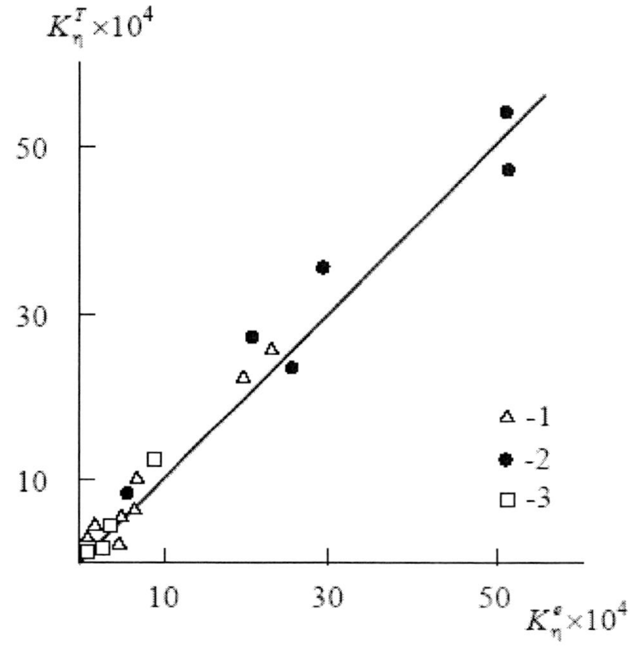

Figure 10. The correlation between experimental (K_η^e) Mark-Kuhn-Houwink constants and those calculated according to the equation (41) for the solutions of polyarylates [16] in symm-tetrachloroethane (1), tetrahydrofurane (2) and 1,4-dioxane (3) [46].

Another method of exponent b_η evaluation is proposed by the authors [49]. It is based on the usage of solubility parameters δ_s and δ_p. From the equation (29) follows, that the value D_f = 1,67 for good solvent is achieved at $\delta_f = 0$ and $D_f = 2,0$ for poor (θ) solvent – at $\delta_f = 1$. At the same time great number of experimental observations are shown the general tendency – the closer the values δ_p and δ_s the better a polymer is dissolved in this solvent [50]. This allows to suppose following form of correlation δ_f, δ_p and δ_s [49]:

$$\delta_f \sim \left|\delta_p - \delta_s\right|. \tag{42}$$

The proportionality coefficient in (42) can be obtained from following simple considerations. As it is known [50], a polymer is not dissolved in a solvent, if $|\delta_p - \delta_s| \geq 2{,}5$ $(cal/cm^3)^{1/2}$. This criterion for poor solvent with $\delta_f = 1$, that allows to adopted mentioned coefficient equal to ~ 0,4 and calculate D_f according to the equation (29) and b_η - according to the equation (4). The comparison of theoretical and experimental b_η values for 18 pairs of polymer-solvent their good agreement is shown [49].

5. Physical Sense of the Structural Parameters of Low-Molecular Solvents

The representation δ_f was used above as the dimension of a molecule of low-molecular solvent [40]. However, another treatment [51], which will be considered below, is possible. At the derivation of the equation (29) the supposition, that the cluster sizes of polymer and solvent are comparable, was made [40]. As in the case of the low-molecular solvents, the macromolecular coil size is much bigger then the solvent molecule size, we should make the conclusion that it is necessary to consider the structure of the solvent molecules set interacting with a macromolecular coil ("swarm" of the solvent molecules), but not a separate molecule. In favour of such a conclusion there exists the fact that the value δ_f for one and the same solvent, but for the various polymers adopts the different magnitudes [10]. Therefore, it should be supposed, that the dimension δ_f characterizes the polymer-solvent interactions. In Fig. 11 the dependence δ_f on Flory-Huggins interaction parameter χ_1 for 4 polymers in the various solvents is shown, which confirms this assumption. The indicated correlation is found approximately linear and it approximates by the following empirical equation [51]:

$$\delta_f \approx 1{,}15\chi_1 - 0{,}3. \tag{43}$$

The most interest feature of the equation (43) is the possibility of the obtaining of δ_f negative values. As it is known [3, 52], the classical fractal analysis assumes only positive fractal dimensions, describing that or another structure. Therefore, the natural question about the physical sense of dimension δ_f appears. The certain similarity can be given by the equation, describing the number N_{is} of the intersections of two fractals with the arbitrary dimensions D_1 and D_2 [40]:

$$N_{is} \sim R^{D_1+D_2-d}, \tag{44}$$

where R is a radius of the biggest fractal.

At $D_1+D_2>d$ the fractals are opaque for each other, at $D_1+D_2<d$ are permeable for each other. It should be noted, that the relationship (44) gives the condition of the permeability (transparency) of macromolecular coils with $D_1=D_2$ at d=3 [51]:

$$2D = 3 \ or \ D_1 = D_2 = 1{,}50. \tag{45}$$

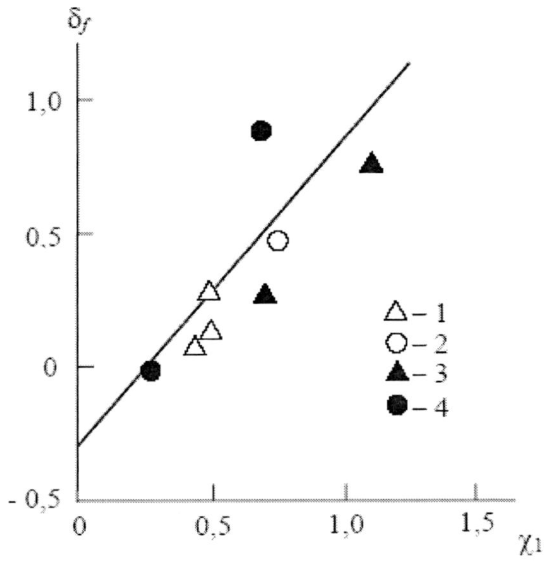

Figure 11. The relationship between the fractal dimension δ_f of "swarm" of solvent molecules and Flory-Huggins interaction parameter χ_1 for polymethylmethacrylate (1), polystyrene (2), polyvynilacetate (3) and polysulfone (4) [51].

For the purposes of the present is discussion is important, that in case of the transparent coils an exponent in the right part of the relationship (44) becomes negative. This allows to suppose, that the dimension δ_f characterizes structure (distribution in space) of the low-molecular solvent molecules and that each contact of the coil and the solvent molecules should be considered as an intersection point of two fractals: coil and "swarm". It is quite obvious, that the solvent molecules must occupy the same space region, that the macromolecular coil with a gyration radius R_g and that's why they cannot form a dense Euclidean object, i.e., $\delta_f<d$. As follows from the equation (29), the maximal value δ_f, at which a compact globule is formed ($D_f=d$) [12], is equal to ~ 2,667. Consequently, the dimension δ_f characterizes the number of the polymer-solvent interactions or the number of the low-molecular solvent molecules n_s in "swarm" of radius R_g according to the general fractal relationship [3]:

$$n_s \sim R_g^{\delta_f}. \tag{46}$$

It is easy to see, that the macromolecular coil will be opaque for the solvent molecules (accounting for the equation (29)) at condition $D_f \geq 2$, $\delta_f \geq 1$. Accounting for the condition of the approximate equality of "swarm" radius and R_g and the known fractal relationship $R_g \sim N^{1/D_f}$ [2], we can obtain the following relationship for n_s [51]:

$$n_s \sim N^{\delta_f / D_f}. \tag{47}$$

From the relationship (47) it follows, that the δ_f decrease, resulting to weaker D_f decrease according to the equation (29), means the reduction n_s at N=const, i.e., the indicated relationship describes the well known effect of the solvent "squeezing" from the macromolecular coil [53].

Using the equations (29), (43) and the literary values χ_1 [13], the limiting magnitudes δ_f can be evaluated. From the equation (29) it follows, that for a coil in a good solvent ($D_f \approx$ 1,667 [2]) the value δ_f=0. It means that the solvent with such a dimension fully screens the interactions of the attraction of macromolecular coil elements. For the permeable coil (D_f=1,50 [2]) realization the value δ_f= - 0,645 is already required, i.e., in this case the solvent must not only screen the noted above interactions, but introduce the repulsive interactions of polymer-solvent. Thus, the value δ_f<0 means the existence of the repulsive interactions of polymer-solvent, δ_f>0 – the attraction interactions and δ_f=0 – the absence of any interactions of polymer-solvent [51]. The cited in [13] values χ_1 for 74 pairs of polymer-solvent vary within limits - 0,30-1,55. According to the equation (43) it means δ_f the variation within limits -0,645-1,483, that in its turn, according to the equation (29) means D_f variation within limits 1,50-2,21. Such a variation includes the full interval of the possible D_f change of real macromolecular coils: from the permeable (D_f=1,50 [20]) to the so-called compensated coil state (D_f=2,28 [12]).

As it is assumed, that δ_f is a measure of polymer-solvent interactions, then it is natural to expect the certain interrelation δ_f with the parameters of the solubility of polymer δ_p and solvent δ_s, characterizing the cohesion energy of these substances [11, 13]. This interrelation is given by the equations (28) and (42) and also by the following relationship [51]:

$$\delta_f = \left(\delta_s^{1/2} - 27/\delta_p\right), \tag{48}$$

where δ_s and δ_p are expressed in $(\text{cal/cm}^3)^{1/2}$.

The equation (48) differs from the classical view on polymer dissolution problem. It is assumed [50] now, that a good solvent is realized at the condition δ_s=δ_p, that corresponds to D_f=1,667 or, according to the equation (29), to δ_f=0. However, according to the equation (48), this criterion will have the other form [51]:

$$\delta_s^{1/2} = \frac{27}{\delta_p}. \tag{49}$$

In paper [11] the great number of polymers and solvents for them are cited and the latter were conditionally divided in 4 groups: good, intermediate, poor (where the polymer is partially dissolved or swollen) and non-solvent. In Table 2 the comparison of the evaluation of a solvent type according to the criteria δ_s=δ_p and (49) in case of polycarbonate, for which the quantitative measures of deviation $|\delta_p-\delta_s|$ and $|\delta_s^{1/2}-27/\delta_p|$ are used, respectively. As one can see from the data of Table 2, the solvent quality is much better described by the criterion (49), then the solvent-non-solvent transition is fully defined by the value $|\delta_p-\delta_s|$.

We should note, that the variation of constant 27 in the equation (48) for the various polymers as a function of their characteristics is possible [51].

It was shown before [16], that for the series of polyarylates (PAr), produced by the high-temperature (equilibrium) and interfacial polycondensation, the different values of exponent b_η in the Mark-Kuhn-Houwink equation for PAr of the same chemical constitution, but produced by the different methods (for the second method the value b_η is higher) are observed. It means the different structure of a macromolecular coil for PAr, produced by the different methods of polycondensation according to the equation (4).

The evaluations according to the equation (31) showed, that in case of the interfacial polycondensation δ_f=0 and in case of the high-temperature polycondensation - $\delta_f > 0$ [54]. This means that the solvent does not take part in the macromolecular coil formation in the first case, but it defines its structure in the second case. Such treatment is assumed by the difference of mechanisms of the indicated methods polycondensation: if the high-temperature polycondensation is carried out in solution, then the interfacial polycondensation – on the phase boundary. Therefore, received in the presence of solvent with $\delta_f > 0$ the macromolecular coil structure is more compact than without its influence (δ_f=0). This effect of the structural memory is preserved at the dissolution of PAr in the same solvent, and it defines the different values of exponent b_η for PAr of the same chemical constitution, but produced by the indicated above methods of polycondensation [54].

Table 2. Characteristic of the solvent quality according to two criteria [51]

Solvent quality	Criterion	
	$\lvert\delta_p-\delta_s\rvert$, (cal/cm^3)$^{1/2}$	$\lvert\delta_s^{1/2}-27/\delta_p\rvert$
Good	0,46-1,82	0,02-0,22
Intermediate	0-3,30	0,28-0,41
Poor	0,10-0,78	0,11-0,79
Non-solvent	1,43-2,35	0,05-0,11

6. Application of the Theory of the Fractional Derivatives for Description of Polymer Solutions

The fractality of macromolecular coil structure allows to use for the description of its characteristics the mathematical theory of the fractional derivation [55-57]. Within the framework of this formalism the possibility of the precise description of such non-linear phenomena as, for example, the spatial correlations, is presented [57]. The authors of papers the [58, 59] used this approach for the calculation of the average end-to-end distance for a chain of polycarbonate (PC) in two different solvents.

There is a number of methods for the calculation of the mean-square end-to-end distance of a polymeric chain $\langle \bar{r}^2 \rangle^{1/2}$, which is an important parameter in the theory of polymeric solutions [13]. So, the following empirical relationships for PC in a solution in methylene chloride (MC) and tetrahydrofuran (THF) were obtained [60]:

$$\left\langle \bar{r}^2 \right\rangle^{1/2} = 0{,}66 M_\eta^{0{,}58}, \text{ for MC,} \tag{50}$$

and

$$\left\langle \bar{r}^2 \right\rangle^{1/2} = 1{,}04 M_\eta^{0{,}53}, \text{ for THF.} \tag{51}$$

Another method is the usage of the following equation [13]:

$$[\eta] = \Phi(\alpha) \frac{\left\langle \bar{r}^2 \right\rangle^{1/2}}{M_\eta}. \tag{52}$$

Besides, the equation for the estimation of value $\left\langle \bar{r}^2 \right\rangle^{1/2}$ can be received within the framework of the fractional derivation as follows [61]. Let $x=x(t)$ be the law of the change of some physical property due time t. The rate of change $x(t)$ has the form:

$$\frac{dx}{dt} = \vartheta(t)\tau_0, \tag{53}$$

where t is dimensionless time, τ_0 is characteristic time of the given process.

The equation (53) can be presented as:

$$D_{O_t}^{\nu} D_{O_t}^{1-\nu} x(t) = \vartheta(t)\tau_0, \tag{54}$$

where

$$D_{0_t}^{\nu} f(t) = \frac{1}{\Gamma(1-\nu)} \frac{d}{dt} \int_0^t \frac{f(\tau)d\tau}{(t-\tau)^{\nu}}, \tag{55}$$

is a fractional derivative of Rimman – Liouville of the order ν; $0 < \nu < 1$.

If we enter the notation $D_{O_t}^{1-\nu} x(t) = r(t)$, the equation (54) becomes:

$$D_{O_t}^{\nu} r(t) = \vartheta(t)\tau_0. \tag{56}$$

The solution of the equation (56) has the form:

$$r(t)=\frac{\tau_0}{\Gamma(\nu)}\int_0^t\frac{\vartheta(\tau)d\tau}{(t-\tau)^\nu}. \quad (57)$$

The integral in the right part (57) is easily calculated at $\vartheta(t)=\vartheta$ =const and, returning to the dimensional t, we receive [61]:

$$r(t)=\frac{\vartheta\tau_0^\nu}{(1-\nu)\Gamma(\nu)}, \quad (58)$$

where $\Gamma(\nu)$ is Euler's gamma function.

Comparing the expression (58) at ν=1/2 with the known Einstein's formula for the mean-square displacement of a particle making Brownian motion, the authors [61] have drawn the conclusion, that the value $r(t)$ can be considered as $r = \langle x^2\rangle^{1/2}$.

As it is known [13], the polymeric macromolecule can be divided in the statistical segments of length l_{st} and one may simulate it (in the simplest case) as Brownian motion of a segment. Then the equation (58) can be used for the description of a polymeric chain. In this case the rate ϑ can be treated as the rate of a segment jump and τ_0 as the time of one jump and supposing $l_{st}=\vartheta\tau_0$, $t=\tau_0N_{st}$ (N_{st} is the number of the statistical segments in a chain), we receive [61]:

$$\left\langle \bar{r}^2\right\rangle^{1/2}=\frac{l_{st}N_{st}^{1-\nu}}{(1-\nu)\Gamma(\nu)}. \quad (59)$$

The authors [58, 59] made the calculation $\left\langle \bar{r}^2\right\rangle^{1/2}$ according to the equations (50) - (52) and (59) with the following comparison of the results on an example of PC solutions in two solvents (MC and THF). For this purpose 5 molecular weights of M for PC were arbitrary chosen: 2,5; 5,0; 7,5; 10,0 and $12{,}5\times10^4$. Then, according to the equations (50) and (51) the empirical values $\left\langle \bar{r}^2\right\rangle^{1/2}$ were calculated for the indicated M. In case of the application of the equation (52) for the determination of values [η], corresponding to the indicated M, the Mark–Kuhn–Houwink equations for PC and MC, THF [60] were used. For the application of the equation (59) one should define the parameters, included in it, l_{st}, N_{st}, ν and $\Gamma(\nu)$. For the determination l_{st}, N_{st} and $\Gamma(\nu)$ the usual techniques, in detail described in [58, 59] were used, and we are going to consider the ν determination more in detail.

As it was shown in paper [55], a fractional exponent ν coincides with the fractal dimension of Cantor's set and indicates a fraction of the system, the status kept for all time of the evolution t. Let's remind that the Cantor's set is considered in one-dimensional Euclidean space (d=1) and consequently its fractal dimension d_f <1 by virtue of the fractal definition. For the fractal objects in Euclidean spaces with the higher dimensions (d>1) as ν one should accept a fractional part d_f or [62, 63]:

$$\nu = d_f - (d - 1), \tag{60}$$

Let's consider the physical sense and the definition of a fractional exponent value ν in the given context. As it is known [21], the transition to the condensed polymeric state occurs at D_f=2,5. It means that the limiting value ν is reached at this value D_f, being equivalent to d, and then we receive [64]:

$$\nu = D_f - (2.5 - 1), \tag{61}$$

or

$$\nu = D_f - 1.5, \tag{62}$$

In Fig. 12 the comparison of the dependences, calculated by three indicated methods, $\left\langle \overline{r}^2 \right\rangle^{1/2}$ on M for the solutions of PC in MC is shown. As one can see, the good correspondence (within the limits of 8 %) of values $\left\langle \overline{r}^2 \right\rangle^{1/2}$ calculated according to the equations (52) and (59) with the usage of a dynamic variant of the estimation C_∞, i.e., the equation (14), is obtained. The application of a static variant C_∞=const=2,4 [65] increases an error of the calculated values up to ~ 12 %. As to the empirical equation (50), it gives the values $\left\langle \overline{r}^2 \right\rangle^{1/2}$ much lower, than the estimation according to the equations (52) and (59). In essence, the similar picture is received for the solutions PC in THF [59]. It is necessary to mark the important feature of the dependences $\left\langle \overline{r}^2 \right\rangle^{1/2}$ on M, calculated according to the equation (59) - at large M they grow faster than the similar dependences calculated according to the equation (52). As it is noted above, at large M the macromolecular coil becomes more compact, that results in the decrease of an exponent b_η in the Mark–Kuhn–Houwink equation and, according to the equation (4), to the increase D_f. The estimations have shown, that in case of PC solutions in MC increase D_f from 1,648 up to 1,681, i.e., on 2 % (or decrease b_η from 0,820 down to 0,785) results in decrease $\left\langle \overline{r}^2 \right\rangle^{1/2}$ from 107,4 nm down to 99,6 nm, i.e., eliminates the indicated difference.

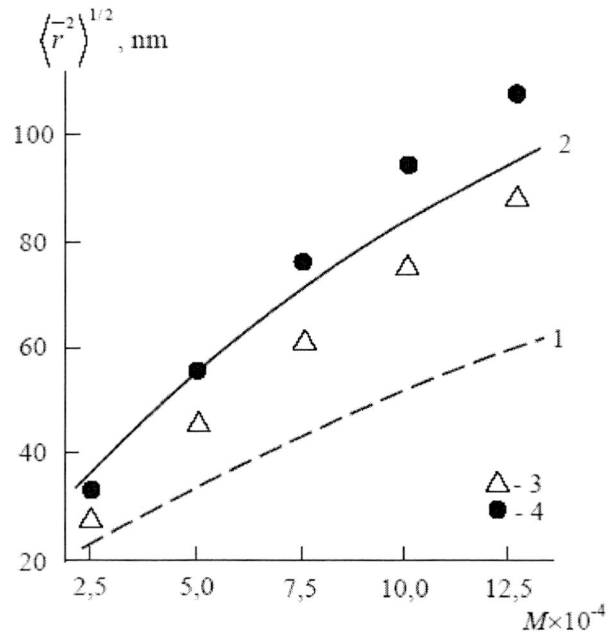

Figure 12. The dependences of mean-square end-to-end distance of polymeric chain $\langle \bar{r}^2 \rangle^{1/2}$ on a molecular weight M for solution of PC in methylene chloride. Calculation according to the equations: (50) (1), (52) (2), (59) at C_∞=const (3) and C_∞, estimated according to the equation (14) (4) [58].

The results of the present chapter allow to make two main conclusions. First, the methods the theory of fractional derivatives (the equation (59)) allows the calculation of the mean-square end-to-end distance of the polymeric chain nor less precisely as the other computational techniques which exist now. Second, the account for the change of the chain molecular characteristics of polymeric is required for the correct calculation of the indicated parameters, i.e., one should take into consideration their dynamics when the structure of macromolecular coil varies [58, 59].

7. Fractal Analysis of the Film Formation from the Polymer Solutions

At first, let's consider the dissolution process of solid polymers, which is essentially opposite in respect to the film formation from a solution. The studies of the heats of dissolution of the amorphous polymers were made rather often. It is supposed that the thermal effect of the system transition from the methastabile (nonequilibrium) glassy state in the quasiequilibrium solution makes the dominant contribution in the value ΔH_d of the heat of dissolution. At the same time the essential contribution can be made by component, determined by the polymer structure, particularly, by the existence ordered regions. The authors [66] made the estimation of the ΔH_d value in terms of a cluster model of the structure of polymer amorphous state [67,

68]. This model assumes the existence in the amorphous polymer structure the regions of a local order (clusters), consisting of some densely packed collinear segments of the different macromolecules (the amorphous analogue of crystallite with the extended chains). These clusters are connected between themselves by "tie" chains, forming by virtue of this the network of the physical entanglements with the density ν_{cl} and surrounded by loosely packed matrix, in which all fluctuation free volume of polymer [68] is concentrated.

Some common features of ΔH_d behaviour attract attention. First, above glass transition temperature of polymer T_g the value ΔH_d does not depend on the dissolution temperature T. In terms of a cluster model the thermofluctuational decay of clusters at T_g is postulated, i.e., we should expect, that the change of the ΔH_d behaviour at T_g is namely due to this effect. Second, the T_g rise results to the increase ΔH_d at the fixed T and in terms of a cluster model, the polymers with the highest T_g have higher values ν_{cl} at the fixed of the tests temperature. Third, the rise of the dissolution temperature results to the ΔH_d reduction and the cluster model assumes the ν_{cl} decrease at the temperature rise. But the linear dependence ΔH_d on the difference of temperatures (T_g-T) the greatest interest represents. Such dependence assumes that ΔH_d does not depend on the features of the amorphous polymer constitution, but it is determined only by the degree of approaching to T_g. The dependences ν_{cl} on the value (T_g-T) are fould similar in form, but different by the absolute values ν_{cl}. Some more general characteristic of the degree of local ordering of the amorphous polymer structure is a relative fraction of clusters φ_{cl} [68, 69]. For three polymers (PC, PAr, PMMA) all data in coordinates φ_{cl}-(T_g-T) lie on one curve. Sequently, as ΔH_d, the degree of local ordering in polymers is defined by approaching T to T_g and this allows to assume the possibility of the correlation ΔH_d and ν_{cl} or φ_{cl}. Let's note that the similar in form dependences of polymer properties on (T_g-T) are often offered in literature [69].

Treating the segments entering in clusters as linear defects (the analogue of dislocations), the mathematical methods of the dislocation theory for the description of the amorphous polymer structure and properties [68] can be used. Within the framework of this treatment the energy U_{dis} of the clusters dissociation was calculated and the comparison of the values U_{dis} and ΔH_d as a function (T_g-T) for three indicated polymers showed their close quantitative agreement, it allows to affirm that the value ΔH_d represents itself the energy which is necessary for the dissociation (decay) of clusters in the process of polymer dissolution. The last phenomenon means the transition of polymer from the methastabile nonequilibrium state in the quasiequilibrium solution [66, 68].

Now it is well known [40], that between the structures of a macromolecular coil in solution and the condensed state of polymers the certain interrelation, expressed by the relationship of the fractal dimensions of these states, exists. The value of fractal dimension of the structure of the condensed state can be determined so [40]:

$$d_f = \frac{d_s(d+2)}{2}. \tag{63}$$

In its turn, the value D_f of a macromolecular coil in the diluted solution is defined according to the equation (27). For the linear polymers d_s=1,0 [41] and the combination of the equations (27) and (63) allows to receive the simple relationship between D_f and d_f for the indicated case (d_s=1,0, d=3) [70-73]:

$$d_f = 1{,}5D_f. \tag{64}$$

The equation (64) clearly demonstrates the genetic interrelation between the structures of a macromolecular coil in solution and the condensed state of polymers and it is basic at the determination of the structure of polymeric films. On using it, the authors [74] gave the quantitative description of the structure formation of polymeric films, obtained from the various solvents, within the framework of the fractal analysis and Witten-Sander model of the irreversible aggregation on the example of the amorphous glassy polyarylatesulfone (PASF).

Let's consider the estimation of dimension d_f within the framework of Witten-Sander model [75]. As it was shown in papers [76, 77], the structure of the amorphous glassy polymers can be modelled as the totality of a large number of Witten-Sander clusters (clusters WS), having the radius R_{cl}. The value R_{cl} is determined so [78]:

$$R_{cl} \sim c_0^{-d}, \tag{65}$$

where c_0 is "seeds" number, around which the clusters WS grow.

The nodes of macromolecular entanglements ("hookings") with the density ν_e play the role of "seeds" at the structure formation of amorphous polymers are played. Then at the high concentration of "seeds" it can be written [78]:

$$R_{cl}^{d_f} \sim c_r / c_0, \tag{66}$$

where c_r is the concentration of particles, forming the cluster WS.

In the considered case under such a particle the statistical segment – "rigidity sector" of a macromolecule with the interindependent orientation of a position in the space [76] is accepted. The length l_{st} of a statistical segment is determined as follows [65]:

$$l_{st} = l_0 C_\infty, \tag{67}$$

where l_0 is skeletal bond length of polymer main chain.

The common length of macromolecules L per unit of polymer volume can be determined so [68]:

$$L = S^{-1}. \tag{68}$$

Thus, the number of the statistical segments per unit of polymer volume is equal to [74]:

$$c_r = \frac{L}{l_{st}} = \frac{1}{Sl_0 C_\infty}. \tag{69}$$

In Fig. 13 the comparison of the fractal dimensions d_f, calculated according to the equations (64) and (66) (at condition c_0=ν_e) is adduced, from which their good correspondence follows. This confirms the correctness of the description of the amorphous polymers structure as the totality of clusters WS and indicates the factors which control it. So, the molecular characteristics of a polymeric chain (l_0, C_∞, S) and the node density of macromolecular hookings network (c_0=ν_e) have the strong influence on the polymer structure. That's why, let's consider the factors defining the value ν_e or c_0. As it is known [40], the number of the intersections of fractals (in our case – a macromolecular coil) N_{is} is determined according to the equation (44) and relationship between R_g and M is given by the formula (39). The combination of the relationships (39) and (44) allows to obtain [74]:

$$N_{is} \sim M^{(2D_f - d)/D_f}. \tag{70}$$

It was shown [74], that at M=const the dependence ν_e on M is linear, i.e., the value ν_e is also determined by the fractal dimension D_f of a macromolecular coil and the polymer molecular weigth (the latter is well known fact [79]).

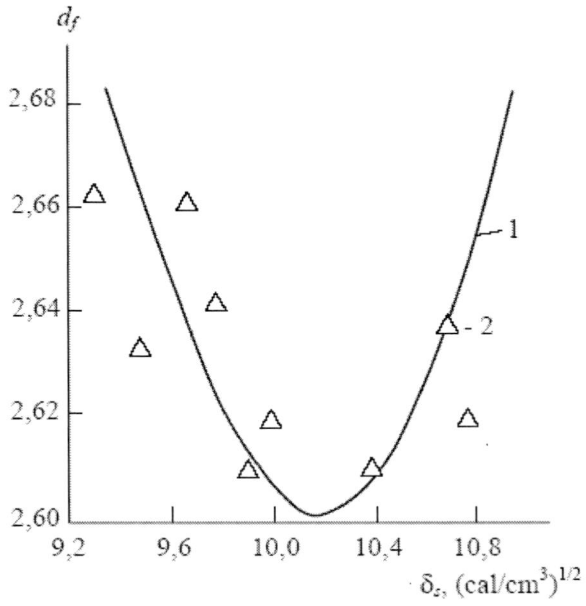

Figure 13. The dependences of the fractal dimension d_f of polymeric films structure on solubility parameter δ_s of solvent for PASF. 1 – the calculation according to the equation (64); 2 – the calculation according to the equation (66) [74].

The authors [80-82] used this approach for the quantitative prediction of the mechanical properties of polymer film samples, obtained from the different solvents. It should be noted that the prediction of the properties on the full extent of stress-strain diagram is carried out within the framework of one approach and with the exactness, that is sufficient for the practical applications. This approach is based on the strict analytical ground of the interrelation between the structures of a macromolecular coil in solution and the condensed state of polymers [73].

8. Some other Problems of Polymer Solutions

The study of the dependences of self-diffusion coefficient D_{sd} of the chain Gaussian macromolecules on the different factors has the great significance both for the development of the ideas about the macromolecule mobility in solution and at the decision of the various problems of the polymer physics and chemistry [13]. However, up to now the main attention was given to D_{sd} dependence on the solution concentration c and the macromolecule molecular weigth M [83]. Nevertheless, another very important factor exists – a macromolecular coil structure [84]. Budtov proposes the following relationship for the evaluation of the value D_{sd} [83]:

$$D_{sd} \approx A_1 \left(\frac{c^*}{c} \right)^{\beta} D_R, \tag{71}$$

where A_1 is a numerical coefficient, c^* is a threshold value of concentration c, D_R is a macromolecule friction coefficient, for which for the intramolecular hydrodynamical interaction between the chain elements is not accounted, β is an exponent, determined as follows [83]:

$$\beta = \frac{2 - \nu_F}{3\nu_F - 1}, \tag{72}$$

where ν_F is Flory exponent in the relationship R_g-N [2, 13].

The value c^* can be determined according to the equation [83]:

$$c^* = 0{,}35c[\eta] = 0{,}35\left[0{,}64\ln^2\left(2\sqrt{N}\right)\right]. \tag{73}$$

The value c^* for polydimethyldiallyl ammonium chloride (PDMDAACh) can be determined at $[\eta]$=0,80 dl/g and N=315 [84]. Besides, c is determined as c_0 $[\eta]$, where c_0 is monomer concentration. From these estimations the authors [84] were obtained the ratio value $c^*/c \approx 1{,}32$.

As it is known [2], between ν_F and D_f the following relationship exists:

$$\nu_F = D_f^{-1}. \tag{74}$$

The combination of the equations (72) and (74) allows to express the exponent β as a function D_f [84]:

$$\beta = \frac{2D_f - 1}{3 - D_f}. \tag{75}$$

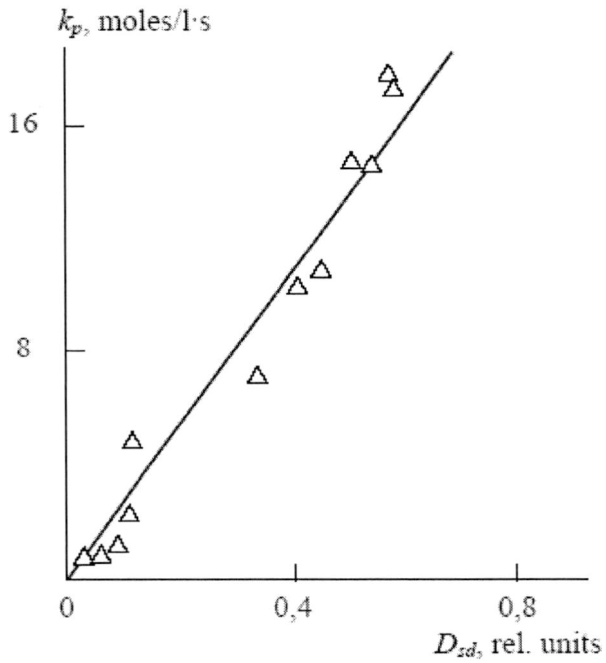

Figure 14. The dependence of reaction rate k_p on self-diffusion coefficient of a macromolecular coils D_{sd} for PDMDAACh [84].

Now the value D_{sd} in the relative units can be estimated, as a function D_f assuming A_1=const and D_R=const and using the equations (71) and (75). In Fig. 14 the dependence of the polymerization rate k_p on D_{sd} for PDMDAACh is shown. As it can be seen, the correlation $k_p(D_{sd})$ is linear and passes through the coordinate origin. Such dependence was expected, as the existing theories of polymer synthesis foresee proportional relationship between k_p and D_{sd}. So, in paper [85] two relationships obtained according to the approximation of the chaotic phases, are cited:

$$k_p = 4\pi D_{sd} R_g \left(\frac{9}{5} + \frac{R_g}{aN} \right)^{-1}, \tag{76}$$

and according to the average density approximation:

$$k_p = 4\pi D_{sd} R_g \left[1 - \left(\frac{R_g}{4\pi a N} \right)^{1/3} \right], \tag{77}$$

where a is a reactive radius of the reacting particles.

For PDMDAACh radical polymerization we can write down [85]:

$$a = 2R_g . \tag{78}$$

In this case the members in brackets of the equations (76) and (77) give very weak correction to the directly proportional dependence k_p on D_{sd}, especially at rather large N [84].

The fractal reactions are admitted to call either the reactions of fractal objects or the reactions in fractal spaces [86]. The typical indication of these reactions is autodeceleration, i.e., the decrease of reaction rate k_p with the time of its duration t [87]. Let's note, that for Euclidean reaction is typical the linear kinetics and, correspondingly, the condition k_p=const [88]. The fractal reactions in wide sense of this term are met very often in the practice (reactions of the synthesis, sorption curves, stress-strain diagrams and so on) [57]. The simplest and visual relationship for the description of this effect is following [87]:

$$k_p \sim t^{-h}, \tag{79}$$

where h is a heterogeneity exponent ($0 \leq h \leq 1$), converted to zero in case of the classical behaviour (reactions in the homogeneous or Euclidean mediums) and then k_p=const.

The relationship (79) allows to give the following treatment of the physical sense of the reactive medium heterogeneity: the higher degree of this heterogeneity (fractality) is, the faster decay k_p at the reaction elapsion is. Within the framework of the fractal kinetics of reactions the authors [89] obtain the following interrelation between h and D_f:

$$h = \frac{D_f - 1}{2}. \tag{80}$$

Let's estimate the limiting values of the fractal dimension of a macromolecular coil in solution for two cases: h=0 (D_f^0) and h=1 (D_f^1). From the equation (80) it follows: D_f^0=1,0 and D_f^1=3,0 [89]. The physical sense of the obtained limiting values D_f in this case is obvious. For the homogeneous reactive medium (h=0) k_p is maximal for all $t > 1$ and D_f^0=1,0 means, that the macromolecule represents itself a fully stretched chain, but not a macromolecular coil. If the latter is a fractal object, the inner regions of which are screened by the surface and inaccessible for the reaction [90], then for the stretched chain all reactive centres are accessible and the steric hindrances for the passing of reaction are absent. Therefore, the value k_p is maximal and constant. The value D_f^1=3,0 means, that the macromolecule is coiled up in a compact globule, only small part of surface of which is accessible to the reaction. That's why, the value k_p is minimal and decreases fastly to zero [89].

And in conclusion let's consider a fractal model of low-molecular admixtures by polymeric flocculator (PDMDAACh) [91]. As it is known [92], the value D_f characterizes the degree of "oppeness" ("looseness") of a fractal object structure. This degree can be more precisely expressed by the object density ρ determined as follows [92]:

$$\rho = \rho_0 R_g^{D_f - d}, \tag{81}$$

where ρ_0 is the density of a three-dimensional Euclidean object.

For a macromolecular coil accounting for the equation (39) we can write down [91]:

$$\rho \sim \left(\frac{M}{m_0}\right)^{-(d-D_f)/D_f}. \tag{82}$$

By the application of the equations (1) and (4) the following results for PDMDAACh were obtained: in aqueous solution D_f=1,44 and in NaCl solution D_f=1,65. So, one of the effects of medium, in which the flocculation takes place, on a macromolecule of flocculator is the change of its conformation (and therefore, D_f) with the variation of the solvent quality [91].

As follows from the relationship (82), for a compact object (D_f=d) it is obtained that ρ=1 (in relative units). One of the most important properties of fractal objects is the reduction of their density (loosening) with the increase R_g or M even at D_f=const. It is obvious that the flocculation degree will be the function not only of the number of active centers, but of the degree of their accessibility. Within the framework of the fractal analysis the last factor can be characterized by the value ρ - the lower ρ is, the easier the trapping of adsorbed substance is (in the considered case – phosphatides [94]) in all the coil volume. In essence, the similar explanation of the observed effect is given in [94], where the decrease of the flocculant (PDMDAACh) concentration c_f with its increase [η] is also connected to the geometrical features of a macromolecular coil, though only on the qualitative level.

In Fig. 15 the relationship $c_f(M^1)$ in double log-log coordinates, plotted according to the data [94] is shown. As it can be seen, this relationship can be approximated by the following scaling expression [94]:

$$c_f \sim M^{-n}, \tag{83}$$

where the exponent n is equal about to 0,80, that is very close on its absolute value to the exponent $(d-D_f)/D_f$ in (82), which for D_f=1,65 is equal to 0,818.

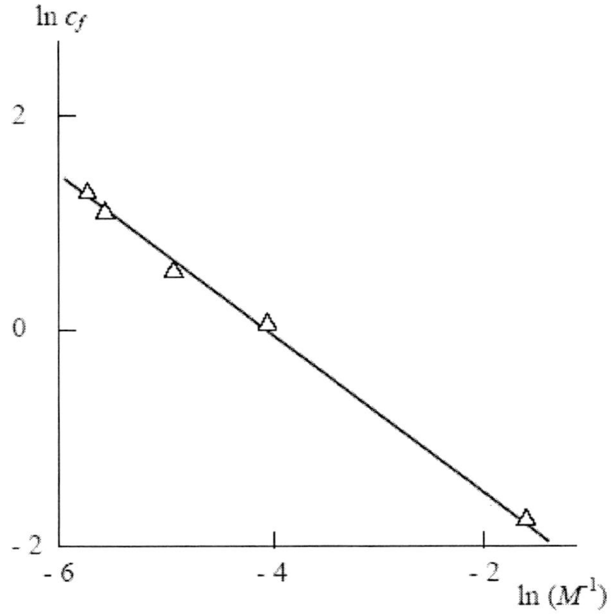

Figure 15. The scaling relationship between a concentration of PDMDAACh c_f and its average viscosity molecular weight M in double log-log coordinates [91].

The physical sense of the equation (83) can be easily understood within the modern ideas about the fractal clusters [86, 95]. The depth of the penetration of a particle in cluster Δr is determined as follows [95]:

$$\Delta r \sim N^{q}. \tag{84}$$

The exponent ratio q/D_f for three-dimensional Euclidean space is equal to ~0,8 [95]. For D_f=1,65 with the appreciation of the condition $q \approx n^{-1}$ we shall obtain n≈0,76 that is close to the value n evaluated from the plot of Fig. 15. Thus, the bigger the depth of the penetration of phosphatides is in a PDMDAACh macromolecular coil, the smaller the optimal concentration of flocculator is [91].

Apparently, one more factor influencing on the efficiency of flocculation process will be the fractal dimension D_f of a flocculator macromolecular coil. So, according to the relationship (83) at the change of D_f it is possible to preserve the value of c_f on previous level by M variation. The authors [93] compared this possibility for PDMDAACh in two solvents: water and NaCl aqueous solution (D_f=1,44 and 1,65, respectively) and they obtained the coefficient k_c which shows how many times it is necessary to decrease M of polymer at the transition from D_f=1,65 to D_f=1,44 in order to preserve the optimal value of c_f. It proves, that at the indicated D_f and initial M=2,99×10^5 it is possible to decrease M almost in one order of the value [93].

Thus, the cited above results show, that between the factors influencing on the efficiency of the flocculation process by polymeric flocculator PDMDAACh there are the molecular weight, the fractal dimension of a macromolecular coil and the medium in which the flocculation takes place. The effect of the medium on the flocculation process can be realized by two ways – by the change of the fractal dimension of a macromolecular coil and the viscosity variation. The last factor can influence on the intensity of diffusive processes. The only factor which will influence in equal degree on the properties of fractal and compact objects is the variation of the medium viscosity and the influence of three remaining factors is defined only by the fractal nature of macromolecular coil of the polymeric flocculator [91, 93].

Let's mark another interesting feature, cited in paper [94], namely, the extreme change of the degree of the removal of phosphatides as a function of PDMDAACh concentration for each molecular weight. Because of its applied and theoretical importance this problem requires the special examination. Therefore, in paper [91] the theoretical description of this effect for PDMDAACh within the framework of the fractal analysis was given.

The increase of the degree of the removal of phosphatides Q up to some polymer concentration c_f (c_{max}), at which the maximal value Q (Q_{max}) is achieved, is due to the simple increase of "active centres" (trapping sites [94]) for the flocculation owing to the increase of a number of macromolecular coils (polymer concentration) in solution. At the concentration c_{max} the touching of macromolecular coils takes place, that corresponds to Q_{max}. At c_f increase the interpenetration of coils is realized and the scale of the mentioned transition r_c is defined by the criterion [78]:

$$r_c \sim c_f^{-1/(d-D_f)}. \tag{85}$$

The value r_c at the condition of coil touching is accepted to be $2R_g$ [91]. The comparison of the calculated according to the relationship (85) theoretical values c_{max} (c_{max}^T) and the experimental magnitudes of this parameter (c_{max}^e) showed the good correspondence of the theory and the experiment for all M, except the smallest out of the used ([η]=0,1 dl/g or $M \approx 4\times10^3$), that proves the made above assumption about the physical sense of the value c_{max}. As for PDMDAACh with $M \approx 4\times10^3$, then the value c_{max}^T is found about 8 times more c_{max}^e, i.e., the efficiency of low-molecular flocculator is found much higher than the expected one. This important question deserves the special consideration.

For its molecular weight, PDMDAACh with $M \approx 4\times10^3$ is closer to olygomer than to polymer. Within the framework of the models of irreversible aggregation it was shown [95], that at low N the small rod-like aggregates are formed, assuming smaller value D_f than the obtained one according to the equation (4) at b_η=0,82 for the high-molecular polymer [96]. Besides, it is known [17], that the value D_f is a monotonously increasing function of M. Thus, for the low-molecular ($M \approx 4\times10^3$) PDMDAACh the value D_f will be much lower than 1,65, obtained for the high-molecular polymers. From the condition $c_{max}^T = c_{max}^e$ and using the relationships (4) and (85), the magnitude D_f for the low-molecular PDMDAACh was estimated, which was found equal to ~ 1,26, that completely agrees with the made above assumptions and results of papers [7, 96]. This example demonstrates very strong effect of the spatial geometric structure of a macromolecule, characterized by dimension D_f, which controls the access of impurities (in our case-phosphatides) into internal regions of flocculator during the flocculation process.

The power dependence of ρ on D_f defines the strong dependence Q on D_f. So, in order to increase Q twofold for PDMDAACh with M=3×10^5 it is necessary to reduce D_f from 1,65 to 1,50 at the constant flocculator concentration. As it is known [45], the value of the exponent b_η in the Mark-Kuhn-Houwink equation depends on the type of solvent used in measurement. Therefore, by the selection of the latter the macromolecular coil conformation can be changed and by this way it is possible to control its ability to flocculate the impurities. The control of this ability can be made by the variation and the measurement of the value b_η accounting for its connection with D_f.

Now let's consider the reasons of the extreme dependence $Q(c)$, experimentally obtained in paper [94], for each value of [η]. As follows from the mentioned above considerations, the value Q should depend, as minimum, on two parameters [91]:

$$Q = k_0 \left(\frac{c}{\rho} \right), \tag{86}$$

where k_0 is a constant, ρ is the density determined according to the equation (82).

For $c < c_{max}$ the coils do not touch each other, the value ρ for them will be constant (at M=const) and, in essence, Q is a function only of c. For $c > c_{max}$ the interpenetration of macromolecular coils on their external regions is realized, i.e., as if they flocculated each other that reduced their ability to flocculate the impurities from the solvent. Let's remind, that the fractal aggregates have one common feature – for rather large N the addition of particles (or the small clusters of particles) to this aggregate can be realized only on narrow enough external "active zone" and their internal regions are screened by the indicated zone and can not add the particles [90]. Therefore, between the effective (non-screened) radius of a macromolecular coil R_{ef} and the polymer concentration c there is a relationship [78]:

$$c \sim R_{ef}^{D_f - d} \tag{87}$$

from which it follows the R_{ef} reduction at $c > c_{max}$. Using the values obtained by that way, the effective density of a coil ρ_{ef} can be estimated according to the relationship (82) and then after calculating the constant k_0 in the equation (86) from the condition of the equality of the theoretical and experimental values Q in the maximum point for PDMDAACh with different M it is possible to estimate theoretically the dependence $Q(c)$. In Fig.16 the comparison of the theoretical (curves) and experimental (points) dependences $Q(c)$ for PDMDAACh with viscosity [η]=0,9; 1,8 and 3,45 dl/g is shown. As it can be seen, the proposed method expresses exactly the qualitative change Q with c, though the quantitative correspondence is worse, especially at $c < c_{max}$. The authors [91] assumed that this discrepancy is not only due to the imperfection of the theoretical calculation though its simplicity gives the certain possibilities for the improvement. So, the obvious extrapolation Q to the zero at c=0 should be expected though the experimental data assume this not always (for example, for [η]=1,8 dl/g [94]). Nevertheless, one from the possible improvements of the proposed methods of calculation can be indicated [91]. All cited up to now equations assume that the aggregate (a macromolecular coil) consists of zero-dimensional objects (points). Apparently, the real polymeric coil consists of the monomer units with finite sizes (for example, of length a_m) and in this case it is necessary to incorporate the dimensional characteristics (scale) in the corresponding relationships. For example, the relationship (39) has the form [97]:

$$\left(\frac{R_g}{a_m}\right)^{D_f} \sim N \, . \tag{88}$$

It is also apparently, that the incorporation of a linear scale is especially important at the comparison of the flocculating ability of the various polymers. However, the authors [91] did not attempt to achieve the full quantitative agreement of the theory and the experiment, but they wanted to demonstrate the principal possibility of the description of such complicated phenomenon as the flocculation within the framework of the models of the irreversible aggregation and the fractal analysis.

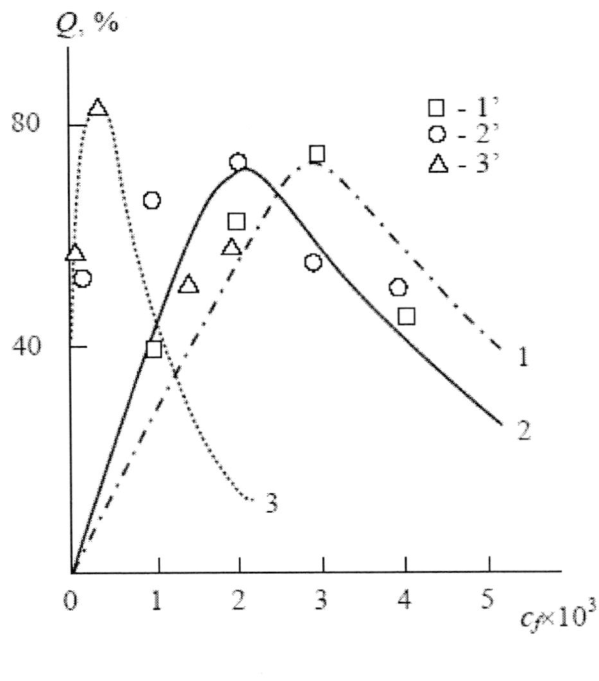

Figure 16. The dependence of the degree of phosphatides removal Q on flocculator concentration c_f for PDMDAACh with molecular weight: 5,8×10^4(1, 1'), 13,5×10^4 (2, 2') and 30×10^4 (3, 3') calculated according to the equation (86) (1, 2, 3) and experimentally obtained (1', 2', 3') [91].

9. Conclusions

Thus, in the present review the main aspects of the application of the fractal analysis for the description of the behaviour of macromolecular coils in the diluted solutions are considered. The coil structure is quantitatively described with the help of its fractal (Hausdorff) dimension, characterizing the distribution of coil elements in space. A number of the experimental methods of the determination of this dimension is offered. The interrelation of the classical and fractal (structural) characteristics of a macromolecular coil is shown. The fractal model is used for the description of such phenomena as film formation from polymer solutions, flocculation and self-diffusion of macromolecular coils. The example of the application of the theory of fractional derivatives for the description of the main parameters of a macromolecular coil in solution is given.

References

[1] Tsallis C. In book: *Fractals in Physics.* Ed. Pietronero L., Tosatti E. Amsterdam, Oxford, New York, Tokyo, North-Holland, 1986, p. 98-105.

[2] Baranov V.G., Frenkel S.Ya., Brestkin Yu.V. *Dokl. Akad. Nauk SSSR,* 1986, v. 290, № 2, p. 369-372.

[3] Mandelbrot B.B. The Fractal Geometry of Nature. San-Francisco, W.H. *Freeman and Comp.,* 1982, 459 p.

[4] Karmanov A.P., Monakov Yu.B. *Vysokomol. Soedin. B,* 1995, v. 37, № 2, p. 328-331.

[5] Pavlov G.M., Korneeva E.V., Mikhailova N.A., Anan'eva E.P. *Biofizika,* 1992, v. 37, № 6, p. 1035-1040.

[6] Pavlov G.M., Korneeva E.V. *Biofizika,* 1995, v. 40, № 6, p. 1227-1233.

[7] Kozlov G.V., Dolbin I.V. *Biofizika,* 2001, v. 46, № 2, p. 216-219.

[8] Kozlov G.V., Dolbin I.V., Zaikov G.E. In book: *Perspectives on Chemical and Biochemical Physics.* Ed. Zaikov G. New York, Nova Science Publishers, Inc., 2002, p. 217-223.

[9] Ptitsyn O.B., Eizner Yu.E. Zh. *Tekh. Fiz.,* 1959, v. 29, № 9, p. 1117-1134.

[10] Kozlov G.V., Dolbin I.V., Mashukov N.I., Burmistr M.V., Korenyako V.A. *Voprosy Khimii i Khim. Technol.,* 2001, № 6, p. 71-77.

[11] Wiehe I.A. *Ind. Engng. Chem. Res.,* 1995, v. 34, № 2, p. 661-673.

[12] Family F. *J. Stat. Phys.,* 1984, v. 36, № 5/6, p. 881-896.

[13] Budtov V.P. *Physical Chemistry of Polymer Solutions* (in Russian). Sankt-Peterburg, Khimiya, 1992, 384 p.

[14] Temiraev K.B., Kozlov G.V., Sozaev V.A. *Vestnik of Kabardino-Balkarian State University,* 1999, № 3, p. 24-27.

[15] Kozlov G.V., Temiraev K.B., Sozaev V.A. Zh. *Fizich. Khimii,* 1999, v. 73, № 4, p. 766-768.

[16] Askadskii A.A. *Physical Chemistry of Polyarylates* (in Russian). Moscow, Khimiya, 1968, 216 p.

[17] Dubrovina L.V., Pavlova S.S.-A., Bragina T.P. *Vysokomol. Soedin.* A, 1988, v. 30, № 5, p. 995-1000.

[18] Kozlov G.V., Temiraev K.B., Kaloev N.I. *Dokl. Russian Akad. Nauk,* 1998, v. 362, № 4, p. 489-492.

[19] Kozlov G.V., Burya A.I., Shustov G.B., Temiraev K.B., Shogenov V.N. *Vestnik of Dniepropetrovskogo University,* 2000, № 5, p. 96-102.

[20] Kozlov G.V., Afaunova Z.I., Temiraev K.B. In book: *Fractals and Local Order in Polymeric Materials*. Ed. Kozlov G., Zaikov G. New York, Nova Science Publishers, Inc., 2001, p. 21-28.
[21] Novikov V.U., Kozlov G.V. *Russian Chemical Reviews,* 2000, v. 69, № 4, p. 378-399.
[22] Kozlov G.V., Shustov G.B., Zaikov G.E. J. *Balkan Tribol. Assoc.,* 2003, v. 9, № 4, p. 467-514.
[23] Novikov V.U., Kozlov G.V. In book: *Prikl. Sinergetika, Fraktaly i Komp'yut. Modelir. Structur*. Ed. Oksogoev A.A. Tomsk, TGU, 2002, p. 268-302.
[24] Dolbin I.V., Kozlov G.V. Mater. Of All-Russian conf. *"Perspectiva-2003",* v. 4, Nal'chik, 2003, p. 108-111.
[25] Dondos A. *Polymer,* 2001, v. 42, № 4, p. 897-901.
[26] Gans C., Schnee J., Scherf U., Staikos G., Pierri E., Dondos A. *Polymer,* 1998, v. 39, № 17, p. 4155-4158.
[27] Papkov S.P. *Vysokomol. Soedin.* B, 1982, v. 24, № 11, p. 869-873.
[28] Kozlov G.V., Temiraev K.B., Mikitaev A.K. *Izvestiya VUZov, Khimiya,* 2002, v. 45, № 2, p. 30-33.
[29] Kozlov G.V., Shustov G.B., Dolbin I.V. Mater. *I Intern. Conf. "Modern Problems of Organic Chemistry, Ecology and Biotechnology"*. Luga, 2001, p. 17-18.
[30] Kozlov G.V., Temiraev K.B., Lipatov Yu.S. In book: *Fractals and Local Order in Polymeric Materials*. Ed. Kozlov G., Zaikov G. New York, Nova Science Publishers, Inc., 2001, p. 1-9.
[31] Okatova O.V., Lavrenko P.N. *Vysokomol. Soedin.* B, 1992, v. 34, № 7, p. 9-18.
[32] Tsvetkov V.N. *Rigidchain Polymer Macromolecules* (in Russian), Leningrad, *Khimiya,* 1986, 328 p.
[33] Ryskina I.I., Zhiganova I.Yu. *Vysokomol. Soedin.* B, 1994, v. 36, № 3, p. 503-506.
[34] Khamrakulov G., Musaev Kh.N., Budtov V.P., Kuleznev V.N. *Vysokomol. Soedin.* A, 1993, v. 35, № 6, p. 705-709.
[35] Korshak V.V., Pavlova S.S.-A., Timofeeva G.I., Kroyan S.A., Krongauz Ye.S., Travnikova A.P., Raubach Ch., Schulz G., Gnauk R. *Vysokomol. Soedin.* A, 1984, v. 26, № 9, p. 1868-1876.
[36] Temiraev K.B., Shustov G.B., Mikitaev A.K. *Vysokomol. Soedin.* B, 1993, v. 35, № 12, p. 2057-2059.
[37] Burya A.I., Kozlov G.V., Temiraev K.B., Malamatov A.Kh. *Voprosy Khimii i Khim. Technol.,* 1999, № 3, p. 26-28.
[38] Shustov G.B., Temiraev K.B., Afaunova Z.I., Kozlov G.V. *Izvestiya KBSC RAS,* 2000, № 2(5), p. 92-94.
[39] Kozlov G.V., Dolbin I.V. *Central Eur. J. Chem.,* 2004 (in press).
[40] Vilgis T.A. *Physica* A, 1988, v. 153, № 2, p. 341-354.
[41] Alexander S., Orbach R. *J. Phys. Lett.* (Paris), 1982, v. 43, № 17, p. L625-L631.
[42] Dolbin I.V., Kozlov G.V. In book: *Book of Young Scientifics Works* (in Russian), Nal'chik, KBSU, 2002, p. 84-89.
[43] Kozlov G.V., Dolbin I.V. Mater. All-Russian Conf. *"Chemistry in Technology and Medicine"*. Makhachkala, DSU, 2002, p. 45-48.
[44] Frenkel S.Ya. *Introduction to the Statistical Theory of Polymerization* (in Russian). Moscow, Nauka, 1965, 268 p.
[45] Kozlov G.V., Dolbin I.V. *Vysokomol. Soedin.* B, 2002, v. 44, № 1, p. 115-118.

[46] Kozlov G.V., Dolbin I.V., Zaikov G.E. In Book: *New Perspectives in Chemistry and Biochemistry.* Ed. Zaikov G. New York, Nova Science Publishers, Inc., 2002, p. 41-48.

[47] Kozlov G.V., Dolbin I.V., Zaikov G.E. In book: *Polymer Yearbook* 18. Ed. Pethrick R., Zaikov G. Shawbyry, Rapra Technology Limited, 2003, p. 393-400.

[48] Kozlov G.V., Malkanduev Yu.A., Ulybina A.S. *Vestnik KBSU, Khimich. Nauki,* Nal'chik, KBSU, 2003, p. 175.

[49] Kozlov G.V., Dolbin I.V., Shustov G.B. *Proceeding of XVII Mendeleev Conf. On General and Applied Chemistry.* Kazan, 2003, v. 1, p. 279.

[50] Askadskii A.A. *Structure and Properties of Thermo-resistant Polymers* (in Russian). Moscow, Khimia, 1981, 320 p.

[51] Kozlov G.V., Dolbin I.V. *Izvestiya VUZov, Severo-Kavkazs. Region, Estestv. Nauki,* 2004, № 3, p. 69-71.

[52] Feder E. *Fractals.* New York, Plenum Press, 1988, 256 p.

[53] Andrianov K.A., Pavlova S.A., Tverdochlebova I.I., Pertsova N.V., Temnikovskii V.A. *Vysokomol. Soedin., Ser.* A, 1972, v. 14, № 8, p. 1816-1821.

[54] Kozlov G.V., Shustov G.B., Dolbin I.V. Proceeding of XVII Mendeleev *Conf. On General and Applied Chemistry. Kazan,* 2003, v. 2, p. 442.

[55] Nigmatullin R.R. *Theoret. i Matemat. Fizika,* 1992, v. 90, № 3, p. 354-367.

[56] Bolotov V.N. *Pis'ma v ZhTF,* 1995, v. 21, № 10, p. 82-84.

[57] Meilanov R.P., Sveshnikova D.A., Shabanov O.M. Izvestiya VUZov, Severo-Kavkazs. Region, *Estestv. Nauki,* 2001, № 1, p. 63-66.

[58] Kozlov G.V., Dolbin I.V., Shogenov V.Kh., Zaikov G.E. In book: *Fractal Analysis of Polymers: From Synthesis to Composites.* Ed. Kozlov G., Zaikov G., Novikov V. New York, Nova Science Publishers, Inc., 2003, p. 123-130.

[59] Kozlov G.V., Dolbin I.V., Shogenov V.Kh., Zaikov G.E. *J. of the Balkan Tribological Association,* 2003, v. 9, № 3, p. 428-434.

[60] Schnell H. *Chemistry and Physics of Polycarbonates.* New York, London, Sidney, Interdcience Publishers, 1965, 229 p.

[61] Shogenov V.Kh., Shkhanukov-Lafishev M.Kh., Beshtoev Kh. M. *Fractional Derivatives: Interpretation and some Applications in Physics.* (Preprint). Dubna, OIYaI, 1997, 14 p.

[62] Kozlov G.V., Batyrova H.M., Zaikov G.E. *J. Appl. Polymer Sci.,* 2003, v. 89, № 7, p. 1764-1767.

[63] Kozlov G.V., Shustov G.V., Zaikov G.E. *J. Appl. Polymer Sci.,* 2003, v. 89, № 9, p. 2378-2381.

[64] Kozlov G.V., Shustov G.V. *Proceeding Intern. Seminar "Fractals and Applied Synergetics".* Moscow, MSOU, 2001, p. 155-157.

[65] Wu S. J. *Polymer Sci.: Part B: Polymer Phys.,* 1989, v. 27, № 4, p. 723-741.

[66] Kozlov G.V., Dolbin I.V., Zaikov G.E. Mater. V All-Russian Seminar *"Modelling of nonequilibrium systems".* Krasnoyarsk, KSU, 2002, p. 96-97.

[67] Kozlov G.V., Beloshenko V.A., Varyukhin V.N., Lipatov Yu.S. *Polymer,* 1999, v. 40, № 4, p. 1045-1051.

[68] Kozlov G.V., Zaikov G.E. *Structure of the Polymer Amorphous State.* Leiden, Brill Academic Publishers, 2004, 465 p.

[69] DiBenedetto A.T., Trachte K.V. *J. Appl. Polymer Sci.,* 1970, v. 14, № 11, p. 2249-2262.

[70] Mashukov N.I., Temiraev K.B., Shustov G.B., Kozlov G.V. *Papers of the 6th Int. Workshop of Polymer Reaction Engng.* Berlin, 1998, v. 134, p. 429-438.

[71] Kozlov G.V., Temiraev K.B., Malamatov A.Kh. *Khimich. Promyshl.,* 1998, № 4, p. 230-232.

[72] Kozlov G.V., Temiraev K.B., Shustov G.B. *Izvestiya VUZov,* Severo-Kavkazs. Region, Estestv. Nauki, 1999, № 3, p. 77-81.

[73] Kozlov G.V., Temiraev K.B., Shustov G.B., Mashukov N.I. *J. Appl. Polymer Sci.,* 2002, v. 85, № 6, p. 1137-1140.

[74] Dolbin I.V., Kozlov G.V. *Izvestiya KBSC RAS,* 2004 (in press).

[75] Witten T.A., Sander L.M. *Phys. Rev. Lett.,* 1981, v. 47, № 19, p. 1400-1403.

[76] Kozlov G.V., Beloshenko V.A., Varyukhin V.N. *Ukrainsk. Fizich. Zhurn.,* 1998, v. 43, № 3, p. 322-323.

[77] Kozlov G.V., Shogenov V.N., Mikitaev A.K. *Inzhenerno-fizich. Zhurn.,* 1998, v. 71, № 6, p. 1012-1015.

[78] Witten T.A., Meakin P. *Phys. Rev.* B, 1983, v. 28, № 10, p. 5632-5642.

[79] Mandelkern L. *Polymer J.,* 1985, v. 17, № 1, p. 337-350.

[80] Novikov V.U., Kozlov G.E., Zaikov G.E. *Russian Polymer News,* 1998, v. 3, № 2, p. 29-34.

[81] Shogenov V.N., Belosshenko V.A., Kozlov G.V., Varyukhin V.N. *Fizika i Technika Vysok. Davl.,* 1999, v. 9, № 3, p. 30-36.

[82] Kozlov G.V., Shogenov V.N., Ozden S., Zaikov G.E. In book: *Perspectives on Chemical and Biochemical Physics.* Ed. Zaikov G. New York, Nova Science Publishers, Inc., 2002, p. 159-168.

[83] Budtov V.P. *Vysokomol. Soedin.* A, 1986, v. 28, № 12, p. 2575-2581.

[84] Kozlov G.V., Malkanduev Yu.A., Burya A.I., Burmistr E.M. *Voprosy Khimii i Khim. Technol.,* 2003, № 2, p. 69-72.

[85] Burlatskii S.F., Oshanin G.S., Likhachev V.N. *Khimichesk. Fizika,* 1988, v. 7, № 7, p. 970-978.

[86] Shogenov V.N., Kozlov G.V. *Fractal Clusters in Physics and Chemistry of Polymers* (in Russian). Nal'chik, Polygraphservice and T, 2002, 268.

[87] Klymko P.W., Kopelman R. *J. Phys. Chem.,* 1983, v. 87, № 23, p. 4565-4567.

[88] Kozlov G.V., Zaikov G/E. *Teoret. Osnovy Khimich. Technol.,* 2003, v. 37, № 5, p. 555-557.

[89] Dolbin I.V., Kozlov G.E. *Dokl. Adygsk. Mezdunar. Akad. Nauk,* 2004, v. 7, № 1, p. 134-137.

[90] Meakin P., Stanley H.E., Coniglio A., Witten T.A. *Phys. Rev.* A, 1985, v. 32, № 4, p. 2364-2369.

[91] Kozlov G.E., Malkanduev Yu.A., Zaikov G.E. In book: *Homolytic and Heterolytic Reactions. Problems and Solytions.* Ed. Zaikov G., Monakov Yu., Jimenez AA. New York, Nova Science Publishers, Inc., 2004, p. 91-120.

[92] Brady R.M., Ball R.C. *Nature,* 1984, v. 309, № 5965, p. 225-229.

[93] Esmursiev A.M., Malkanduev Yu.A., Kozlov G.E. Vestnik KBSU, *Khimich. Nauki,* Nal'chik, KBSU, 2001, № 4, p. 121-123.

[94] Shnaider M.A., Kolganova I.V., Ter-Minasyan R.I., Topchiev D.A. *Maslozhirov. Promyshl.,* 1990, № 10, p. 17-21.

[95] Smirnov B.M. *Physics of Fractal Cluster* (in Russian). Moscow, Nauka, 1991, 136 p.

[96] Kolb M., Botet R., Jullien R. *Phys. Rev. Lett.*, 1983, v. 51, № 13, p. 1123-1126.
[97] Hentschel H.G.E., Deutch J.M., Meakin P. *J. Chem. Phys.,* 1984, v. 81, № 5, p. 2496-2502.

INDEX

A

D

E

H

I

N

O

P

S

V

W

X

Y

Z